Martin Braun

Differential Equations and Their Applications

Short Version

Springer-Verlag
New York Heidelberg Berlin

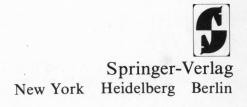

Martin Braun
Department of Mathematics
Queens College
City University of New York
Flushing, NY 11367
USA

AMS Subject Classifications: 34-01

This book is an abridged version of *Differential Equations and Their Applications: An Introduction to Applied Mathematics* by Martin Braun, © 1975, 1978 by Springer-Verlag New York Inc.

Printed in the United States of America.

9 8 7 6 5 4 3 2 1

ISBN 0-387-90289-9 Springer-Verlag New York
ISBN 3-540-90289-9 Springer-Verlag Berlin Heidelberg

To four beautiful people:

Zelda Lee

Adeena Rachelle, I. Nasanayl, and Shulamit

Preface

This textbook is a unique blend of the theory of differential equations and their exciting application to "real world" problems. First, and foremost, it is a rigorous study of ordinary differential equations and can be fully understood by anyone who has completed one year of calculus. However, in addition to the traditional applications, it also contains many exciting "real life" problems. These applications are completely self contained. First, the problem to be solved is outlined clearly, and one or more differential equations are derived as a model for this problem. These equations are then solved, and the results are compared with real world data. The following applications are covered in this text.

1. In Section 1.3 we prove that the beautiful painting "Disciples at Emmaus" which was bought by the Rembrandt Society of Belgium for $170,000 was a modern forgery.

2. In Section 1.5 we derive differential equations which govern the population growth of various species, and compare the results predicted by our models with the known values of the populations.

3. In Section 1.6 we try to determine whether tightly sealed drums filled with concentrated waste material will crack upon impact with the ocean floor. In this section we also describe several tricks for obtaining information about solutions of a differential equation that cannot be solved explicitly.

4. In Section 2.7 we derive a very simple model of the blood glucose regulatory system and obtain a fairly reliable criterion for the diagnosis of diabetes.

5. In Section 4.3 we derive two Lanchestrian combat models, and fit one of these models, with astonishing accuracy, to the battle of Iwo Jima in World War II.

This textbook also contains the following important, and often unique features.

1. In Section 1.9 we give a complete proof of the existence–uniqueness theorem for solutions of first-order equations. Our proof is based on the method of Picard iterates, and can be fully understood by anyone who has completed one year of calculus.

2. Modesty aside, Section 2.12 contains an absolutely super and unique treatment of the Dirac delta function. We are very proud of this section because it eliminates all the ambiguities which are inherent in the traditional exposition of this topic.

3. All the linear algebra pertinent to the study of systems of equations is presented in Sections 3.1–3.5. One advantage of our approach is that the reader gets a concrete feeling for the very important but extremely abstract properties of linear independence, spanning, and dimension. Indeed, many linear algebra students sit in on our course to find out what's really going on in their course.

I greatly appreciate the help of the following people in the preparation of this manuscript: Eleanor Addison who drew the original figures, and Kate MacDougall, Sandra Spinacci, and Miriam Green who typed portions of this manuscript.

I am grateful to Walter Kaufmann-Bühler, the mathematics editor at Springer-Verlag, and Elizabeth Kaplan, the production editor, for their extensive assistance and courtesy during the preparation of this manuscript. It is a pleasure to work with these true professionals.

Finally, I am especially grateful to Joseph P. LaSalle for the encouragement and help he gave me. Thanks again, Joe.

New York City
October, 1977

Martin Braun

Contents

Chapter 1

First-order differential equations 1

1.1 Introduction 1
1.2 First-order linear differential equations 2
1.3 The van Meegeren art forgeries 11
1.4 Separable equations 20
1.5 Population models 27
1.6 An atomic waste disposal problem 37
1.7 The dynamics of tumor growth, mixing problems, and orthogonal trajectories 42
1.8 Exact equations, and why we cannot solve very many differential equations 48
1.9 The existence–uniqueness theorem; Picard iteration 57
1.10 Difference equations, and how to compute the interest due on your student loans 71

Chapter 2

Second-order linear differential equations 76

2.1 Algebraic properties of solutions 76
2.2 Linear equations with constant coefficients 87
  2.2.1 Complex roots 90
  2.2.2 Equal roots; reduction of order 94
2.3 The nonhomogeneous equation 100
2.4 The method of variation of parameters 102
2.5 The method of judicious guessing 106
2.6 Mechanical vibrations 114
  2.6.1 The Tacoma Bridge disaster 122
  2.6.2 Electrical networks 124

2.7 A model for the detection of diabetes 127
2.8 Series solutions 134
2.8.1 Singular points; the method of Frobenius 146
2.9 The method of Laplace transforms 154
2.10 Some useful properties of Laplace transforms 163
2.11 Differential equations with discontinuous right-hand sides 168
2.12 The Dirac delta function 173
2.13 The convolution integral 181
2.14 The method of elimination for systems 186
2.15 A few words about higher-order equations 188

Chapter 3
Systems of differential equations 194

3.1 Algebraic properties of solutions of linear systems 194
3.2 Vector spaces 203
3.3 Dimension of a vector space 209
3.4 Applications of linear algebra to differential equations 221
3.5 The theory of determinants 227
3.6 The eigenvalue–eigenvector method of finding solutions 240
3.7 Complex roots 249
3.8 Equal roots 253
3.9 Fundamental matrix solutions; $e^{\mathbf{A}t}$ 263
3.10 The nonhomogeneous equation; variation of parameters 268
3.11 Solving systems by Laplace transforms 276

Chapter 4
Qualitative theory of differential equations 280

4.1 Introduction 280
4.2 The phase-plane 286
4.3 Lanchester's combat models and the battle of Iwo Jima 291

Appendix A
Some simple facts concerning functions of several variables 300

Appendix B
Sequences and series 302

Answers to odd-numbered exercises 305

Index 317

1 First-order differential equations

1.1 Introduction

This book is a study of differential equations and their applications. A differential equation is a relationship between a function of time and its derivatives. The equations

$$\frac{dy}{dt} = 3y^2 \sin(t+y) \tag{i}$$

and

$$\frac{d^3y}{dt^3} = e^{-y} + t + \frac{d^2y}{dt^2} \tag{ii}$$

are both examples of differential equations. The order of a differential equation is the order of the highest derivative of the function y that appears in the equation. Thus (i) is a first-order differential equation and (ii) is a third-order differential equation. By a solution of a differential equation we will mean a continuous function $y(t)$ which together with its derivatives satisfies the relationship. For example, the function

$$y(t) = 2\sin t - \tfrac{1}{3}\cos 2t$$

is a solution of the second-order differential equation

$$\frac{d^2y}{dt^2} + y = \cos 2t$$

since

$$\frac{d^2}{dt^2}\left(2\sin t - \tfrac{1}{3}\cos 2t\right) + \left(2\sin t - \tfrac{1}{3}\cos 2t\right)$$

$$= \left(-2\sin t + \tfrac{4}{3}\cos 2t\right) + 2\sin t - \tfrac{1}{3}\cos 2t = \cos 2t.$$

Differential equations appear naturally in many areas of science and the humanities. In this book, we will present serious discussions of the applications of differential equations to such diverse and fascinating problems as the detection of art forgeries, the diagnosis of diabetes, the growth of cancerous tumor cells, the battle of Iwo Jima during World War II, and the growth of various populations. Our purpose is to show how researchers have used differential equations to solve, or try to solve, *real life* problems. And while we will discuss some of the great success stories of differential equations, we will also point out their limitations and document some of their failures.

1.2 First-order linear differential equations

We begin by studying first-order differential equations and we will assume that our equation is, or can be put, in the form

$$\frac{dy}{dt}=f(t,y). \tag{1}$$

The problem before us is this: Given $f(t,y)$ find all functions $y(t)$ which satisfy the differential equation (1). We approach this problem in the following manner. A fundamental principle of mathematics is that the way to solve a new problem is to reduce it, in some manner, to a problem that we have already solved. In practice this usually entails successively simplifying the problem until it resembles one we have already solved. Since we are presently in the business of solving differential equations, it is advisable for us to take inventory and list all the differential equations we can solve. If we assume that our mathematical background consists of just elementary calculus then the very sad fact is that the only first-order differential equation we can solve at present is

$$\frac{dy}{dt}=g(t) \tag{2}$$

where g is any integrable function of time. To solve Equation (2) simply integrate both sides with respect to t, which yields

$$y(t)=\int g(t)\,dt+c.$$

Here c is an arbitrary constant of integration, and by $\int g(t)\,dt$ we mean an anti-derivative of g, that is, a function whose derivative is g. Thus, to solve any other differential equation we must somehow reduce it to the form (2). As we will see in Section 1.8, this is impossible to do in most cases. Hence, we will not be able, without the aid of a computer, to solve most differential equations. It stands to reason, therefore, that to find those differential equations that we *can* solve, we should start with very simple equations and not ones like

$$\frac{dy}{dt} = e^{\sin(t-37\sqrt{|y|})}$$

(which incidentally, cannot be solved exactly). Experience has taught us that the "simplest" equations are those which are *linear* in the dependent variable y.

Definition. The general first-order linear differential equation is

$$\frac{dy}{dt} + a(t)y = b(t). \tag{3}$$

Unless otherwise stated, the functions $a(t)$ and $b(t)$ are assumed to be continuous functions of time. We single out this equation and call it linear because the dependent variable y appears by itself, that is, no terms such as e^{-y}, y^3 or $\sin y$ etc. appear in the equation. For example $dy/dt = y^2 + \sin t$ and $dy/dt = \cos y + t$ are both *nonlinear* equations because of the y^2 and $\cos y$ terms respectively.

Now it is not immediately apparent how to solve Equation (3). Thus, we simplify it even further by setting $b(t) = 0$.

Definition. The equation

$$\frac{dy}{dt} + a(t)y = 0 \tag{4}$$

is called the *homogeneous* first-order linear differential equation, and Equation (3) is called the *nonhomogeneous* first-order linear differential equation for $b(t)$ not identically zero.

Fortunately, the homogeneous equation (4) can be solved quite easily. First, divide both sides of the equation by y and rewrite it in the form

$$\frac{\frac{dy}{dt}}{y} = -a(t).$$

Second, observe that

$$\frac{\frac{dy}{dt}}{y} \equiv \frac{d}{dt}\ln|y(t)|$$

where by $\ln|y(t)|$ we mean the natural logarithm of $|y(t)|$. Hence Equation (4) can be written in the form

$$\frac{d}{dt}\ln|y(t)| = -a(t). \tag{5}$$

But this is Equation (2) "essentially" since we can integrate both sides of (5) to obtain that

$$\ln|y(t)| = -\int a(t)\,dt + c_1$$

where c_1 is an arbitrary constant of integration. Taking exponentials of both sides yields

$$|y(t)| = \exp\left(-\int a(t)\,dt + c_1\right) = c\exp\left(-\int a(t)\,dt\right)$$

or

$$\left|y(t)\exp\left(\int a(t)\,dt\right)\right| = c. \tag{6}$$

Now, $y(t)\exp\left(\int a(t)\,dt\right)$ is a continuous function of time and Equation (6) states that its absolute value is constant. But if the absolute value of a continuous function $g(t)$ is constant then g itself must be constant. To prove this observe that if g is not constant, then there exist two different times t_1 and t_2 for which $g(t_1) = c$ and $g(t_2) = -c$. By the intermediate value theorem of calculus g must achieve all values between $-c$ and $+c$ which is impossible if $|g(t)| = c$. Hence, we obtain the equation $y(t)\exp\left(\int a(t)\,dt\right) = c$ or

$$\boxed{y(t) = c\exp\left(-\int a(t)\,dt\right).} \tag{7}$$

Equation (7) is said to be the *general solution* of the homogeneous equation since every solution of (4) must be of this form. Observe that an arbitrary constant c appears in (7). This should not be too surprising. Indeed, we will always expect an arbitrary constant to appear in the general solution of any first-order differential equation. To wit, if we are given dy/dt and we want to recover $y(t)$, then we must perform an integration, and this, of necessity, yields an arbitrary constant. Observe also that Equation (4) has infinitely many solutions; for each value of c we obtain a distinct solution $y(t)$.

Example 1. Find the general solution of the equation $(dy/dt) + 2ty = 0$.

Solution. Here $a(t) = 2t$ so that $y(t) = c\exp\left(-\int 2t\,dt\right) = ce^{-t^2}$.

Example 2. Determine the behavior, as $t \to \infty$, of all solutions of the equation $(dy/dt) + ay = 0$, a constant.

Solution. The general solution is $y(t) = c\exp\left(-\int a\,dt\right) = ce^{-at}$. Hence if $a < 0$, all solutions, with the exception of $y = 0$, approach infinity, and if $a > 0$, all solutions approach zero as $t \to \infty$.

In applications, we are usually not interested in all solutions of (4). Rather, we are looking for the *specific* solution $y(t)$ which at some initial time t_0 has the value y_0. Thus, we want to determine a function $y(t)$ such that

an initial value problem is one with a predetermined condition.

$$\frac{dy}{dt}+a(t)y=0, \qquad y(t_0)=y_0. \tag{8}$$

Equation (8) is referred to as an initial-value problem for the obvious reason that of the totality of all solutions of the differential equation, we are looking for the one solution which initially (at time t_0) has the value y_0. To find this solution we integrate both sides of (5) between t_0 and t. Thus

$$\int_{t_0}^{t}\frac{d}{ds}\ln|y(s)|\,ds=-\int_{t_0}^{t}a(s)\,ds$$

and, therefore

$$\ln|y(t)|-\ln|y(t_0)|=\ln\left|\frac{y(t)}{y(t_0)}\right|=-\int_{t_0}^{t}a(s)\,ds.$$

Taking exponentials of both sides of this equation we obtain that

$$\left|\frac{y(t)}{y(t_0)}\right|=\exp\left(-\int_{t_0}^{t}a(s)\,ds\right)$$

or

$$\left|\frac{y(t)}{y(t_0)}\exp\left(\int_{t_0}^{t}a(s)\,ds\right)\right|=1.$$

The function inside the absolute value sign is a continuous function of time. Thus, by the argument given previously, it is either identically $+1$ or identically -1. To determine which one it is, evaluate it at the point t_0; since

$$\frac{y(t_0)}{y(t_0)}\exp\left(\int_{t_0}^{t_0}a(s)\,ds\right)=1$$

we see that

$$\frac{y(t)}{y(t_0)}\exp\left(\int_{t_0}^{t}a(s)\,ds\right)=1.$$

Hence

$$y(t)=y(t_0)\exp\left(-\int_{t_0}^{t}a(s)\,ds\right)=y_0\exp\left(-\int_{t_0}^{t}a(s)\,ds\right).$$

Example 3. Find the solution of the initial-value problem

$$\frac{dy}{dt}+(\sin t)y=0, \qquad y(0)=\tfrac{3}{2}.$$

Solution. Here $a(t)=\sin t$ so that

$$y(t)=\tfrac{3}{2}\exp\left(-\int_0^t \sin s\,ds\right)=\tfrac{3}{2}e^{(\cos t)-1}.$$

Example 4. Find the solution of the initial-value problem

$$\frac{dy}{dt}+e^{t^2}y=0, \qquad y(1)=2.$$

Solution. Here $a(t)=e^{t^2}$ so that

$$y(t)=2\exp\left(-\int_1^t e^{s^2}ds\right).$$

Now, at first glance this problem would seem to present a very serious difficulty in that we cannot integrate the function e^{s^2} directly. However, this solution is equally as valid and equally as useful as the solution to Example 3. The reason for this is twofold. First, there are very simple numerical schemes to evaluate the above integral to any degree of accuracy with the aid of a computer. Second, even though the solution to Example 3 is given explicitly, we still cannot evaluate it at any time t without the aid of a table of trigonometric functions and some sort of calculating aid, such as a slide rule, electronic calculator or digital computer.

We return now to the nonhomogeneous equation

$$\frac{dy}{dt}+a(t)y=b(t).$$

It should be clear from our analysis of the homogeneous equation that the way to solve the nonhomogeneous equation is to express it in the form

$$\frac{d}{dt}(\text{“something”})=b(t)$$

and then to integrate both sides to solve for "something". However, the expression $(dy/dt)+a(t)y$ does not appear to be the derivative of some simple expression. The next logical step in our analysis therefore should be the following: Can we make the left hand side of the equation to be d/dt of "something"? More precisely, we can multiply both sides of (3) by any continuous function $\mu(t)$ to obtain the equivalent equation

$$\mu(t)\frac{dy}{dt}+a(t)\,\mu(t)y=\mu(t)b(t). \tag{9}$$

(By equivalent equations we mean that every solution of (9) is a solution of (3) and vice-versa.) Thus, can we *choose* $\mu(t)$ so that $\mu(t)(dy/dt)+a(t)\mu(t)y$ is the derivative of some simple expression? The answer to this question is yes, and is obtained by observing that

$$\frac{d}{dt}\mu(t)y=\mu(t)\frac{dy}{dt}+\frac{d\mu}{dt}y.$$

Hence, $\mu(t)(dy/dt)+a(t)\mu(t)y$ will be equal to the derivative of $\mu(t)y$ if and only if $d\mu(t)/dt=a(t)\mu(t)$. But this is a first-order linear homogeneous equation for $\mu(t)$, i.e. $(d\mu/dt)-a(t)\mu=0$ which we already know how to solve, and since we only need one such function $\mu(t)$ we set the constant c in (7) equal to one and take

$$\mu(t)=\exp\left(\int a(t)\,dt\right).$$

For this $\mu(t)$, Equation (9) can be written as

$$\frac{d}{dt}\mu(t)y=\mu(t)b(t). \tag{10}$$

To obtain the general solution of the nonhomogeneous equation (3), that is, to find all solutions of the nonhomogeneous equation, we take the indefinite integral (anti-derivative) of both sides of (10) which yields

$$\mu(t)y=\int\mu(t)b(t)\,dt+c$$

or

$$y=\frac{1}{\mu(t)}\left(\int\mu(t)b(t)\,dt+c\right)=\exp\left(-\int a(t)\,dt\right)\left(\int\mu(t)b(t)\,dt+c\right). \tag{11}$$

Alternately, if we are interested in the specific solution of (3) satisfying the initial condition $y(t_0)=y_0$, that is, if we want to solve the initial-value problem

$$\frac{dy}{dt}+a(t)y=b(t),\qquad y(t_0)=y_0$$

then we can take the definite integral of both sides of (10) between t_0 and t to obtain that

$$\mu(t)y-\mu(t_0)y_0=\int_{t_0}^{t}\mu(s)b(s)\,ds$$

or

$$y=\frac{1}{\mu(t)}\left(\mu(t_0)y_0+\int_{t_0}^{t}\mu(s)b(s)\,ds\right). \tag{12}$$

Remark 1. Notice how we used our knowledge of the solution of the homogeneous equation to find the function $\mu(t)$ which enables us to solve the nonhomogeneous equation. This is an excellent illustration of how we use our knowledge of the solution of a simpler problem to solve a harder problem.

Remark 2. The function $\mu(t)=\exp\left(\int a(t)\,dt\right)$ is called an *integrating factor* for the nonhomogeneous equation since after multiplying both sides by $\mu(t)$ we can immediately integrate the equation to find all solutions.

Remark 3. The reader should not memorize formulae (11) and (12). Rather, we will solve all nonhomogeneous equations by first multiplying both sides by $\mu(t)$, by writing the new left-hand side as the derivative of $\mu(t)y(t)$, and then by integrating both sides of the equation.

Remark 4. An alternative way of solving the initial-value problem $(dy/dt)+a(t)y=b(t)$, $y(t_0)=y_0$ is to find the general solution (11) of (3) and then use the initial condition $y(t_0)=y_0$ to evaluate the constant c. If the function $\mu(t)b(t)$ cannot be integrated directly, though, then we must take the definite integral of (10) to obtain (12), and this equation is then approximated numerically.

Example 5. Find the general solution of the equation $(dy/dt)-2ty=t$.
Solution. Here $a(t)=-2t$ so that

$$\mu(t)=\exp\left(\int a(t)\,dt\right)=\exp\left(-\int 2t\,dt\right)=e^{-t^2}.$$

Multiplying both sides of the equation by $\mu(t)$ we obtain the equivalent equation

$$e^{-t^2}\left(\frac{dy}{dt}-2ty\right)=te^{-t^2} \quad \text{or} \quad \frac{d}{dt}e^{-t^2}y=te^{-t^2}.$$

Hence,

$$e^{-t^2}y=\int te^{-t^2}\,dt+c=\frac{-e^{-t^2}}{2}+c$$

and

$$y(t)=-\tfrac{1}{2}+ce^{t^2}.$$

Example 6. Find the solution of the initial-value problem

$$\frac{dy}{dt}+2ty=t, \qquad y(1)=2.$$

Solution. Here $a(t)=2t$ so that

$$\mu(t)=\exp\left(\int a(t)\,dt\right)=\exp\left(\int 2t\,dt\right)=e^{t^2}.$$

Multiplying both sides of the equation by $\mu(t)$ we obtain that

$$e^{t^2}\left(\frac{dy}{dt}+2ty\right)=te^{t^2} \quad \text{or} \quad \frac{d}{dt}(e^{t^2}y)=te^{t^2}.$$

Hence,

$$\int_1^t \frac{d}{ds} e^{s^2} y(s)\,ds = \int_1^t s e^{s^2}\,ds$$

so that

$$e^{s^2} y(s)\Big|_1^t = \frac{e^{s^2}}{2}\Big|_1^t.$$

Consequently,

$$e^{t^2} y - 2e = \frac{e^{t^2}}{2} - \frac{e}{2}$$

and

$$y = \frac{1}{2} + \frac{3e}{2} e^{-t^2} = \tfrac{1}{2} + \tfrac{3}{2} e^{1-t^2}.$$

Example 7. Find the solution of the initial-value problem

$$\frac{dy}{dt} + y = \frac{1}{1+t^2}, \quad y(2) = 3.$$

Solution. Here $a(t) = 1$, so that

$$\mu(t) = \exp\left(\int a(t)\,dt\right) = \exp\left(\int 1\,dt\right) = e^t.$$

Multiplying both sides of the equation by $\mu(t)$ we obtain that

$$e^t\left(\frac{dy}{dt} + y\right) = \frac{e^t}{1+t^2} \quad \text{or} \quad \frac{d}{dt} e^t y = \frac{e^t}{1+t^2}.$$

Hence

$$\int_2^t \frac{d}{ds} e^s y(s)\,ds = \int_0^t \frac{e^s}{1+s^2}\,ds,$$

so that

$$e^t y - 3e^2 = \int_2^t \frac{e^s}{1+s^2}\,ds$$

and

$$y = e^{-t}\left[3e^2 + \int_2^t \frac{e^s}{1+s^2}\,ds\right].$$

EXERCISES

In each of Problems 1–7 find the general solution of the given differential equation.

1. $\frac{dy}{dt} + y\cos t = 0$

2. $\frac{dy}{dt} + y\sqrt{t}\,\sin t = 0$

3. $\frac{dy}{dt}+\frac{2t}{1+t^2}y=\frac{1}{1+t^2}$

4. $\frac{dy}{dt}+y=te^t$

5. $\frac{dy}{dt}+t^2y=1$

6. $\frac{dy}{dt}+t^2y=t^2$

7. $\frac{dy}{dt}+\frac{t}{1+t^2}y=1-\frac{t^3}{1+t^4}y$

In each of Problems 8–14, find the solution of the given initial-value problem.

8. $\frac{dy}{dt}+\sqrt{1+t^2}\,y=0, \quad y(0)=\sqrt{5}$

9. $\frac{dy}{dt}+\sqrt{1+t^2}\,e^{-t}y=0, \quad y(0)=1$

10. $\frac{dy}{dt}+\sqrt{1+t^2}\,e^{-t}y=0, \quad y(0)=0$

11. $\frac{dy}{dt}-2ty=t, \quad y(0)=1$

12. $\frac{dy}{dt}+ty=1+t, \quad y(\frac{3}{2})=0$

13. $\frac{dy}{dt}+y=\frac{1}{1+t^2}, \quad y(1)=2$

14. $\frac{dy}{dt}-2ty=1, \quad y(0)=1$

15. Find the general solution of the equation

$$(1+t^2)\frac{dy}{dt}+ty=(1+t^2)^{5/2}.$$

(*Hint*: Divide both sides of the equation by $1+t^2$.)

16. Find the solution of the initial-value problem

$$(1+t^2)\frac{dy}{dt}+4ty=t, \qquad y(1)=\tfrac{1}{4}.$$

17. Find a continuous solution of the initial-value problem

$$y'+y=g(t), \qquad y(0)=0$$

where

$$g(t)=\begin{cases}2, & 0\leqslant t\leqslant 1\\ 0, & t>1\end{cases}.$$

18. Show that every solution of the equation $(dy/dt)+ay=be^{-ct}$ where a and c are positive constants and b is any real number approaches zero as t approaches infinity.

19. Given the differential equation $(dy/dt)+a(t)y=f(t)$ with $a(t)$ and $f(t)$ continuous for $-\infty<t<\infty$, $a(t)\geqslant c>0$, and $\lim_{t\to\infty}f(t)=0$, show that every solution tends to zero as t approaches infinity.

When we derived the solution of the nonhomogeneous equation we tacitly assumed that the functions $a(t)$ and $b(t)$ were continuous so that we could perform the necessary integrations. If either of these functions was discontinuous at a point t_1, then we would expect that our solutions might be discontinuous at $t=t_1$. Problems 20–23 illustrate the variety of things that may happen. In Problems 20–22 determine the behavior of all solutions of the given differential equation as $t \to 0$, and in Problem 23 determine the behavior of all solutions as $t \to \pi/2$.

20. $\dfrac{dy}{dt} + \dfrac{1}{t} y = \dfrac{1}{t^2}$

21. $\dfrac{dy}{dt} + \dfrac{1}{\sqrt{t}} y = e^{\sqrt{t}/2}$

22. $\dfrac{dy}{dt} + \dfrac{1}{t} y = \cos t + \dfrac{\sin t}{t}$

23. $\dfrac{dy}{dt} + y \tan t = \sin t \cos t.$

1.3 The Van Meegeren art forgeries

After the liberation of Belgium in World War II, the Dutch Field Security began its hunt for Nazi collaborators. They discovered, in the records of a firm which had sold numerous works of art to the Germans, the name of a banker who had acted as an intermediary in the sale to Goering of the painting "Woman Taken in Adultery" by the famed 17th century Dutch painter Jan Vermeer. The banker in turn revealed that he was acting on behalf of a third rate Dutch painter H. A. Van Meegeren, and on May 29, 1945 Van Meegeren was arrested on the charge of collaborating with the enemy. On July 12, 1945 Van Meegeren startled the world by announcing from his prison cell that he had never sold "Woman Taken in Adultery" to Goering. Moreover, he stated that this painting and the very famous and beautiful "Disciples at Emmaus", as well as four other presumed Vermeers and two de Hooghs (a 17th century Dutch painter) were his own work. Many people, however, thought that Van Meegeren was only lying to save himself from the charge of treason. To prove his point, Van Meegeren began, while in prison, to forge the Vermeer painting "Jesus Amongst the Doctors" to demonstrate to the skeptics just how good a forger of Vermeer he was. The work was nearly completed when Van Meegeren learned that a charge of forgery had been substituted for that of collaboration. He, therefore, refused to finish and age the painting so that hopefully investigators would not uncover his secret of aging his forgeries. To settle the question an international panel of distinguished chemists, physicists and art historians was appointed to investigate the matter. The panel took x-rays of the paintings to determine whether other paintings were underneath them. In addition, they analyzed the pigments (coloring materials) used in the paint, and examined the paintings for certain signs of old age.

Now, Van Meegeren was well aware of these methods. To avoid detection, he scraped the paint from old paintings that were not worth much, just to get the canvas, and he tried to use pigments that Vermeer would have used. Van Meegeren also knew that old paint was extremely hard, and impossible to dissolve. Therefore, he very cleverly mixed a chemical, phenoformaldehyde, into the paint, and this hardened into bakelite when the finished painting was heated in an oven.

However, Van Meegeren was careless with several of his forgeries, and the panel of experts found traces of the modern pigment cobalt blue. In addition, they also detected the phenoformaldehyde, which was not discovered until the turn of the 19th century, in several of the paintings. On the basis of this evidence Van Meegeren was convicted, of forgery, on October 12, 1947 and sentenced to one year in prison. While in prison he suffered a heart attack and died on December 30, 1947.

However, even following the evidence gathered by the panel of experts, many people still refused to believe that the famed "Disciples at Emmaus" was forged by Van Meegeren. Their contention was based on the fact that the other alleged forgeries and Van Meegeren's nearly completed "Jesus Amongst the Doctors" were of a very inferior quality. Surely, they said, the creator of the beautiful "Disciples at Emmaus" could not produce such inferior pictures. Indeed, the "Disciples at Emmaus" was certified as an authentic Vermeer by the noted art historian A. Bredius and was bought by the Rembrandt Society for $170,000. The answer of the panel to these skeptics was that because Van Meegeren was keenly disappointed by his lack of status in the art world, he worked on the "Disciples at Emmaus" with the fierce determination of proving that he was better than a third rate painter. After producing such a masterpiece his determination was gone. Moreover, after seeing how easy it was to dispose of the "Disciples at Emmaus" he devoted less effort to his subsequent forgeries. This explanation failed to satisfy the skeptics. They demanded a thoroughly scientific and conclusive proof that the "Disciples at Emmaus" was indeed a forgery. This was done recently in 1967 by scientists at Carnegie Mellon University, and we would now like to describe their work.

The key to the dating of paintings and other materials such as rocks and fossils lies in the phenomenon of radioactivity discovered at the turn of the century. The physicist Rutherford and his colleagues showed that the atoms of certain "radioactive" elements are unstable and that within a given time period a fixed proportion of the atoms spontaneously disintegrates to form atoms of a new element. Because radioactivity is a property of the atom, Rutherford showed that the radioactivity of a substance is directly proportional to the number of atoms of the substance present. Thus, if $N(t)$ denotes the number of atoms present at time t, then dN/dt, the number of atoms that disintegrate per unit time is proportional to N, that is,

$$\frac{dN}{dt} = -\lambda N. \tag{1}$$

The constant λ which is positive, is known as the decay constant of the substance. The larger λ is, of course, the faster the substance decays. One measure of the rate of disintegration of a substance is its *half-life* which is defined as the time required for half of a given quantity of radioactive atoms to decay. To compute the half-life of a substance in terms of λ, assume that at time t_0, $N(t_0) = N_0$. Then, the solution of the initial-value problem $dN/dt = -\lambda N$, $N(t_0) = N_0$ is

$$N(t) = N_0 \exp\left(-\lambda \int_{t_0}^{t} ds\right) = N_0 e^{-\lambda(t-t_0)}$$

or $N/N_0 = \exp(-\lambda(t-t_0))$. Taking logarithms of both sides we obtain that

$$-\lambda(t-t_0) = \ln\frac{N}{N_0}. \tag{2}$$

Now, if $N/N_0 = \frac{1}{2}$ then $-\lambda(t-t_0) = \ln\frac{1}{2}$ so that

$$(t-t_0) = \frac{\ln 2}{\lambda} = \frac{0.6931}{\lambda}. \tag{3}$$

Thus, the half-life of a substance is $\ln 2$ divided by the decay constant λ. The dimension of λ, which we suppress for simplicity of writing, is reciprocal time. If t is measured in years then λ has the dimension of reciprocal years, and if t is measured in minutes, then λ has the dimension of reciprocal minutes. The half-lives of many substances have been determined and recorded. For example, the half-life of carbon-14 is 5568 years and the half-life of uranium-238 is 4.5 billion years.

Now the basis of "radioactive dating" is essentially the following. From Equation (2) we can solve for $t - t_0 = 1/\lambda \ln(N_0/N)$. If t_0 is the time the substance was initially formed or manufactured, then the age of the substance is $1/\lambda \ln(N_0/N)$. The decay constant λ is known or can be computed, in most instances. Moreover, we can usually evaluate N quite easily. Thus, if we knew N_0 we could determine the age of the substance. But this is the real difficulty of course, since we usually do not know N_0. In some instances though, we can either determine N_0 indirectly, or else determine certain suitable ranges for N_0, and such is the case for the forgeries of Van Meegeren.

We begin with the following well-known facts of elementary chemistry. Almost all rocks in the earth's crust contain a small quantity of uranium. The uranium in the rock decays to another radioactive element, and that one decays to another and another, and so forth (see Figure 1) in a series of elements that results in lead, which is not radioactive. The uranium (whose half-life is over four billion years) keeps feeding the elements following it in the series, so that as fast as they decay, they are replaced by the elements before them.

Now, all paintings contain a small amount of the radioactive element lead-210 (^{210}Pb), and an even smaller amount of radium-226 (^{226}Ra), since these elements are contained in white lead (lead oxide), which is a pigment that artists have used for over 2000 years. For the analysis which follows, it is important to note that white lead is made from lead metal, which, in turn, is extracted from a rock called lead ore, in a process called smelting. In this process, the lead-210 in the ore goes along with the lead metal. However, 90–95% of the radium and its descendants are removed with

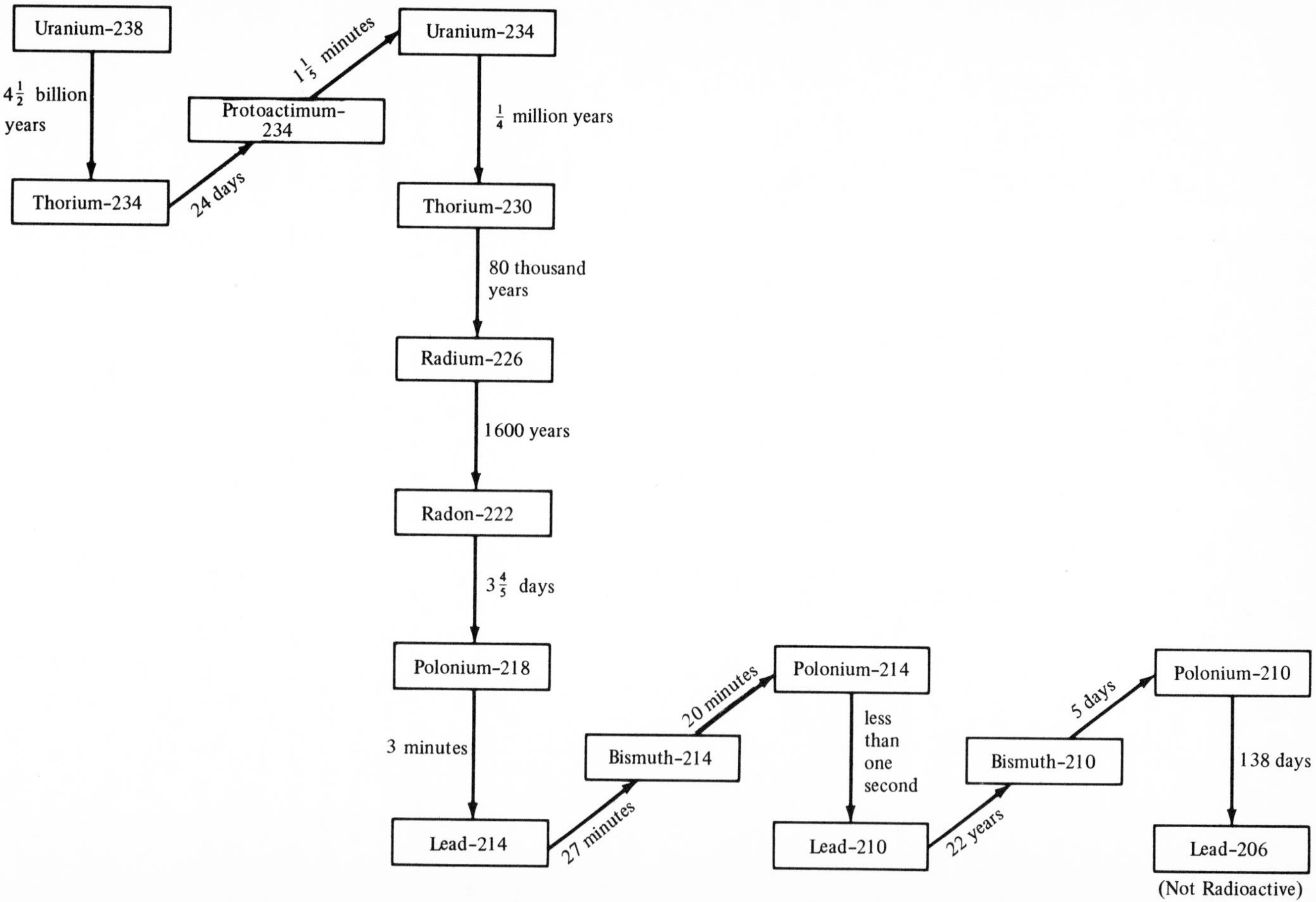

Figure 1. The Uranium series. (The times shown on the arrows are the half-lives of each step.)

other waste products in a material called slag. Thus, most of the supply of lead-210 is cut off and it begins to decay very rapidly, with a half-life of 22 years. This process continues until the lead-210 in the white lead is once more in radioactive equilibrium with the small amount of radium present, i.e. the disintegration of the lead-210 is exactly balanced by the disintegration of the radium.

Let us now use this information to compute the amount of lead-210 present in a sample in terms of the amount originally present at the time of manufacture. Let $y(t)$ be the amount of lead-210 per gram of white lead at time t, y_0 the amount of lead-210 per gram of white lead present at the time of manufacture t_0, and $r(t)$ the number of disintegrations of radium-226 per minute per gram of white lead, at time t. If λ is the decay constant for lead-210, then

$$\frac{dy}{dt} = -\lambda y + r(t), \qquad y(t_0) = y_0. \tag{4}$$

Since we are only interested in a time period of at most 300 years we may assume that the radium-226, whose half-life is 1600 years, remains constant, so that $r(t)$ is a constant r. Multiplying both sides of the differential equation by the integrating factor $\mu(t) = e^{\lambda t}$ we obtain that

$$\frac{d}{dt} e^{\lambda t} y = r e^{\lambda t}.$$

Hence

$$e^{\lambda t} y(t) - e^{\lambda t_0} y_0 = \frac{r}{\lambda}\left(e^{\lambda t} - e^{\lambda t_0}\right)$$

or

$$y(t) = \frac{r}{\lambda}\left(1 - e^{-\lambda(t - t_0)}\right) + y_0 e^{-\lambda(t - t_0)}. \tag{5}$$

Now $y(t)$ and r can be easily measured. Thus, if we knew y_0 we could use Equation (5) to compute $(t - t_0)$ and consequently, we could determine the age of the painting. As we pointed out, though, we cannot measure y_0 directly. One possible way out of this difficulty is to use the fact that the original quantity of lead-210 was in radioactive equilibrium with the larger amount of radium-226 in the ore from which the metal was extracted. Let us, therefore, take samples of different ores and count the number of disintegrations of the radium-226 in the ores. This was done for a variety of ores and the results are given in Table 1 below. These numbers vary from 0.18 to 140. Consequently, the number of disintegrations of the lead-210 per minute per gram of white lead at the time of manufacture will vary from 0.18 to 140. This implies that y_0 will also vary over a very large interval, since the number of disintegrations of lead-210 is proportional to the amount present. Thus, we cannot use Equation (5) to obtain an accurate, or even a crude estimate, of the age of a painting.

Table 1. Ore and ore concentrate samples. All disintegration rates are per gram of white lead.

Description and Source		Disintegrations per minute of ^{226}Ra
Ore concentrate	(Oklahoma–Kansas)	4.5
Crushed raw ore	(S.E. Missouri)	2.4
Ore concentrate	(S.E. Missouri)	0.7
Ore concentrate	(Idaho)	2.2
Ore concentrate	(Idaho)	0.18
Ore concentrate	(Washington)	140.0
Ore concentrate	(British Columbia)	1.9
Ore concentrate	(British Columbia)	0.4
Ore concentrate	(Bolivia)	1.6
Ore concentrate	(Australia)	1.1

However, we can still use Equation (5) to distinguish between a 17th century painting and a modern forgery. The basis for this statement is the simple observation that if the paint is very old compared to the 22 year half-life of lead, then the amount of radioactivity from the lead-210 in the paint will be nearly equal to the amount of radioactivity from the radium in the paint. On the other hand, if the painting is modern (approximately 20 years old, or so) then the amount of radioactivity from the lead-210 will be much greater than the amount of radioactivity from the radium.

We make this argument precise in the following manner. Let us assume that the painting in question is either very new or about 300 years old. Set $t-t_0=300$ years in (5). Then, after some simple algebra, we see that

$$\lambda y_0=\lambda y(t)e^{300\lambda}-r(e^{300\lambda}-1). \tag{6}$$

If the painting is indeed a modern forgery, then λy_0 will be absurdly large. To determine what is an absurdly high disintegration rate we observe (see Exercise 1) that if the lead-210 decayed originally (at the time of manufacture) at the rate of 100 disintegrations per minute per gram of white lead, then the ore from which it was extracted had a uranium content of approximately 0.014 per cent. This is a fairly high concentration of uranium since the average amount of uranium in rocks of the earth's crust is about 2.7 parts per million. On the other hand, there are some very rare ores in the Western Hemisphere whose uranium content is 2-3 per cent. To be on the safe side, we will say that a disintegration rate of lead-210 is certainly absurd if it exceeds 30,000 disintegrations per minute per gram of white lead.

To evaluate λy_0, we must evaluate the present disintegration rate, $\lambda y(t)$, of the lead-210, the disintegration rate r of the radium-226, and $e^{300\lambda}$. Since the disintegration rate of polonium-210 (^{210}Po) equals that of lead-210 after several years, and since it is easier to measure the disintegration rate of polonium-210, we substitute these values for those of lead-210. To compute

$e^{300\lambda}$, we observe from (3) that $\lambda = (\ln 2/22)$. Hence

$$e^{300\lambda} = e^{(300/22)\ln 2} = 2^{(150/11)}.$$

The disintegration rates of polonium-210 and radium-226 were measured for the "Disciples at Emmaus" and various other alleged forgeries and are given in Table 2 below.

Table 2. Paintings of questioned authorship. All disintegration rates are per minute, per gram of white lead.

Description	^{210}Po disintegration	^{226}Ra disintegration
"Disciples at Emmaus"	8.5	0.8
"Washing of Feet"	12.6	0.26
"Woman Reading Music"	10.3	0.3
"Woman Playing Mandolin"	8.2	0.17
"Lace Maker"	1.5	1.4
"Laughing Girl"	5.2	6.0

If we now evaluate λy_0 from (6) for the white lead in the painting "Disciples at Emmaus" we obtain that

$$\begin{aligned}\lambda y_0 &= (8.5)2^{150/11} - 0.8(2^{150/11} - 1)\\ &= 98{,}050\end{aligned}$$

which is unacceptably large. Thus, this painting must be a modern forgery. By a similar analysis, (see Exercises 2–4) the paintings "Washing of Feet", "Woman Reading Music" and "Woman Playing Mandolin" were indisputably shown to be faked Vermeers. On the other hand, the paintings "Lace Maker" and "Laughing Girl" cannot be recently forged Vermeers, as claimed by some experts, since for these two paintings, the polonium-210 is very nearly in radioactive equilibrium with the radium-226, and no such equilibrium has been observed in any samples from 19th or 20th century paintings.

References

Coremans, P., Van Meegeren's Faked Vermeers and De Hooghs, Meulenhoff, Amsterdam, 1949.

Keisch, B., Feller, R. L., Levine, A. S., Edwards, P. R., Dating and Authenticating Works of Art by Measurement of Natural Alpha Emitters, Science (155), 1238–1241, March 1967.

Keisch, B., Dating Works of Art through Their Natural Radioactivity: Improvements and Applications, Science, 160, 413–415, April 1968.

EXERCISES

1. In this exercise we show how to compute the concentration of uranium in an ore from the dpm/(g of Pb) of the lead-210 in the ore.

(a) The half-life of uranium-238 is 4.51×10^9 years. Since this half-life is so large, we may assume that the amount of uranium in the ore is constant over a period of two to three hundred years. Let $N(t)$ denote the number of atoms of ^{238}U per gram of ordinary lead in the ore at time t. Since the lead-210 is in radioactive equilibrium with the uranium-238 in the ore, we know that $dN/dt = -\lambda N = -100$ dpm/g of Pb at time t_0. Show that there are 3.42×10^{17} atoms of uranium-238 per gram of ordinary lead in the ore at time t_0. (Hint: 1 year = 525,600 minutes.)

(b) Using the fact that one mole of uranium-238 weighs 238 grams, and that there are 6.02×10^{23} atoms in a mole, show that the concentration of uranium in the ore is approximately 0.014 percent.

For each of the paintings 2, 3, and 4 use the data in Table 2 to compute the disintegrations per minute of the original amount of lead-210 per gram of white lead, and conclude that each of these paintings is a forged Vermeer.

2. "Washing of Feet"

3. "Woman Reading Music"

4. "Woman Playing Mandolin"

5. The following problem describes a very accurate derivation of the age of uranium.

(a) Let $N_{238}(t)$ and $N_{235}(t)$ denote the number of atoms of ^{238}U and ^{235}U at time t in a given sample of uranium, and let $t=0$ be the time this sample was created. By the radioactive decay law,

$$\frac{d}{dt}N_{238}(t) = \frac{-\ln 2}{(4.5)10^9}N_{238}(t),$$

$$\frac{d}{dt}N_{235}(t) = \frac{-\ln 2}{0.707(10)^9}N_{235}(t).$$

Solve these equations for $N_{238}(t)$ and $N_{235}(t)$ in terms of their original numbers $N_{238}(0)$ and $N_{235}(0)$.

(b) In 1946 the ratio of $^{238}U/^{235}U$ in any sample was 137.8. Assuming that equal amounts of ^{238}U and ^{235}U appeared in any sample at the time of its creation, show that the age of uranium is 5.96×10^9 years. This figure is universally accepted as the age of uranium.

6. In a samarskite sample discovered recently, there was 3 grams of Thorium (^{232}TH). Thorium decays to lead-208 (^{208}Pb) through the reaction $^{232}Th \rightarrow {}^{208}Pb + 6(4\,{}^4He)$. It was determined that 0.0376 of a gram of lead-208 was produced by the disintegration of the original Thorium in the sample. Given that the half-life of Thorium is 13.9 billion years, derive the age of this samarskite sample. (Hint: 0.0376 grams of ^{208}Pb is the product of the decay of $(232/208) \times$ 0.0376 grams of Thorium.)

One of the most accurate ways of dating archaeological finds is the method of carbon-14 (^{14}C) dating discovered by Walter Libby around 1949. The basis of this method is delightfully simple: The atmosphere of the earth is continuously bombarded by cosmic rays. These cosmic rays produce neutrons in the earth's atmosphere, and these neutrons combine with nitrogen to produce ^{14}C, which is usually called radiocarbon, since it decays radioactively. Now, this radiocarbon is incorporated in carbon dioxide and thus moves through the atmosphere to be absorbed by plants. Animals, in turn, build radiocarbon into their tissues by eating the plants. In living tissue, the rate of ingestion of ^{14}C exactly balances the rate of disintegration of ^{14}C. When an organism dies, though, it ceases to ingest carbon-14 and thus its ^{14}C concentration begins to decrease through disintegration of the ^{14}C present. Now, it is a fundamental assumption of physics that the rate of bombardment of the earth's atmosphere by cosmic rays has always been constant. This implies that the original rate of disintegration of the ^{14}C in a sample such as charcoal is the same as the rate measured today.* This assumption enables us to determine the age of a sample of charcoal. Let $N(t)$ denote the amount of carbon-14 present in a sample at time t, and N_0 the amount present at time $t=0$ when the sample was formed. If λ denotes the decay constant of ^{14}C (the half-life of carbon-14 is 5568 years) then $dN(t)/dt = -\lambda N(t)$, $N(0)=N_0$. Consequently, $N(t)=N_0e^{-\lambda t}$. Now the present rate $R(t)$ of disintegration of the ^{14}C in the sample is given by $R(t)=\lambda N(t)=\lambda N_0e^{-\lambda t}$ and the original rate of disintegration is $R(0)=\lambda N_0$. Thus, $R(t)/R(0)=e^{-\lambda t}$ so that $t=(1/\lambda)\ln[R(0)/R(t)]$. Hence if we measure $R(t)$, the present rate of disintegration of the ^{14}C in the charcoal, and observe that $R(0)$ must equal the rate of disintegration of the ^{14}C in a comparable amount of living wood, then we can compute the age t of the charcoal. The following two problems are real life illustrations of this method.

7. Charcoal from the occupation level of the famous Lascaux Cave in France gave an average count in 1950 of 0.97 disintegrations per minute per gram. Living wood gave 6.68 disintegrations. Estimate the date of occupation and hence the probable date of the remarkable paintings in the Lascaux Cave.

8. In the 1950 excavation at Nippur, a city of Babylonia, charcoal from a roof beam gave a count of 4.09 disintegrations per minute per gram. Living wood gave 6.68 disintegrations. Assuming that this charcoal was formed during the time of Hammurabi's reign, find an estimate for the likely time of Hamurabi's succession.

9. Many savings banks now advertise continuous compounding of interest. This means that the amount of money $P(t)$ on deposit at time t, satisfies the differential equation $dP(t)/dt=rP(t)$ where r is the annual interest rate and t is measured in years. Let P_0 denote the original principal.

*Since the mid 1950's the testing of nuclear weapons has significantly increased the amount of radioactive carbon in our atmosphere. Ironically this unfortunate state of affairs provides us with yet another extremely powerful method of detecting art forgeries. To wit, many artists' materials, such as linseed oil and canvas paper, come from plants and animals, and so will contain the same concentration of carbon-14 as the atmosphere at the time the plant or animal dies. Thus linseed oil (which is derived from the flax plant) that was produced during the last few years will contain a much greater concentration of carbon-14 than linseed oil produced before 1950.

(a) Show that $P(1)=P_0e^r$.
(b) Let $r=0.0575$, 0.065, 0.0675, and 0.075. Show that $e^r=1.05919$, 1.06716, 1.06983, and 1.07788, respectively. Thus, the effective annual yield on interest rates of $5\frac{3}{4}$, $6\frac{1}{2}$, $6\frac{3}{4}$, and $7\frac{1}{2}\%$ should be 5.919, 6.716, 6.983, and 7.788%, respectively. Most banks, however, advertise effective annual yields of 6, 6.81, 7.08, and 7.9%, respectively. The reason for this discrepancy is that banks calculate a daily rate of interest based on 360 days, and they pay interest for each day money is on deposit. For a year, one gets five extra days. Thus, we must multiply the annual yields of 5.919, 6.716, 6.983, and 7.788% by 365/360, and then we obtain the advertised values.
(c) It is interesting to note that the Old Colony Cooperative Bank in Rhode Island advertises an effective annual yield of 6.72% on an annual interest rate of $6\frac{1}{2}\%$ (the lower value), and an effective annual yield of 7.9% on an annual interest rate of $7\frac{1}{2}\%$. Thus they are inconsistent.

10. The presence of toxins in a certain medium destroys a strain of bacteria at a rate jointly proportional to the number of bacteria present and to the amount of toxin. Call the constant of proportionality a. If there were no toxins present, the bacteria would grow at a rate proportional to the amount present. Call this constant of proportionality b. Assume that the amount T of toxin is increasing at a constant rate c, that is, $dT/dt=c$, and that the production of toxins begins at time $t=0$. Let $y(t)$ denote the number of living bacteria present at time t.
(a) Find a first-order differential equation satisfied by $y(t)$.
(b) Solve this differential equation to obtain $y(t)$. What happens to $y(t)$ as t approaches ∞?

1.4 Separable equations

We solved the first-order linear homogeneous equation

$$\frac{dy}{dt}+a(t)y=0 \tag{1}$$

by dividing both sides of the equation by $y(t)$ to obtain the equivalent equation

$$\frac{1}{y(t)}\frac{dy(t)}{dt}=-a(t) \tag{2}$$

and observing that Equation (2) can be written in the form

$$\frac{d}{dt}\ln|y(t)|=-a(t). \tag{3}$$

We then found $\ln|y(t)|$, and consequently $y(t)$, by integrating both sides of (3). In an exactly analogous manner, we can solve the more general differential equation

$$\frac{dy}{dt}=\frac{g(t)}{f(y)} \tag{4}$$

where f and g are continuous functions of y and t. This equation, and any other equation which can be put into this form, is said to be separable. To solve (4), we first multiply both sides by $f(y)$ to obtain the equivalent equation

$$f(y)\frac{dy}{dt}=g(t). \tag{5}$$

Then, we observe that (5) can be written in the form

$$\frac{d}{dt}F(y(t))=g(t) \tag{6}$$

where $F(y)$ is any anti-derivative of $f(y)$; i.e., $F(y)=\int f(y)dy$. Consequently,

$$F(y(t))=\int g(t)dt+c \tag{7}$$

where c is an arbitrary constant of integration, and we solve for $y=y(t)$ from (7) to find the general solution of (4).

Example 1. Find the general solution of the equation $dy/dt=t^2/y^2$.
Solution. Multiplying both sides of this equation by y^2 gives

$$y^2\frac{dy}{dt}=t^2, \quad \text{or} \quad \frac{d}{dt}\frac{y^3(t)}{3}=t^2.$$

Hence, $y^3(t)=t^3+c$ where c is an arbitrary constant, and $y(t)=(t^3+c)^{1/3}$.

Example 2. Find the general solution of the equation

$$e^y\frac{dy}{dt}-t-t^3=0.$$

Solution. This equation can be written in the form

$$\frac{d}{dt}e^{y(t)}=t+t^3$$

and thus $e^{y(t)}=t^2/2+t^4/4+c$. Taking logarithms of both sides of this equation gives $y(t)=\ln(t^2/2+t^4/4+c)$.

In addition to the differential equation (4), we will often impose an initial condition on $y(t)$ of the form $y(t_0)=y_0$. The differential equation (4) together with the initial condition $y(t_0)=y_0$ is called an initial-value problem. We can solve an initial-value problem two different ways. Either we use the initial condition $y(t_0)=y_0$ to solve for the constant c in (7), or else we integrate both sides of (6) between t_0 and t to obtain that

$$F(y(t))-F(y_0)=\int_{t_0}^{t}g(s)\,ds. \tag{8}$$

If we now observe that

$$F(y)-F(y_0)=\int_{y_0}^{y}f(r)\,dr, \tag{9}$$

then we can rewrite (8) in the simpler form

$$\int_{y_0}^{y} f(r)\,dr = \int_{t_0}^{t} g(s)\,ds. \tag{10}$$

Example 3. Find the solution $y(t)$ of the initial-value problem

$$e^y\frac{dy}{dt} - (t + t^3) = 0, \qquad y(1) = 1.$$

Solution. Method (i). From Example 2, we know that the general solution of this equation is $y = \ln(t^2/2 + t^4/4 + c)$. Setting $t = 1$ and $y = 1$ gives $1 = \ln(3/4 + c)$, or $c = e - 3/4$. Hence, $y(t) = \ln(e - 3/4 + t^2/2 + t^4/4)$.
Method (ii). From (10),

$$\int_1^y e^r\,dr = \int_1^t (s + s^3)\,ds.$$

Consequently,

$$e^y - e = \frac{t^2}{2} + \frac{t^4}{4} - \frac{1}{2} - \frac{1}{4}, \quad \text{and} \quad y(t) = \ln(e - 3/4 + t^2/2 + t^4/4).$$

Example 4. Solve the initial-value problem $dy/dt = 1 + y^2$, $y(0) = 0$.
Solution. Divide both sides of the differential equation by $1 + y^2$ to obtain the equivalent equation $1/(1 + y^2)dy/dt = 1$. Then, from (10)

$$\int_0^y \frac{dr}{1 + r^2} = \int_0^t ds.$$

Consequently, $\arctan y = t$, and $y = \tan t$.

The solution $y = \tan t$ of the above problem has the disturbing property that it goes to $\pm\infty$ at $t = \pm\pi/2$. And what's even more disturbing is the fact that there is nothing at all in this initial-value problem which even hints to us that there is any trouble at $t = \pm\pi/2$. The sad fact of life is that solutions of perfectly nice differential equations can go to infinity in finite time. Thus, solutions will usually exist only on a finite open interval $a < t < b$, rather than for all time. Moreover, as the following example shows, different solutions of the same differential equation usually go to infinity at different times.

Example 5. Solve the initial-value problem $dy/dt = 1 + y^2$, $y(0) = 1$.
Solution. From (10)

$$\int_1^y \frac{dr}{1 + r^2} = \int_0^t ds.$$

Consequently, $\arctan y - \arctan 1 = t$, and $y = \tan(t + \pi/4)$. This solution exists on the open interval $-3\pi/4 < t < \pi/4$.

Example 6. Find the solution $y(t)$ of the initial-value problem

$$y\frac{dy}{dt}+(1+y^2)\sin t=0, \qquad y(0)=1.$$

Solution. Dividing both sides of the differential equation by $1+y^2$ gives

$$\frac{y}{1+y^2}\frac{dy}{dt}=-\sin t.$$

Consequently,

$$\int_1^y \frac{r\,dr}{1+r^2}=\int_0^t -\sin s\,ds,$$

so that

$$\tfrac{1}{2}\ln(1+y^2)-\tfrac{1}{2}\ln 2=\cos t-1.$$

Solving this equation for $y(t)$ gives

$$y(t)=\pm\left(2e^{-4\sin^2 t/2}-1\right)^{1/2}$$

To determine whether we take the plus or minus branch of the square root, we note that $y(0)$ is positive. Hence,

$$y(t)=\left(2e^{-4\sin^2 t/2}-1\right)^{1/2}.$$

This solution is only defined when

$$2e^{-4\sin^2 t/2}\geqslant 1$$

or

$$e^{4\sin^2 t/2}\leqslant 2. \tag{11}$$

Since the logarithm function is monotonic increasing, we may take logarithms of both sides of (11) and still preserve the inequality. Thus, $4\sin^2 t/2 \leqslant \ln 2$, which implies that

$$\left|\frac{t}{2}\right|\leqslant \arcsin\frac{\sqrt{\ln 2}}{2}$$

Therefore, $y(t)$ only exists on the open interval $(-a,a)$ where

$$a=2\arcsin\left[\sqrt{\ln 2}\;/2\right].$$

Now, this appears to be a new difficulty associated with nonlinear equations, since $y(t)$ just "disappears" at $t=\pm a$, without going to infinity. However, this apparent difficulty can be explained quite easily, and moreover, can even be anticipated, if we rewrite the differential equation above in the standard form

$$\frac{dy}{dt}=-\frac{(1+y^2)\sin t}{y}$$

Notice that this differential equation is not defined when $y=0$. Therefore, if a solution $y(t)$ achieves the value zero at some time $t=t^*$, then we cannot expect it to be defined for $t>t^*$. This is exactly what happens here, since $y(\pm a)=0$.

Example 7. Solve the initial-value problem $dy/dt=(1+y)t$, $y(0)=-1$.
Solution. In this case, we cannot divide both sides of the differential equation by $1+y$, since $y(0)=-1$. However, it is easily seen that $y(t)=-1$ is one solution of this initial-value problem, and in Section 1.9 we show that it is the only solution. More generally, consider the initial-value problem $dy/dt=f(y)g(t)$, $y(t_0)=y_0$, where $f(y_0)=0$. Certainly, $y(t)=y_0$ is one solution of this initial-value problem, and in Section 1.9 we show that it is the only solution if $\partial f/\partial y$ exists and is continuous.

Example 8. Solve the initial-value problem

$$(1+e^y)dy/dt=\cos t, \qquad y(\pi/2)=3.$$

Solution. From (10),

$$\int_3^y (1+e^r)\,dr=\int_{\pi/2}^t \cos s\,ds$$

so that $y+e^y=2+e^3+\sin t$. This equation cannot be solved explicitly for y as a function of t. Indeed, most separable equations cannot be solved explicitly for y as a function of t. Thus, when we say that

$$y+e^y=2+e^3+\sin t$$

is the solution of this initial-value problem, we really mean that it is an implicit, rather than an explicit solution. This does not present us with any difficulties in applications, since we can always find $y(t)$ numerically with the aid of a digital computer.

Example 9. Find all solutions of the differential equation $dy/dt=-t/y$.
Solution. Multiplying both sides of the differential equation by y gives $y\,dy/dt=-t$. Hence

$$y^2+t^2=c^2. \tag{12}$$

Now, the curves (12) are *closed*, and we cannot solve for y as a *single-valued* function of t. The reason for this difficulty, of course, is that the differential equation is not defined when $y=0$. Nevertheless, the circles $t^2+y^2=c^2$ are perfectly well defined, even when $y=0$. Thus, we will call the circles $t^2+y^2=c^2$ *solution curves* of the differential equation

$$dy/dt=-t/y.$$

More generally, we will say that any curve defined by (7) is a solution curve of (4).

Exercises

In each of Problems 1–5, find the general solution of the given differential equation.

1. $(1+t^2)\dfrac{dy}{dt}=1+y^2$. *Hint*: $\tan(x+y)=\dfrac{\tan x+\tan y}{1-\tan x\tan y}$.

2. $\dfrac{dy}{dt}=(1+t)(1+y)$

3. $\dfrac{dy}{dt}=1-t+y^2-ty^2$

4. $\dfrac{dy}{dt}=e^{t+y+3}$

5. $\cos y\sin t\dfrac{dy}{dt}=\sin y\cos t$

In each of Problems 6–12, solve the given initial-value problem, and determine the interval of existence of each solution.

6. $t^2(1+y^2)+2y\dfrac{dy}{dt}=0,\quad y(0)=1$

7. $\dfrac{dy}{dt}=\dfrac{2t}{y+yt^2},\quad y(2)=3$

8. $(1+t^2)^{1/2}\dfrac{dy}{dt}=ty^3(1+t^2)^{-1/2},\quad y(0)=1$

9. $\dfrac{dy}{dt}=\dfrac{3t^2+4t+2}{2(y-1)}\quad y(0)=-1$

10. $\cos y\dfrac{dy}{dt}=\dfrac{-t\sin y}{1+t^2},\quad y(1)=\pi/2$

11. $\dfrac{dy}{dt}=k(a-y)(b-y),\quad y(0)=0,\ a,b>0$

12. $3t\dfrac{dy}{dt}=y\cos t,\quad y(1)=0$

13. Any differential equation of the form $dy/dt=f(y)$ is separable. Thus, we can solve all those first-order differential equations in which time does not appear explicitly. Now, suppose we have a differential equation of the form $dy/dt=f(y/t)$, such as, for example, the equation $dy/dt=\sin(y/t)$. Differential equations of this form are called homogeneous equations. Since the right-hand side only depends on the single variable y/t, it suggests itself to make the substitution $y/t=v$ or $y=tv$.
(a) Show that this substitution replaces the equation $dy/dt=f(y/t)$ by the equivalent equation $t\,dv/dt+v=f(v)$, which is separable.
(b) Find the general solution of the equation $dy/dt=2(y/t)+(y/t)^2$.

14. Determine whether each of the following functions of t and y can be expressed as a function of the single variable y/t.

(a) $\dfrac{y^2+2ty}{y^2}$ (b) $\dfrac{y^3+t^3}{yt^2+y^3}$

(c) $\dfrac{y^3+t^3}{t^2+y^3}$ (d) $\ln y-\ln t+\dfrac{t+y}{t-y}$

(e) $\dfrac{e^{t+y}}{e^{t-y}}$ (f) $\ln\sqrt{t+y}-\ln\sqrt{t-y}$

(g) $\sin\dfrac{t+y}{t-y}$ (h) $\dfrac{(t^2+7ty+9y^2)^{1/2}}{3t+5y}$.

15. Solve the initial-value problem $t(dy/dt)=y+\sqrt{t^2+y^2}$, $y(1)=0$.

Find the general solution of the following differential equations.

16. $2ty\dfrac{dy}{dt}=3y^2-t^2$ **17.** $(t-\sqrt{ty}\,)\dfrac{dy}{dt}=y$

18. $\dfrac{dy}{dt}=\dfrac{t+y}{t-y}$.

19. $e^{t/y}(y-t)\dfrac{dy}{dt}+y(1+e^{t/y})=0$

$\left[\textit{Hint}: \displaystyle\int\frac{v-1}{ve^{-1/v}+v^2}\,dv=\ln(1+ve^{1/v})\right]$

20. Consider the differential equation

$$\frac{dy}{dt}=\frac{t+y+1}{t-y+3}. \qquad (*)$$

We could solve this equation if the constants 1 and 3 were not present. To eliminate these constants, we make the substitution $t=T+h$, $y=Y+k$.

(a) Determine h and k so that (*) can be written in the form $dY/dT=(T+Y)/T-Y$.

(b) Find the general solution of (*). (See Exercise 18).

21. (a) Prove that the differential equation

$$\frac{dy}{dt}=\frac{at+by+m}{ct+dy+n}$$

where a, b, c, d, m, and n are constants, can always be reduced to $dy/dt=(at+by)/(ct+dy)$ if $ad-bc\neq0$.

(b) Solve the above equation in the special case that $ad=bc$.

Find the general solution of the following equations.

22. $(1+t-2y)+(4t-3y-6)dy/dt=0$

23. $(t+2y+3)+(2t+4y-1)dy/dt=0$

1.5 Population models

In this section we will study first-order differential equations which govern the growth of various species. At first glance it would seem impossible to model the growth of a species by a differential equation since the population of any species always changes by integer amounts. Hence the population of any species can never be a differentiable function of time. However, if a given population is very large and it is suddenly increased by one, then the change is very small compared to the given population. Thus, we make the approximation that large populations change continuously and even differentiably with time.

Let $p(t)$ denote the population of a given species at time t and let $r(t,p)$ denote the difference between its birth rate and its death rate. If this population is isolated, that is, there is no net immigration or emigration, then dp/dt, the rate of change of the population, equals $rp(t)$. In the most simplistic model we assume that r is constant, that is, it does not change with either time or population. Then, we can write down the following differential equation governing population growth:

$$\frac{dp(t)}{dt}=ap(t), \qquad a=\text{constant}.$$

This is a linear equation and is known as the Malthusian law of population growth. If the population of the given species is p_0 at time t_0, then $p(t)$ satisfies the initial-value problem $dp(t)/dt=ap(t)$, $p(t_0)=p_0$. The solution of this initial-value problem is $p(t)=p_0e^{a(t-t_0)}$. Hence any species satisfying the Malthusian law of population growth grows exponentially with time.

Now, we have just formulated a very simple model for population growth; so simple, in fact, that we have been able to solve it completely in a few lines. It is important, therefore, to see if this model, with its simplicity, has any relationship at all with reality. Let $p(t)$ denote the human population of the earth at time t. It was estimated that the earth's human population in 1961 was 3,060,000,000 and that during the past decade the population was increasing at a rate of 2% per year. Thus, $t_0=1961$, $p_0=(3.06)10^9$, and $a=0.02$ so that

$$p(t)=(3.06)10^9e^{0.02(t-1961)}.$$

Now, we can certainly check this formula out as far as past populations. *Result*: It reflects with surprising accuracy the population estimate for the period 1700–1961. The population of the earth has been doubling about every 35 years and our equation predicts a doubling of the earth's population every 34.6 years. To prove this, observe that the human population of the earth doubles in a time $T=t-t_0$ where $e^{0.02T}=2$. Taking logarithms of both sides of this equation gives $0.02T=\ln 2$ so that $T=50\ln 2\cong 34.6$.

However, let us look into the distant future. Our equation predicts that the earth's population will be 200,000 billion in the year 2510, 1,800,000 billion in the year 2635, and 3,600,000 billion in the year 2670. These are astronomical numbers whose significance is difficult to gauge. The total surface of this planet is approximately 1,860,000 billion square feet. Eighty percent of this surface is covered by water. Assuming that we are willing to live on boats as well as land, it is easy to see that by the year 2510 there will be only 9.3 square feet per person; by 2635 each person will have only one square foot on which to stand; and by 2670 we will be standing two deep on each other's shoulders.

It would seem therefore, that this model is unreasonable and should be thrown out. However, we cannot ignore the fact that it offers exceptional agreement in the past. Moreover, we have additional evidence that populations do grow exponentially. Consider the Microtus Arvallis Pall, a small rodent which reproduces very rapidly. We take the unit of time to be a month, and assume that the population is increasing at the rate of 40% per month. If there are two rodents present initially at time $t=0$, then $p(t)$, the number of rodents at time t, satisfies the initial-value problem

$$dp(t)/dt=0.4p, \qquad p(0)=2.$$

Consequently,

$$p(t)=2e^{0.4t}. \tag{1}$$

Table 1 compares the observed population with the population calculated from Equation (1).

Table 1. The growth of Microtus Arvallis Pall.

Months	0	2	6	10
p Observed	2	5	20	109
p Calculated	2	4.5	22	109.1

As one can see, there is excellent agreement.

Remark. In the case of the Microtus Arvallis Pall, p observed is very accurate since the pregnancy period is three weeks and the time required for the census taking is considerably less. If the pregnancy period were very short then p observed could not be accurate since many of the pregnant rodents would have given birth before the census was completed.

The way out of our dilemma is to observe that linear models for population growth are satisfactory *as long as* the population is not too large. When the population gets extremely large though, these models cannot be very accurate, since they do not reflect the fact that individual members are now competing with each other for the limited living space, natural resources and food available. Thus, we must add a competition term to our linear differential equation. A suitable choice of a competition term is $-bp^2$, where b is a constant, since the statistical average of the number of encounters of two members per unit time is proportional to p^2. We consider, therefore, the modified equation

$$\frac{dp}{dt} = ap - bp^2.$$

This equation is known as the logistic law of population growth and the numbers a, b are called the vital coefficients of the population. It was first introduced in 1837 by the Dutch mathematical-biologist Verhulst. Now, the constant b, in general, will be very small compared to a, so that if p is not too large then the term $-bp^2$ will be negligible compared to ap and the population will grow exponentially. However, when p is very large, the term $-bp^2$ is no longer negligible, and thus serves to slow down the rapid rate of increase of the population. Needless to say, the more industrialized a nation is, the more living space it has, and the more food it has, the smaller the coefficient b is.

Let us now use the logistic equation to predict the future growth of an isolated population. If p_0 is the population at time t_0, then $p(t)$, the population at time t, satisfies the initial-value problem

$$\frac{dp}{dt} = ap - bp^2, \qquad p(t_0) = p_0.$$

This is a separable differential equation, and from Equation (10), Section 1.4,

$$\int_{p_0}^{p} \frac{dr}{ar - br^2} = \int_{t_0}^{t} ds = t - t_0.$$

To integrate the function $1/(ar - br^2)$ we resort to partial fractions. Let

$$\frac{1}{ar - br^2} \equiv \frac{1}{r(a - br)} = \frac{A}{r} + \frac{B}{a - br}.$$

To find A and B, observe that

$$\frac{A}{r} + \frac{B}{a - br} = \frac{A(a - br) + Br}{r(a - br)} = \frac{Aa + (B - bA)r}{r(a - br)}.$$

Therefore, $Aa + (B - bA)r = 1$. Since this equation is true for all values of r, we see that $Aa = 1$ and $B - bA = 0$. Consequently, $A = 1/a$, $B = b/a$, and

$$\int_{p_0}^{p} \frac{dr}{r(a - br)} = \frac{1}{a} \int_{p_0}^{p} \left(\frac{1}{r} + \frac{b}{a - br} \right) dr$$

$$= \frac{1}{a}\left[\ln \frac{p}{p_0} + \ln \left| \frac{a - bp_0}{a - bp} \right| \right] = \frac{1}{a} \ln \frac{p}{p_0} \left| \frac{a - bp_0}{a - bp} \right|.$$

Thus,

$$a(t-t_0)=\ln\frac{p}{p_0}\left|\frac{a-bp_0}{a-bp}\right|. \tag{2}$$

Now, it is a simple matter to show (see Exercise 1) that

$$\frac{a-bp_0}{a-bp(t)}$$

is always positive. Hence,

$$a(t-t_0)=\ln\frac{p}{p_0}\frac{(a-bp_0)}{a-bp}.$$

Taking exponentials of both sides of this equation gives

$$e^{a(t-t_0)}=\frac{p}{p_0}\frac{a-bp_0}{a-bp},$$

or

$$p_0(a-bp)e^{a(t-t_0)}=(a-bp_0)p.$$

Bringing all terms involving p to the left-hand side of this equation, we see that

$$\left[a-bp_0+bp_0e^{a(t-t_0)}\right]p(t)=ap_0e^{a(t-t_0)}.$$

Consequently,

$$p(t)=\frac{ap_0e^{a(t-t_0)}}{a-bp_0+bp_0e^{a(t-t_0)}}=\frac{ap_0}{bp_0+(a-bp_0)e^{-a(t-t_0)}}. \tag{3}$$

Let us now examine Equation (3) to see what kind of population it predicts. Observe that as $t\to\infty$,

$$p(t)\to\frac{ap_0}{bp_0}=\frac{a}{b}.$$

Thus, *regardless of its initial value, the population always approaches the limiting value a/b*. Next, observe that $p(t)$ is a monotonically increasing function of time if $0<p_0<a/b$. Moreover, since

$$\frac{d^2p}{dt^2}=a\frac{dp}{dt}-2bp\frac{dp}{dt}=(a-2bp)p(a-bp),$$

we see that dp/dt is increasing if $p(t)<a/2b$, and that dp/dt is decreasing if $p(t)>a/2b$. Hence, if $p_0<a/2b$, the graph of $p(t)$ must have the form given in Figure 1. Such a curve is called a logistic, or S-shaped curve. From its shape we conclude that the time period before the population reaches half its limiting value is a period of accelerated growth. After this point, the rate of growth decreases and in time reaches zero. This is a period of diminishing growth.

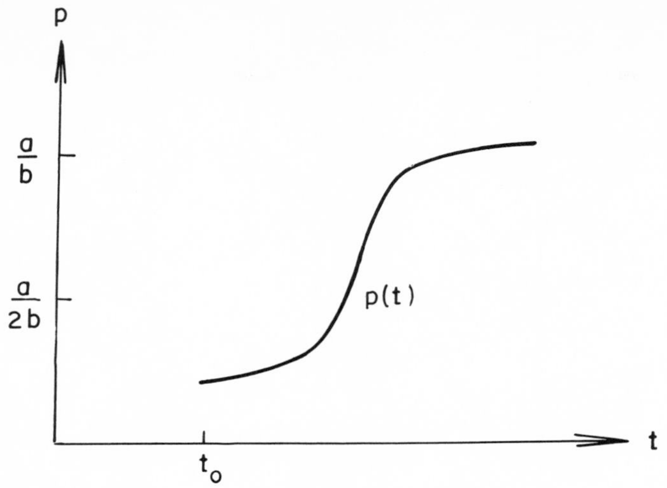

Figure 1. Graph of $p(t)$

These predictions are borne out by an experiment on the protozoa Paramecium caudatum performed by the mathematical biologist G. F. Gause. Five individuals of Paramecium were placed in a small test tube containing 0.5 cm^3 of a nutritive medium, and for six days the number of individuals in every tube was counted daily. The Paramecium were found to increase at a rate of 230.9% per day when their numbers were low. The number of individuals increased rapidly at first, and then more slowly, until towards the fourth day it attained a maximum level of 375, saturating the test tube. From this data we conclude that if the Paramecium caudatum grow according to the logistic law $dp/dt = ap - bp^2$, then $a = 2.309$ and $b = 2.309/375$. Consequently, the logistic law predicts that

$$p(t) = \frac{(2.309)5}{\frac{(2.309)5}{375} + \left(2.309 - \frac{(2.309)5}{375}\right)e^{-2.309t}}$$

$$= \frac{375}{1 + 74e^{-2.309t}}. \tag{4}$$

(We have taken the initial time t_0 to be 0.) Figure 2 compares the graph of $p(t)$ predicted by Equation (4) with the actual measurements, which are denoted by o. As can be seen, the agreement is remarkably good.

In order to apply our results to predict the future human population of the earth, we must estimate the vital coefficients a and b in the logistic equation governing its growth. Some ecologists have estimated that the natural value of a is 0.029. We also know that the human population was increasing at the rate of 2% per year when the population was $(3.06)10^9$.

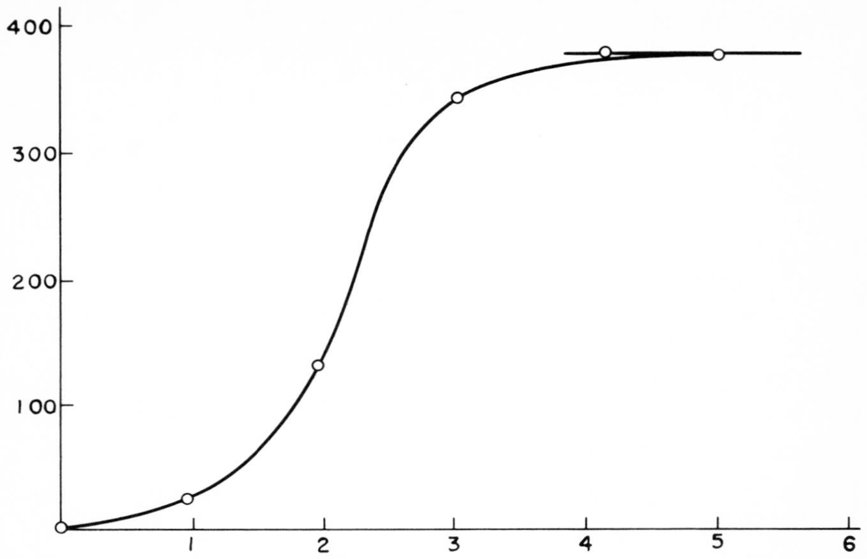

Figure 2. The growth of paramecium

Since $(1/p)(dp/dt) = a - bp$, we see that

$$0.02 = a - b(3.06)10^9.$$

Consequently, $b = 2.941 \times 10^{-12}$. Thus, according to the logistic law of population growth, the human population of the earth will tend to the limiting value of

$$\frac{a}{b} = \frac{0.029}{2.941 \times 10^{-12}} = 9.86 \text{ billion people.}$$

Note that according to this prediction, we were still on the accelerated growth portion of the logistic curve in 1961, since we had not yet attained half the limiting population predicted for us.

As another verification of the validity of the logistic law of population growth, we consider the equation

$$p(t) = \frac{197{,}273{,}000}{1 + e^{-0.03134(t - 1913.25)}} \tag{5}$$

which was introduced by Pearl and Reed as a model of the population growth of the United States. First, using the census figures for the years 1790, 1850, and 1910, Pearl and Reed found from (3) (see Exercise 2a) that $a = 0.03134$ and $b = (1.5887)10^{-10}$. Then (see Exercise 2b), Pearl and Reed calculated that the population of the United States reached half its limiting population of $a/b = 197{,}273{,}000$ in April 1913. Consequently (see Exercise 2c), we can rewrite (3) in the simpler form (5).

Table 2 below compares Pearl and Reed's predictions with the observed values of the population of the United States. These results are remarkable,

Table 2. Population of the U.S. from 1790–1950. (The last 4 entries were added by the Dartmouth College Writing Group.)

	Actual	Predicted	Error	%
1790	3,929,000	3,929,000	0	0.0
1800	5,308,000	5,336,000	28,000	0.5
1810	7,240,000	7,228,000	− 12,000	−0.2
1820	9,638,000	9,757,000	119,000	1.2
1830	12,866,000	13,109,000	243,000	1.9
1840	17,069,000	17,506,000	437,000	2.6
1850	23,192,000	23,192,000	0	0.0
1860	31,443,000	30,412,000	− 1,031,000	−3.3
1870	38,558,000	39,372,000	814,000	2.1
1880	50,156,000	50,177,000	21,000	0.0
1890	62,948,000	62,769,000	− 179,000	−0.3
1900	75,995,000	76,870,000	875,000	1.2
1910	91,972,000	91,972,000	0	0.0
1920	105,711,000	107,559,000	1,848,000	1.7
1930	122,775,000	123,124,000	349,000	0.3
1940	131,669,000	136,653,000	4,984,000	3.8
1950	150,697,000	149,053,000	− 1,644,000	− 1.1

especially since we have not taken into account the large waves of immigration into the United States, and the fact that the United States was involved in five wars during this period.

In 1845 Verhulst prophesied a maximum population for Belgium of 6,600,000, and a maximum population for France of 40,000,000. Now, the population of Belgium in 1930 was already 8,092,000. This large discrepancy would seem to indicate that the logistic law of population growth is very inaccurate, at least as far as the population of Belgium is concerned. However, this discrepancy can be explained by the astonishing rise of industry in Belgium, and by the acquisition of the Congo which secured for the country sufficient additional wealth to support the extra population. Thus, after the acquisition of the Congo, and the astonishing rise of industry in Belgium, Verhulst should have lowered the vital coefficient b.

On the other hand, the population of France in 1930 was in remarkable agreement with Verhulst's forecast. Indeed, we can now answer the following tantalizing paradox: Why was the population of France increasing extremely slowly in 1930 while the French population of Canada was increasing very rapidly? After all, they are the same people! The answer to this paradox, of course, is that the population of France in 1930 was very near its limiting value and thus was far into the period of diminishing growth, while the population of Canada in 1930 was still in the period of accelerated growth.

Remark 1. It is clear that technological developments, pollution considerations and sociological trends have significant influence on the vital coefficients a and b. Therefore, they must be re-evaluated every few years.

Remark 2. To derive more accurate models of population growth, we should not consider the population as made up of one homogeneous group of individuals. Rather, we should subdivide it into different age groups. We should also subdivide the population into males and females, since the reproduction rate in a population usually depends more on the number of females than on the number of males.

Remark 3. Perhaps the severest criticism leveled at the logistic law of population growth is that some populations have been observed to fluctuate periodically between two values, and any type of fluctuation is ruled out in a logistic curve. However, some of these fluctuations can be explained by the fact that when certain populations reach a sufficiently high density, they become susceptible to epidemics. The epidemic brings the population down to a lower value where it again begins to increase, until when it is large enough, the epidemic strikes again. In Exercise 7 we derive a model to describe this phenomenon, and we apply this model in Exercise 8 to explain the sudden appearance and disappearance of hordes of small rodents.

References

1. Gause, G. F., *The Struggle for Existence*, Dover Publications, New York, 1964.
2. Pearl and Reed, *Proceedings of the National Academy of Sciences*, 1920, p. 275.

EXERCISES

1. Prove that $(a-bp_0)/(a-bp(t))$ is positive for $t_0<t<\infty$. *Hint*: Use Equation (2) to show that $p(t)$ can never equal a/b if $p_0\neq a/b$.

2. (a) Choose 3 times t_0, t_1, and t_2, with $t_1-t_0=t_2-t_1$. Show that (3) determines a and b uniquely in terms of t_0, $p(t_0)$, t_1, $p(t_1)$, t_2, and $p(t_2)$.
(b) Show that the period of accelerated growth for the United States ended in April, 1913.
(c) Let a population $p(t)$ grow according to the logistic law (3), and let $\bar{t}$ be the time at which half the limiting population is achieved. Show that

$$p(t)=\frac{a/b}{1+e^{-a(t-\bar{t})}}.$$

3. In 1879 and 1881 a number of yearling bass were seined in New Jersey, taken across the continent in tanks by train, and planted in San Francisco Bay. A total of only 435 Striped Bass survived the rigors of these two trips. Yet, in 1899, the commercial net catch alone was 1,234,000 pounds. Since the growth of this population was so fast, it is reasonable to assume that it obeyed the Malthusian law $dp/dt = ap$. Assuming that the average weight of a bass fish is three pounds, and that in 1899 every tenth bass fish was caught, find a lower bound for a.

4. A population grows according to the logistic law, with a limiting population of 5×10^8 individuals. When the population is low it doubles every 40 minutes. What will the population be after two hours if initially it is (a) 10^8, (b) 10^9?

5. A family of salmon fish living off the Alaskan Coast obeys the Malthusian law of population growth $dp(t)/dt = 0.003p(t)$, where t is measured in minutes. At time $t=0$ a group of sharks establishes residence in these waters and begins attacking the salmon. The rate at which salmon are killed by the sharks is $0.001p^2(t)$, where $p(t)$ is the population of salmon at time t. Moreover, since an undesirable element has moved into their neighborhood, 0.002 salmon per minute leave the Alaskan waters.

(a) Modify the Malthusian law of population growth to take these two factors into account.

(b) Assume that at time $t=0$ there are one million salmon. Find the population $p(t)$. What happens as $t\to\infty$?

6. The population of New York City would satisfy the logistic law

$$\frac{dp}{dt} = \frac{1}{25}p - \frac{1}{(25)10^6}p^2,$$

where t is measured in years, if we neglected the high emigration and homicide rates.

(a) Modify this equation to take into account the fact that 6,000 people per year move from the city, and 4,000 people per year are murdered.

(b) Assume that the population of New York City was 8,000,000 in 1970. Find the population for all future time. What happens as $t\to\infty$?

7. We can model a population which becomes susceptible to epidemics in the following manner. Assume that our population is originally governed by the logistic law

$$\frac{dp}{dt} = ap - bp^2 \tag{i}$$

and that an epidemic strikes as soon as p reaches a certain value Q, with Q less than the limiting population a/b. At this stage the vital coefficients become $A < a$, $B < b$, and Equation (i) is replaced by

$$\frac{dp}{dt} = Ap - Bp^2. \tag{ii}$$

Suppose that $Q > A/B$. The population then starts decreasing. A point is reached when the population falls below a certain value $q > A/B$. At this moment the epidemic ceases and the population again begins to grow following

Equation (i), until the incidence of a fresh epidemic. In this way there are periodic fluctuations of p between q and Q. We now indicate how to calculate the period T of these fluctuations.

(a) Show that the time T_1 taken by the first part of the cycle, when p increases from q to Q is given by

$$T_1 = \frac{1}{a}\ln\frac{Q(a-bq)}{q(a-bQ)}.$$

(b) Show that the time T_2 taken by the second part of the cycle, when p decreases from Q to q is given by

$$T_2 = \frac{1}{A}\ln\frac{q(QB-A)}{Q(qB-A)}.$$

Thus, the time for the entire cycle is $T_1 + T_2$.

8. It has been observed that plagues appear in mice populations whenever the population becomes too large. Further, a local increase of density attracts predators in large numbers. These two factors will succeed in destroying 97–98% of a population of small rodents in two or three weeks, and the density then falls to a level at which the disease cannot spread. The population, reduced to 2% of its maximum, finds its refuges from the predators sufficient, and its food abundant. The population therefore begins to grow again until it reaches a level favorable to another wave of disease and predation. Now, the speed of reproduction in mice is so great that we may set $b=0$ in Equation (i) of Exercise 7. In the second part of the cycle, on the contrary, A is very small in comparison with B, and it may be neglected therefore in Equation (ii).

(a) Under these assumptions, show that

$$T_1 = \frac{1}{a}\ln\frac{Q}{q} \quad \text{and} \quad T_2 = \frac{Q-q}{qQB}.$$

(b) Assuming that T_1 is approximately four years, and Q/q is approximately fifty, show that a is approximately one. This value of a, incidentally, corresponds very well with the rate of multiplication of mice in natural circumstances.

9. There are many important classes of organisms whose birth rate is *not* proportional to the population size. Suppose, for example, that each member of the population requires a partner for reproduction, and that each member relies on chance encounters for meeting a mate. If the expected number of encounters is proportional to the product of the numbers of males and females, and if these are equally distributed in the population, then the number of encounters, and hence the birthrate too, is proportional to p^2. The death rate is still proportional to p. Consequently, the population size $p(t)$ satisfies the differential equation

$$\frac{dp}{dt} = bp^2 - ap, \quad a,b>0.$$

Show that $p(t)$ approaches 0 as $t\to\infty$ if $p_0 < a/b$. Thus, once the population size drops below the critical size a/b, the population tends to extinction. Thus, a species is classified endangered if its current size is perilously close to its critical size.

1.6 An atomic waste disposal problem

For several years the Atomic Energy Commission (now known as the Nuclear Regulatory Commission) had disposed of concentrated radioactive waste material by placing it in tightly sealed drums which were then dumped at sea in fifty fathoms (300 feet) of water. When concerned ecologists and scientists questioned this practice, they were assured by the A.E.C. that the drums would never develop leaks. Exhaustive tests on the drums proved the A.E.C. right. However, several engineers then raised the question of whether the drums could crack from the impact of hitting the ocean floor. "Never," said the A.E.C. "We'll see about that," said the engineers. After performing numerous experiments, the engineers found that the drums could crack on impact if their velocity exceeded forty feet per second. The problem before us, therefore, is to compute the velocity of the drums upon impact with the ocean floor. To this end, we digress briefly to study elementary Newtonian mechanics.

Newtonian mechanics is the study of Newton's famous laws of motion and their consequences. Newton's first law of motion states that an object will remain at rest, or move with constant velocity, if no force is acting on it. A force should be thought of as a push or pull. This push or pull can be exerted directly by something in contact with the object, or it can be exerted indirectly, as the earth's pull of gravity is.

Newton's second law of motion is concerned with describing the motion of an object which is acted upon by several forces. Let $y(t)$ denote the position of the center of gravity of the object. (We assume that the object moves in only one direction.) Those forces acting on the object, which tend to increase y, are considered positive, while those forces tending to decrease y are considered negative. The resultant force F acting on an object is defined to be the sum of all positive forces minus the sum of all negative forces. Newton's second law of motion states that the acceleration d^2y/dt^2 of an object is proportional to the resultant force F acting on it; i.e.,

$$\frac{d^2y}{dt^2} = \frac{1}{m}F. \tag{1}$$

The constant m is the mass of the object. It is related to the weight W of the object by the relation $W = mg$, where g is the acceleration of gravity. Unless otherwise stated, we assume that the weight of an object and the acceleration of gravity are constant. We will also adopt the English system of units, so that t is measured in seconds, y is measured in feet, and F is measured in pounds. The units of m are then slugs, and the gravitational acceleration g equals 32.2 ft/s^2.

We return now to our atomic waste disposal problem. As a drum descends through the water, it is acted upon by three forces W, B, and D. The force W is the weight of the drum pulling it down, and in magnitude,

$W=527.436$ lb. The force B is the buoyancy force of the water acting on the drum. This force pushes the drum up, and its magnitude is the weight of the water displaced by the drum. Now, the Atomic Energy Commission used 55 gallon drums, whose volume is 7.35 ft^3. The weight of one cubic foot of salt water is 63.99 lb. Hence $B=(63.99)(7.35)=470.327$ lb.

The force D is the drag force of the water acting on the drum; it resists the motion of the drum through the water. Experiments have shown that any medium such as water, oil, and air resists the motion of an object through it. This resisting force acts in the direction opposite the motion, and is usually directly proportional to the velocity V of the object. Thus, $D=cV$, for some positive constant c. Notice that the drag force increases as V increases, and decreases as V decreases. To calculate D, the engineers conducted numerous towing experiments. They concluded that the orientation of the drum had little effect on the drag force, and that

$$D=0.08\,V\frac{(\text{lb})(\text{s})}{\text{ft}}.$$

Now, set $y=0$ at sea level, and let the direction of increasing y be downwards. Then, W is a positive force, and B and D are negative forces. Consequently, from (1),

$$\frac{d^2y}{dt^2}=\frac{1}{m}(W-B-cV)=\frac{g}{W}(W-B-cV).$$

We can rewrite this equation as a first-order linear differential equation for $V=dy/dt$; i.e.,

$$\frac{dV}{dt}+\frac{cg}{W}V=\frac{g}{W}(W-B). \tag{2}$$

Initially, when the drum is released in the ocean, its velocity is zero. Thus, $V(t)$, the velocity of the drum, satisfies the initial-value problem

$$\frac{dV}{dt}+\frac{cg}{W}V=\frac{g}{W}(W-B), \qquad V(0)=0, \tag{3}$$

and this implies that

$$V(t)=\frac{W-B}{c}\left[1-e^{(-cg/W)t}\right]. \tag{4}$$

Equation (4) expresses the velocity of the drum as a function of time. In order to determine the impact velocity of the drum, we must compute the time t at which the drum hits the ocean floor. Unfortunately, though, it is impossible to find t as an explicit function of y (see Exercise 2). Therefore, we cannot use Equation (4) to find the velocity of the drum when it hits the ocean floor. However, the A.E.C. can use this equation to try and prove that the drums do not crack on impact. To wit, observe from (4) that $V(t)$ is a monotonic increasing function of time which approaches the limiting value

$$V_T = \frac{W-B}{c}$$

as t approaches infinity. The quantity V_T is called the terminal velocity of the drum. Clearly, $V(t) \leqslant V_T$, so that the velocity of the drum when it hits the ocean floor is certainly less than $(W-B)/c$. Now, if this terminal velocity is less than 40 ft/s, then the drums could not possibly break on impact. However,

$$\frac{W-B}{c} = \frac{527.436-470.327}{0.08} = 713.86 \text{ ft/s},$$

and this is way too large.

It should be clear now that the only way we can resolve the dispute between the A.E.C. and the engineers is to find $v(y)$, the velocity of the drum as a function of position. The function $v(y)$ is very different from the function $V(t)$, which is the velocity of the drum as a function of time. However, these two functions are related through the equation

$$V(t) = v(y(t))$$

if we express y as a function of t. By the chain rule of differentiation, $dV/dt = (dv/dy)(dy/dt)$. Hence

$$\frac{W}{g}\frac{dv}{dy}\frac{dy}{dt} = W - B - cV.$$

But $dy/dt = V(t) = v(y(t))$. Thus, suppressing the dependence of y on t, we see that $v(y)$ satisfies the first-order differential equation

$$\frac{W}{g} v \frac{dv}{dy} = W - B - cv, \quad \text{or} \quad \frac{v}{W-B-cv}\frac{dv}{dy} = \frac{g}{W}.$$

Moreover,

$$v(0) = v(y(0)) = V(0) = 0.$$

Hence,

$$\int_0^v \frac{r\,dr}{W-B-cr} = \int_0^y \frac{g}{W}\,ds = \frac{gy}{W}.$$

Now,

$$\begin{aligned}\int_0^v \frac{r\,dr}{W-B-cr} &= \int_0^v \frac{r-(W-B)/c}{W-B-cr}\,dr + \frac{W-B}{c}\int_0^v \frac{dr}{W-B-cr}\\ &= -\frac{1}{c}\int_0^v dr + \frac{W-B}{c}\int_0^v \frac{dr}{W-B-cr}\\ &= -\frac{v}{c} - \frac{(W-B)}{c^2}\ln\frac{|W-B-cv|}{W-B}.\end{aligned}$$

We know already that $v<(W-B)/c$. Consequently, $W-B-cv$ is always positive, and

$$\frac{gy}{W}=-\frac{v}{c}-\frac{(W-B)}{c^2}\ln\frac{W-B-cv}{W-B}. \tag{5}$$

At this point, we are ready to scream in despair since we cannot find v as an explicit function of y from (5). This is not an insurmountable difficulty, though, for it is quite simple, with the aid of a digital computer, to find $v(300)$ from (5). We need only supply the computer with a good approximation of $v(300)$ and this is obtained in the following manner. The velocity $v(y)$ of the drum satisfies the initial-value problem

$$\frac{W}{g}v\frac{dv}{dy}=W-B-cv, \qquad v(0)=0. \tag{6}$$

Let us, for the moment, set $c=0$ in (6) to obtain the new initial-value problem

$$\frac{W}{g}u\frac{du}{dy}=W-B, \qquad u(0)=0. \tag{6'}$$

(We have replaced v by u to avoid confusion later.) We can integrate (6′) immediately to obtain that

$$\frac{W}{g}\frac{u^2}{2}=(W-B)y, \quad \text{or} \quad u(y)=\left[\frac{2g}{W}(W-B)y\right]^{1/2}.$$

In particular,

$$u(300)=\left[\frac{2g}{W}(W-B)300\right]^{1/2}=\left[\frac{2(32.2)(57.109)(300)}{527.436}\right]^{1/2}$$

$$\cong\sqrt{2092}\cong 45.7 \text{ ft/s}.$$

We claim, now, that $u(300)$ is a very good approximation of $v(300)$. The proof of this is as follows. First, observe that the velocity of the drum is always greater if there is no drag force opposing the motion. Hence,

$$v(300)<u(300).$$

Second, the velocity v increases as y increases, so that $v(y)\leqslant v(300)$ for $y\leqslant 300$. Consequently, the drag force D of the water acting on the drum is always less than $0.08\times u(300)\cong 3.7$ lb. Now, the resultant force $W-B$ pulling the drum down is approximately 57.1 lb, which is very large compared to D. It stands to reason, therefore, that $u(y)$ should be a very good approximation of $v(y)$. And indeed, this is the case, since we find numerically that $v(300)=45.1$ ft/s. Thus, the drums can break upon impact, and the engineers were right.

Epilog. The rules of the Atomic Energy Commission now expressly forbid the dumping of low level atomic waste at sea. This author is uncertain though, as to whether Western Europe has also forbidden this practice.

Remark. The methods introduced in this section can also be used to find the velocity of any object which is moving through a medium that resists the motion. We just disregard the buoyancy force if the medium is not water. For example, let $V(t)$ denote the velocity of a parachutist falling to earth under the influence of gravity. Then,

$$\frac{W}{g}\frac{dV}{dt} = W - D$$

where W is the weight of the man and the parachute, and D is the drag force exerted by the atmosphere on the falling parachutist. The drag force on a bluff object in air, or in any fluid of small viscosity is usually very nearly proportional to V^2. Proportionality to V is the exceptional case, and occurs only at very low speeds. The criterion as to whether the square or the linear law applies is the "Reynolds number"

$$R = \rho VL/\mu.$$

L is a representative length dimension of the object, and ρ and μ are the density and viscosity of the fluid. If $R < 10$, then $D \sim V$, and if $R > 10^3$, $D \sim V^2$. For $10 < R < 10^3$, neither law is accurate.

Exercises

1. Solve the initial-value problem (3).

2. Solve for $y = y(t)$ from (4), and then show that the equation $y = y(t)$ cannot be solved explicitly for $t = t(y)$.

3. Show that the drums of atomic waste will not crack upon impact if they are dropped into L feet of water with $(2g(W-B)L/W)^{1/2} < 40$.

4. Fat Richie, an enormous underworld hoodlum weighing 400 lb, was pushed out of a penthouse window 2800 feet above the ground in New York City. Neglecting air resistance find (a) the velocity with which Fat Richie hit the ground; (b) the time elapsed before Fat Richie hit the ground.

5. An object weighing 300 lb is dropped into a river 150 feet deep. The volume of the object is 2 ft^3, and the drag force exerted by the water on it is 0.05 times its velocity. The drag force may be considered negligible if it does not exceed 5% of the resultant force pulling the drum down. Prove that the drag force is negligible in this case. (Here $B = 2(62.4) = 124.8$.)

6. A 400 lb sphere of volume $4\pi/3$ and a 300 lb cylinder of volume π are simultaneously released from rest into a river. The drag force exerted by the water on the falling sphere and cylinder is λV_s and λV_c, respectively, where V_s and V_c are the velocities of the sphere and cylinder, and λ is a positive constant. Determine which object reaches the bottom of the river first.

7. A parachutist falls from rest toward earth. The combined weight of man and parachute is 161 lb. Before the parachute opens, the air resistance equals $V/2$.

The parachute opens 5 seconds after the fall begins; and the air resistance is then $V^2/2$. Find the velocity $V(t)$ of the parachutist after the parachute opens.

8. A man wearing a parachute jumps from a great height. The combined weight of man and parachute is 161 lb. Let $V(t)$ denote his speed at time t seconds after the fall begins. During the first 10 seconds, the air resistance is $V/2$. Thereafter, while the parachute is open, the air resistance is $10V$. Find an explicit formula for $V(t)$ at any time t greater than 10 seconds.

9. An object of mass m is projected vertically downward with initial velocity V_0 in a medium offering resistance proportional to the square root of the magnitude of the velocity.
(a) Find a relation between the velocity V and the time t if the drag force equals $c\sqrt{V}$.
(b) Find the terminal velocity of the object. *Hint*: You can find the terminal velocity even though you cannot solve for $V(t)$.

10. A body of mass m falls from rest in a medium offering resistance proportional to the square of the velocity; that is, $D = cV^2$. Find $V(t)$ and compute the terminal velocity V_T.

11. A body of mass m is projected upward from the earth's surface with an initial velocity V_0. Take the y-axis to be positive upward, with the origin on the surface of the earth. Assuming there is no air resistance, but taking into account the variation of the earth's gravitational field with altitude, we obtain that

$$m\frac{dV}{dt} = -\frac{mgR^2}{(y+R)^2}$$

where R is the radius of the earth.
(a) Let $V(t) = v(y(t))$. Find a differential equation satisfied by $v(y)$.
(b) Find the smallest initial velocity V_0 for which the body will not return to earth. This is the so-called escape velocity. *Hint*: The escape velocity is found by requiring that $v(y)$ remain strictly positive.

12. It is not really necessary to find $v(y)$ explicitly in order to prove that $v(300)$ exceeds 40 ft/s. Here is an alternate proof. Observe first that $v(y)$ increases as y increases. This implies that y is a monotonic increasing function of v. Therefore, if y is less than 300 ft when v is 40 ft/s, then v must be greater than 40 ft/s when y is 300 ft. Substitute $v = 40$ ft/s in Equation (5), and show that y is less than 300 ft. Conclude, therefore, that the drums can break upon impact.

1.7 The dynamics of tumor growth, mixing problems and orthogonal trajectories

In this section we present three very simple but extremely useful applications of first-order equations. The first application concerns the growth of solid tumors; the second application is concerned with "mixing problems" or "compartment analysis"; and the third application shows how to find a family of curves which is orthogonal to a given family of curves.

(a) *The dynamics of tumor growth*

It has been observed experimentally, that "free living" dividing cells, such as bacteria cells, grow at a rate proportional to the volume of dividing cells at that moment. Let $V(t)$ denote the volume of dividing cells at time t. Then,

$$\frac{dV}{dt} = \lambda V \tag{1}$$

for some positive constant λ. The solution of (1) is

$$V(t) = V_0 e^{\lambda(t-t_0)} \tag{2}$$

where V_0 is the volume of dividing cells at the initial time t_0. Thus, free living dividing cells grow *exponentially* with time. One important consequence of (2) is that the volume of cells keeps doubling (see Exercise 1) every time interval of length $\ln 2/\lambda$.

On the other hand, solid tumors do not grow exponentially with time. As the tumor becomes larger, the doubling time of the total tumor volume continuously increases. Various researchers have shown that the data for many solid tumors is fitted remarkably well, over almost a 1000 fold increase in tumor volume, by the equation

$$V(t) = V_0 \exp\left(\frac{\lambda}{\alpha}(1 - \exp(-\alpha t))\right) \tag{3}$$

where $\exp(x) \equiv e^x$, and λ and α are positive constants.

Equation (3) is usually referred to as a Gompertzian relation. It says that the tumor grows more and more slowly with the passage of time, and that it ultimately approaches the limiting volume $V_0 e^{\lambda/\alpha}$. Medical researchers have long been concerned with explaining this deviation from simple exponential growth. A great deal of insight into this problem can be gained by finding a differential equation satisfied by $V(t)$. Differentiating (3) gives

$$\begin{aligned}\frac{dV}{dt} &= V_0 \lambda \exp(-\alpha t) \exp\left(\frac{\lambda}{\alpha}(1 - \exp(-\alpha(t)))\right) \\ &= \lambda e^{-\alpha t} V. \end{aligned} \tag{4}$$

Two conflicting theories have been advanced for the dynamics of tumor growth. They correspond to the two arrangements

$$\frac{dV}{dt} = (\lambda e^{-\alpha t}) V \tag{4a}$$

$$\frac{dV}{dt} = \lambda (e^{-\alpha t} V) \tag{4b}$$

of the differential equation (4). According to the first theory, the retarding effect of tumor growth is due to an increase in the mean generation time of the cells, without a change in the proportion of reproducing cells. As time goes on, the reproducing cells mature, or age, and thus divide more slowly. This theory corresponds to the bracketing (a).

The bracketing (b) suggests that the mean generation time of the dividing cells remains constant, and the retardation of growth is due to a loss in reproductive cells in the tumor. One possible explanation for this is that a *necrotic region* develops in the center of the tumor. This necrosis appears at a critical size for a particular type of tumor, and thereafter the necrotic "core" increases rapidly as the total tumor mass increases. According to this theory, a necrotic core develops because in many tumors the supply of blood, and thus of oxygen and nutrients, is almost completely confined to the surface of the tumor and a short distance beneath it. As the tumor grows, the supply of oxygen to the central core by diffusion becomes more and more difficult resulting in the formation of a necrotic core.

(b) *Mixing problems*

Many important problems in biology and engineering can be put into the following framework. A solution containing a fixed concentration of substance x flows into a tank, or compartment, containing the substance x and possibly other substances, at a specified rate. The mixture is stirred together very rapidly, and then leaves the tank, again at a specified rate. Find the concentration of substance x in the tank at any time t.

Problems of this type fall under the general heading of "mixing problems," or compartment analysis. The following example illustrates how to solve these problems.

Example 1. A tank contains S_0 lb of salt dissolved in 200 gallons of water. Starting at time $t=0$, water containing $\frac{1}{2}$ lb of salt per gallon enters the tank at the rate of 4 gal/min, and the well stirred solution leaves the tank at the same rate. Find the concentration of salt in the tank at any time $t>0$.

Solution. Let $S(t)$ denote the amount of salt in the tank at time t. Then, $S'(t)$, which is the rate of change of salt in the tank at time t, must equal the rate at which salt enters the tank minus the rate at which it leaves the tank. Obviously, the rate at which salt enters the tank is

$$\tfrac{1}{2}\ \text{lb/gal times } 4\ \text{gal/min} = 2\ \text{lb/min}.$$

After a moment's reflection, it is also obvious that the rate at which salt leaves the tank is

$$4\ \text{gal/min times } \frac{S(t)}{200}.$$

Thus

$$S'(t) = 2 - \frac{S(t)}{50}, \qquad S(0) = S_0,$$

and this implies that

$$S(t) = S_0 e^{-0.02t} + 100(1 - e^{-0.02t}). \tag{5}$$

Hence, the concentration $c(t)$ of salt in the tank is given by

$$c(t) = \frac{S(t)}{200} = \frac{S_0}{200} e^{-0.02t} + \tfrac{1}{2}(1 - e^{-0.02t}). \tag{6}$$

Remark. The first term on the right-hand side of (5) represents the portion of the original amount of salt remaining in the tank at time t. This term becomes smaller and smaller with the passage of time as the original solution is drained from the tank. The second term on the right-hand side of (5) represents the amount of salt in the tank at time t due to the action of the flow process. Clearly, the amount of salt in the tank must ultimately approach the limiting value of 100 lb, and this is easily verified by letting t approach ∞ in (5).

(c) *Orthogonal trajectories*

In many physical applications, it is often necessary to find the orthogonal trajectories of a given family of curves. (A curve which intersects each member of a family of curves at right angles is called an orthogonal trajectory of the given family.) For example, a charged particle moving under the influence of a magnetic field always travels on a curve which is perpendicular to each of the magnetic field lines. The problem of computing orthogonal trajectories of a family of curves can be solved in the following manner. Let the given family of curves be described by the relation

$$F(x,y,c) = 0. \tag{7}$$

Differentiating this equation yields

$$F_x + F_y y' = 0, \quad \text{or} \quad y' = -\frac{F_x}{F_y}. \tag{8}$$

Next, we solve for $c = c(x,y)$ from (7) and replace every c in (8) by this value $c(x,y)$. Finally, since the slopes of curves which intersect orthogonally are negative reciprocals of each other, we see that the orthogonal trajectories of (7) are the solution curves of the equation

$$y' = \frac{F_y}{F_x}. \tag{9}$$

Example 2. Find the orthogonal trajectories of the family of parabolas

$$x = cy^2.$$

Solution. Differentiating the equation $x = cy^2$ gives $1 = 2cyy'$. Since $c = x/y^2$, we see that $y' = y/2x$. Thus, the orthogonal trajectories of the family of parabolas $x = cy^2$ are the solution curves of the equation

$$y' = -\frac{2x}{y}. \tag{10}$$

This equation is separable, and its solution is

$$y^2 + 2x^2 = k^2. \tag{11}$$

Thus, the family of ellipses (11) (see Figure 1) are the orthogonal trajectories of the family of parabolas $x = cy^2$.

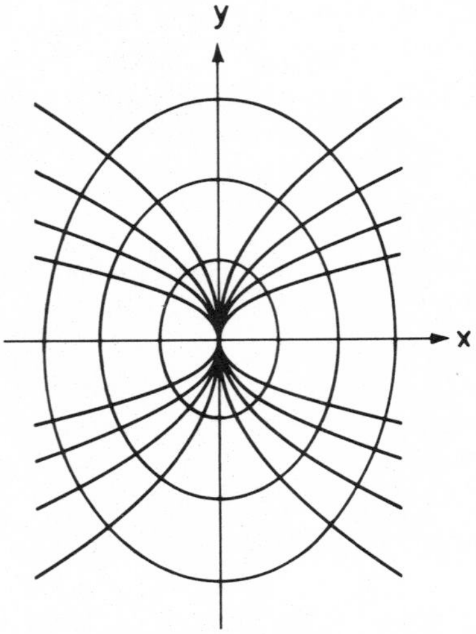

Figure 1. The parabolas $x = cy^2$ and their orthogonal trajectories

Reference

Burton, Alan C., Rate of growth of solid tumors as a problem of diffusion, *Growth*, 1966, vol. 30, pp. 157–176.

Exercises

1. A given substance satisfies the exponential growth law (1). Show that the graph of $\ln V$ versus t is a straight line.

2. A substance x multiplies exponentially, and a given quantity of the substance doubles every 20 years. If we have 3 lb of substance x at the present time, how many lb will we have 7 years from now?

3. A substance x decays exponentially, and only half of the given quantity of x remains after 2 years. How long does it take for 5 lb of x to decay to 1 lb?

4. The equation $p' = ap^{\alpha}$, $\alpha > 1$, is proposed as a model of the population growth of a certain species. Show that $p(t) \to \infty$ in finite time. Conclude, therefore, that this model is not accurate over a reasonable length of time.

5. A cancerous tumor satisfies the Gompertzian relation (3). Originally, when it contained 10^4 cells, the tumor was increasing at the rate of 20% per unit time. The numerical value of the retarding constant α is 0.02. What is the limiting number of cells in this tumor?

6. A *tracer dose* of radioactive iodine ^{131}I is injected into the blood stream at time $t = 0$. Assume that the original amount Q_0 of iodine is distributed evenly in the entire blood stream before any loss occurs. Let $Q(t)$ denote the amount of iodine in the blood at time $t > 0$. Part of the iodine leaves the blood and enters the urine at the rate $k_1 Q$. Another part of the iodine enters the thyroid gland at the rate $k_2 Q$. Find $Q(t)$.

7. Industrial waste is pumped into a tank containing 1000 gallons of water at the rate of 1 gal/min, and the well-stirred mixture leaves the tank at the same rate. (a) Find the concentration of waste in the tank at time t. (b) How long does it take for the concentration to reach 20%?

8. A tank contains 300 gallons of water and 100 gallons of pollutants. Fresh water is pumped into the tank at the rate of 2 gal/min, and the well-stirred mixture leaves at the same rate. How long does it take for the concentration of pollutants in the tank to decrease to 1/10 of its original value?

9. Consider a tank containing, at time $t = 0$, Q_0 lb of salt dissolved in 150 gallons of water. Assume that water containing $\frac{1}{2}$ lb of salt per gallon is entering the tank at a rate of 3 gal/min, and that the well-stirred solution is leaving the tank at the same rate. Find an expression for the concentration of salt in the tank at time t.

10. A room containing 1000 cubic feet of air is originally free of carbon monoxide. Beginning at time $t = 0$, cigarette smoke containing 4 percent carbon monoxide is blown into the room at the rate of 0.1 ft^3/min, and the well-circulated mixture leaves the room at the same rate. Find the time when the concentration of carbon monoxide in the room reaches 0.012 percent. (Extended exposure to this concentration of carbon monoxide is dangerous.)

11. A 500 gallon tank originally contains 100 gallons of fresh water. Beginning at time $t = 0$, water containing 50 percent pollutants flows into the tank at the rate of 2 gal/min, and the well-stirred mixture leaves at the rate of 1 gal/min. Find the concentration of pollutants in the tank at the moment it overflows.

In Exercises 12–17, find the orthogonal trajectories of the given family of curves.

12. $y = cx^2$

13. $y^2 - x^2 = c$

14. $y = c \sin x$

15. $x^2 + y^2 = cx$ (see Exercise 13 of Section 1.4)

16. $y = ce^x$

17. $y = e^{cx}$

1.8 Exact equations, and why we cannot solve very many differential equations

When we began our study of differential equations, the only equation we could solve was $dy/dt = g(t)$. We then enlarged our inventory to include all linear and separable equations. More generally, we can solve all differential equations which are, or can be put, in the form

$$\frac{d}{dt}\phi(t,y)=0 \tag{1}$$

for some function $\phi(t,y)$. To wit, we can integrate both sides of (1) to obtain that

$$\phi(t,y)=\text{constant} \tag{2}$$

and then solve for y as a function of t from (2).

Example 1. The equation $1+\cos(t+y)+\cos(t+y)(dy/dt)=0$ can be written in the form $(d/dt)[t+\sin(t+y)]=0$. Hence,

$$\phi(t,y)=t+\sin(t+y)=c, \quad \text{and} \quad y=-t+\arcsin(c-t).$$

Example 2. The equation $\cos(t+y)+[1+\cos(t+y)]dy/dt=0$ can be written in the form $(d/dt)[y+\sin(t+y)]=0$. Hence,

$$\phi(t,y)=y+\sin(t+y)=c.$$

We must leave the solution in this form though, since we cannot solve for y explicitly as a function of time.

Equation (1) is clearly the most general first-order differential equation that we can solve. Thus, it is important for us to be able to recognize when a differential equation can be put in this form. This is not as simple as one might expect. For example, it is certainly not obvious that the differential equation

$$2t+y-\sin t+(3y^2+\cos y+t)\frac{dy}{dt}=0$$

can be written in the form $(d/dt)(y^3+t^2+ty+\sin y+\cos t)=0$. To find all those differential equations which can be written in the form (1), observe, from the chain rule of partial differentiation, that

$$\frac{d}{dt}\phi(t,y(t))=\frac{\partial\phi}{\partial t}+\frac{\partial\phi}{\partial y}\frac{dy}{dt}.$$

Hence, the differential equation $M(t,y)+N(t,y)(dy/dt)=0$ can be written in the form $(d/dt)\phi(t,y)=0$ if and only if there exists a function $\phi(t,y)$ such that $M(t,y)=\partial\phi/\partial t$ and $N(t,y)=\partial\phi/\partial y$.

This now leads us to the following question. Given two functions $M(t,y)$ and $N(t,y)$, does there exist a function $\phi(t,y)$ such that $M(t,y) = \partial\phi/\partial t$ and $N(t,y)=\partial\phi/\partial y$? Unfortunately, the answer to this question is almost always no as the following theorem shows.

Theorem 1. *Let $M(t,y)$ and $N(t,y)$ be continuous and have continuous partial derivatives with respect to t and y in the rectangle R consisting of those points (t,y) with $a<t<b$ and $c<y<d$. There exists a function $\phi(t,y)$ such that $M(t,y)=\partial\phi/\partial t$ and $N(t,y)=\partial\phi/\partial y$ if, and only if,*

$$\partial M/\partial y = \partial N/\partial t$$

in R.

PROOF. Observe that $M(t,y)=\partial\phi/\partial t$ for some function $\phi(t,y)$ if, and only if,

$$\phi(t,y)=\int M(t,y)\,dt+h(y) \tag{3}$$

where $h(y)$ is an arbitrary function of y. Taking partial derivatives of both sides of (3) with respect to y, we obtain that

$$\frac{\partial\phi}{\partial y}=\int\frac{\partial M(t,y)}{\partial y}dt+h'(y).$$

Hence, $\partial\phi/\partial y$ will be equal to $N(t,y)$ if, and only if,

$$N(t,y)=\int\frac{\partial M(t,y)}{\partial y}dt+h'(y)$$

or

$$h'(y)=N(t,y)-\int\frac{\partial M(t,y)}{\partial y}dt. \tag{4}$$

Now $h'(y)$ is a function of y alone, while the right-hand side of (4) appears to be a function of both t and y. But a function of y alone cannot be equal to a function of both t and y. Thus Equation (4) makes sense only if the right-hand side is a function of y alone, and this is the case if, and only if,

$$\frac{\partial}{\partial t}\left[N(t,y)-\int\frac{\partial M(t,y)}{\partial y}dt\right]=\frac{\partial N}{\partial t}-\frac{\partial M}{\partial y}=0.$$

Hence, if $\partial N/\partial t\neq\partial M/\partial y$, then there is no function $\phi(t,y)$ such that $M =\partial\phi/\partial t$, $N=\partial\phi/\partial y$. On the other hand, if $\partial N/\partial t=\partial M/\partial y$ then we can solve for

$$h(y)=\int\left[N(t,y)-\int\frac{\partial M(t,y)}{\partial y}dt\right]dy.$$

Consequently, $M=\partial\phi/\partial t$, and $N=\partial\phi/\partial y$ with

$$\phi(t,y)=\int M(t,y)\,dt+\int\left[N(t,y)-\int\frac{\partial M(t,y)}{\partial y}dt\right]dy. \qquad \square \tag{5}$$

Definition. The differential equation

$$M(t,y)+N(t,y)\frac{dy}{dt}=0 \tag{6}$$

is said to be *exact* if $\partial M/\partial y=\partial N/\partial t$.

The reason for this definition, of course, is that the left-hand side of (6) is the exact derivative of a known function of t and y if $\partial M/\partial y=\partial N/\partial t$.

Remark 1. It is not essential, in the statement of Theorem 1, that $\partial M/\partial y=\partial N/\partial t$ in a rectangle. It is sufficient if $\partial M/\partial y=\partial N/\partial t$ in any region R which contains no "holes". That is to say, if C is any closed curve lying entirely in R, then its interior also lies entirely in R.

Remark 2. The differential equation $dy/dt=f(t,y)$ can always be written in the form $M(t,y)+N(t,y)(dy/dt)=0$ by setting $M(t,y)=-f(t,y)$ and $N(t,y)=1$.

Remark 3. It is customary to say that the solution of an exact differential equation is given by $\phi(t,y)=$ constant. What we really mean is that the equation $\phi(t,y)=c$ is to be solved for y as a function of t and c. Unfortunately, most exact differential equations cannot be solved explicitly for y as a function of t. While this may appear to be very disappointing, we wish to point out that it is quite simple, with the aid of a computer, to compute $y(t)$ to any desired accuracy.

In practice, we do not recommend memorizing Equation (5). Rather, we will follow one of three different methods to obtain $\phi(t,y)$.

First Method: The equation $M(t,y)=\partial\phi/\partial t$ determines $\phi(t,y)$ up to an arbitrary function of y alone, that is,

$$\phi(t,y)=\int M(t,y)\,dt+h(y).$$

The function $h(y)$ is then determined from the equation

$$h'(y)=N(t,y)-\int\frac{\partial M(t,y)}{\partial y}dt.$$

Second Method: If $N(t,y)=\partial\phi/\partial y$, then, of necessity,

$$\phi(t,y)=\int N(t,y)\,dy+k(t)$$

where $k(t)$ is an arbitrary function of t alone. Since

$$M(t,y)=\frac{\partial\phi}{\partial t}=\int\frac{\partial N(t,y)}{\partial t}dy+k'(t)$$

we see that $k(t)$ is determined from the equation

$$k'(t)=M(t,y)-\int\frac{\partial N(t,y)}{\partial t}dy.$$

Note that the right-hand side of this equation (see Exercise 2) is a function of t alone if $\partial M/\partial y=\partial N/\partial t$.

Third Method: The equations $\partial\phi/\partial t=M(t,y)$ and $\partial\phi/\partial y=N(t,y)$ imply that

$$\phi(t,y)=\int M(t,y)\,dt+h(y)\quad\text{and}\quad\phi(t,y)=\int N(t,y)\,dy+k(t).$$

Usually, we can determine $h(y)$ and $k(t)$ just by inspection.

Example 3. Find the general solution of the differential equation

$$3y+e^t+(3t+\cos y)\frac{dy}{dt}=0.$$

Solution. Here $M(t,y)=3y+e^t$ and $N(t,y)=3t+\cos y$. This equation is exact since $\partial M/\partial y=3$ and $\partial N/\partial t=3$. Hence, there exists a function $\phi(t,y)$ such that

$$\text{(i)}\ 3y+e^t=\frac{\partial\phi}{\partial t}\quad\text{and}\quad\text{(ii)}\ 3t+\cos y=\frac{\partial\phi}{\partial y}.$$

We will find $\phi(t,y)$ by each of the three methods outlined above.

First Method: From (i), $\phi(t,y)=e^t+3ty+h(y)$. Differentiating this equation with respect to y and using (ii) we obtain that

$$h'(y)+3t=3t+\cos y.$$

Thus, $h(y)=\sin y$ and $\phi(t,y)=e^t+3ty+\sin y$. (Strictly speaking, $h(y)=\sin y+\text{constant}$. However, we already incorporate this constant of integration into the solution when we write $\phi(t,y)=c$.) The general solution of the differential equation must be left in the form $e^t+3ty+\sin y=c$ since we cannot find y explicitly as a function of t from this equation.

Second Method: From (ii), $\phi(t,y)=3ty+\sin y+k(t)$. Differentiating this expression with respect to t, and using (i) we obtain that

$$3y+k'(t)=3y+e^t.$$

Thus, $k(t)=e^t$ and $\phi(t,y)=3ty+\sin y+e^t$.

Third Method: From (i) and (ii)

$$\phi(t,y)=e^t+3ty+h(y)\quad\text{and}\quad\phi(t,y)=3ty+\sin y+k(t).$$

Comparing these two expressions for the *same* function $\phi(t,y)$ it is obvious that $h(y)=\sin y$ and $k(t)=e^t$. Hence

$$\phi(t,y)=e^t+3ty+\sin y.$$

Example 4. Find the solution of the initial-value problem

$$3t^2y+8ty^2+(t^3+8t^2y+12y^2)\frac{dy}{dt}=0, \qquad y(2)=1.$$

Solution. Here $M(t,y)=3t^2y+8ty^2$ and $N(t,y)=t^3+8t^2y+12y^2$. This equation is exact since

$$\frac{\partial M}{\partial y}=3t^2+16ty \quad \text{and} \quad \frac{\partial N}{\partial t}=3t^2+16ty.$$

Hence, there exists a function $\phi(t,y)$ such that

$$\text{(i)}\ 3t^2y+8ty^2=\frac{\partial \phi}{\partial t} \quad \text{and} \quad \text{(ii)}\ t^3+8t^2y+12y^2=\frac{\partial \phi}{\partial y}.$$

Again, we will find $\phi(t,y)$ by each of three methods.

First Method: From (i), $\phi(t,y)=t^3y+4t^2y^2+h(y)$. Differentiating this equation with respect to y and using (ii) we obtain that

$$t^3+8t^2y+h'(y)=t^3+8t^2y+12y^2.$$

Hence, $h(y)=4y^3$ and the general solution of the differential equation is $\phi(t,y)=t^3y+4t^2y^2+4y^3=c$. Setting $t=2$ and $y=1$ in this equation, we see that $c=28$. Thus, the solution of our initial-value problem is defined implicitly by the equation $t^3y+4t^2y^2+4y^3=28$.

Second Method: From (ii), $\phi(t,y)=t^3y+4t^2y^2+4y^3+k(t)$. Differentiating this expression with respect to t and using (i) we obtain that

$$3t^2y+8ty^2+k'(t)=3t^2y+8ty^2.$$

Thus $k(t)=0$ and $\phi(t,y)=t^3y+4t^2y^2+4y^3$.

Third Method: From (i) and (ii)

$$\phi(t,y)=t^3y+4t^2y^2+h(y) \quad \text{and} \quad \phi(t,y)=t^3y+4t^2y^2+4y^3+k(t).$$

Comparing these two expressions for the same function $\phi(t,y)$ we see that $h(y)=4y^3$ and $k(t)=0$. Hence, $\phi(t,y)=t^3y+4t^2y^2+4y^3$.

In most instances, as Examples 3 and 4 illustrate, the third method is the simplest to use. However, if it is much easier to integrate N with respect to y than it is to integrate M with respect to t, we should use the second method, and vice-versa.

Example 5. Find the solution of the initial-value problem

$$4t^3e^{t+y}+t^4e^{t+y}+2t+(t^4e^{t+y}+2y)\frac{dy}{dt}=0, \qquad y(0)=1.$$

Solution. This equation is exact since

$$\frac{\partial}{\partial y}(4t^3e^{t+y}+t^4e^{t+y}+2t)=(t^4+4t^3)e^{t+y}=\frac{\partial}{\partial t}(t^4e^{t+y}+2y).$$

Hence, there exists a function $\phi(t,y)$ such that

$$\text{(i)}\ \ 4t^3e^{t+y}+t^4e^{t+y}+2t=\frac{\partial\phi}{\partial t}$$

and

$$\text{(ii)}\ \ t^4e^{t+y}+2y=\frac{\partial\phi}{\partial y}.$$

Since it is much simpler to integrate $t^4e^{t+y}+2y$ with respect to y than it is to integrate $4t^3e^{t+y}+t^4e^{t+y}+2t$ with respect to t, we use the second method. From (ii), $\phi(t,y)=t^4e^{t+y}+y^2+k(t)$. Differentiating this expression with respect to t and using (i) we obtain

$$(t^4+4t^3)e^{t+y}+k'(t)=4t^3e^{t+y}+t^4e^{t+y}+2t.$$

Thus, $k(t)=t^2$ and the general solution of the differential equation is $\phi(t,y)=t^4e^{t+y}+y^2+t^2=c$. Setting $t=0$ and $y=1$ in this equation yields $c=1$. Thus, the solution of our initial-value problem is defined implicitly by the equation $t^4e^{t+y}+t^2+y^2=1$.

Suppose now that we are given a differential equation

$$M(t,y)+N(t,y)\frac{dy}{dt}=0 \tag{7}$$

which is not exact. Can we make it exact? More precisely, can we find a function $\mu(t,y)$ such that the equivalent differential equation

$$\mu(t,y)M(t,y)+\mu(t,y)N(t,y)\frac{dy}{dt}=0 \tag{8}$$

is exact? This question is simple, in principle, to answer. The condition that (8) be exact is that

$$\frac{\partial}{\partial y}(\mu(t,y)M(t,y))=\frac{\partial}{\partial t}(\mu(t,y)N(t,y))$$

or

$$M\frac{\partial\mu}{\partial y}+\mu\frac{\partial M}{\partial y}=N\frac{\partial\mu}{\partial t}+\mu\frac{\partial N}{\partial t}. \tag{9}$$

(For simplicity of writing, we have suppressed the dependence of μ, M and N on t and y in (9).) Thus, Equation (8) is exact if and only if $\mu(t,y)$ satisfies Equation (9).

Definition. A function $\mu(t,y)$ satisfying Equation (9) is called an *integrating factor* for the differential equation (7).

The reason for this definition, of course, is that if μ satisfies (9) then we can write (8) in the form $(d/dt)\phi(t,y)=0$ and this equation can be integrated immediately to yield the solution $\phi(t,y)=c$. Unfortunately, though, there are only two special cases where we can find an explicit solution of (9). These occur when the differential equation (7) has an integrating factor which is either a function of t alone, or a function of y alone. Observe that if μ is a function of t alone, then Equation (9) reduces to

$$N\frac{d\mu}{dt}=\mu\left(\frac{\partial M}{\partial y}-\frac{\partial N}{\partial t}\right)\quad\text{or}\quad\frac{d\mu}{dt}=\frac{\left(\frac{\partial M}{\partial y}-\frac{\partial N}{\partial t}\right)}{N}\mu.$$

But this equation is meaningless unless the expression

$$\frac{\frac{\partial M}{\partial y}-\frac{\partial N}{\partial t}}{N}$$

is a function of t alone, that is,

$$\frac{\frac{\partial M}{\partial y}-\frac{\partial N}{\partial t}}{N}=R(t).$$

If this is the case then $\mu(t)=\exp\left(\int R(t)\,dt\right)$ is an integrating factor for the differential equation (7).

Remark. It should be noted that the expression

$$\frac{\frac{\partial M}{\partial y}-\frac{\partial N}{\partial t}}{N}$$

is almost always a function of both t and y. Only for very special pairs of functions $M(t,y)$ and $N(t,y)$ is it a function of t alone. A similar situation occurs if μ is a function of y alone (see Exercise 17). It is for this reason that we cannot solve very many differential equations.

Example 6. Find the general solution of the differential equation

$$\frac{y^2}{2}+2ye^t+(y+e^t)\frac{dy}{dt}=0.$$

Solution. Here $M(t,y)=(y^2/2)+2ye^t$ and $N(t,y)=y+e^t$. This equation is not exact since $\partial M/\partial y=y+2e^t$ and $\partial N/\partial t=e^t$. However,

$$\frac{1}{N}\left(\frac{\partial M}{\partial y}-\frac{\partial N}{\partial t}\right)=\frac{y+e^t}{y+e^t}=1.$$

Hence, this equation has $\mu(t)=\exp\left(\int 1\,dt\right)=e^t$ as an integrating factor. This means, of course, that the equivalent differential equation

$$\frac{y^2}{2}e^t+2ye^{2t}+(ye^t+e^{2t})\frac{dy}{dt}=0$$

is exact. Therefore, there exists a function $\phi(t,y)$ such that

$$\text{(i)}\ \frac{y^2}{2}e^t+2ye^{2t}=\frac{\partial\phi}{\partial t}$$

and

$$\text{(ii)}\ ye^t+e^{2t}=\frac{\partial\phi}{\partial y}.$$

From Equations (i) and (ii),

$$\phi(t,y)=\frac{y^2}{2}e^t+ye^{2t}+h(y)$$

and

$$\phi(t,y)=\frac{y^2}{2}e^t+ye^{2t}+k(t).$$

Thus, $h(y)=0$, $k(t)=0$ and the general solution of the differential equation is

$$\phi(t,y)=\frac{y^2}{2}e^t+ye^{2t}=c.$$

Solving this equation for y as a function of t we see that

$$y(t)=-e^t\pm\left[e^{2t}+2ce^{-t}\right]^{1/2}.$$

Example 7. Use the methods of this section to find the general solution of the linear equation $(dy/dt)+a(t)y=b(t)$.

Solution. We write this equation in the form $M(t,y)+N(t,y)(dy/dt)=0$ with $M(t,y)=a(t)y-b(t)$ and $N(t,y)=1$. This equation is not exact since $\partial M/\partial y=a(t)$ and $\partial N/\partial t=0$. However, $((\partial M/\partial y)-(\partial N/\partial t))/N=a(t)$. Hence, $\mu(t)=\exp\left(\int a(t)\,dt\right)$ is an integrating factor for the first-order linear equation. Therefore, there exists a function $\phi(t,y)$ such that

$$\text{(i)}\ \mu(t)\left[a(t)y-b(t)\right]=\frac{\partial\phi}{\partial t}$$

and

$$\text{(ii)}\ \mu(t)=\frac{\partial\phi}{\partial y}.$$

Now, observe from (ii) that $\phi(t,y)=\mu(t)y+k(t)$. Differentiating this equation with respect to t and using (i) we see that

$$\mu'(t)y+k'(t)=\mu(t)a(t)y-\mu(t)b(t).$$

But, $\mu'(t)=a(t)\mu(t)$. Consequently, $k'(t)=-\mu(t)b(t)$ and

$$\phi(t,y)=\mu(t)y-\int\mu(t)b(t)\,dt.$$

Hence, the general solution of the first-order linear equation is

$$\mu(t)y-\int u(t)b(t)\,dt=c,$$

and this is the result we obtained in Section 1.2.

Exercises

1. Use the theorem of equality of mixed partial derivatives to show that $\partial M/\partial y = \partial N/\partial t$ if the equation $M(t,y)+N(t,y)(dy/dt)=0$ is exact.

2. Show that the expression $M(t,y)-\int(\partial N(t,y)/\partial t)dy$ is a function of t alone if $\partial M/\partial y=\partial N/\partial t$.

In each of Problems 3–6 find the general solution of the given differential equation.

3. $2t\sin y+y^3e^t+(t^2\cos y+3y^2e^t)\dfrac{dy}{dt}=0$

4. $1+(1+ty)e^{ty}+(1+t^2e^{ty})\dfrac{dy}{dt}=0$

5. $y\sec^2 t+\sec t\tan t+(2y+\tan t)\dfrac{dy}{dt}=0$

6. $\dfrac{y^2}{2}-2ye^t+(y-e^t)\dfrac{dy}{dt}=0$

In each of Problems 7–11, solve the given initial-value problem.

7. $2ty^3+3t^2y^2\dfrac{dy}{dt}=0, \qquad y(1)=1$

8. $2t\cos y+3t^2y+(t^3-t^2\sin y-y)\dfrac{dy}{dt}=0, \qquad y(0)=2$

9. $3t^2+4ty+(2y+2t^2)\dfrac{dy}{dt}=0, \qquad y(0)=1$

10. $y(\cos 2t)e^{ty}-2(\sin 2t)e^{ty}+2t+(t(\cos 2t)e^{ty}-3)\dfrac{dy}{dt}=0, \qquad y(0)=0$

11. $3ty+y^2+(t^2+ty)\dfrac{dy}{dt}=0, \qquad y(2)=1$

In each of Problems 12–14, determine the constant a so that the equation is exact, and then solve the resulting equation.

12. $t+ye^{2ty}+ate^{2ty}\dfrac{dy}{dt}=0$

13. $\dfrac{1}{t^2}+\dfrac{1}{y^2}+\dfrac{(at+1)}{y^3}\dfrac{dy}{dt}=0$

14. $e^{at+y}+3t^2y^2+(2yt^3+e^{at+y})\dfrac{dy}{dt}=0$

15. Show that every separable equation of the form $M(t)+N(y)dy/dt=0$ is exact.

16. Find all functions $f(t)$ such that the differential equation

$$y^2\sin t+yf(t)(dy/dt)=0$$

is exact. Solve the differential equation for these $f(t)$.

17. Show that if $((\partial N/\partial t)-(\partial M/\partial y))/M=Q(y)$, then the differential equation $M(t,y)+N(t,y)dy/dt=0$ has an integrating factor $\mu(y)=\exp\left(\int Q(y)\,dy\right)$.

18. The differential equation $f(t)(dy/dt)+t^2+y=0$ is known to have an integrating factor $\mu(t)=t$. Find all possible functions $f(t)$.

19. The differential equation $e^t\sec y-\tan y+(dy/dt)=0$ has an integrating factor of the form $e^{-at}\cos y$ for some constant a. Find a, and then solve the differential equation.

20. The Bernoulli differential equation is $(dy/dt)+a(t)y=b(t)y^n$. Multiplying through by $\mu(t)=\exp\left(\int a(t)\,dt\right)$, we can rewrite this equation in the form $d/dt(\mu(t)y)=b(t)\mu(t)y^n$. Find the general solution of this equation by finding an appropriate integrating factor. *Hint*: Divide both sides of the equation by an appropriate function of y.

1.9 The existence–uniqueness theorem; Picard iteration

Consider the initial-value problem

$$\frac{dy}{dt}=f(t,y),\qquad y(t_0)=y_0 \tag{1}$$

where f is a given function of t and y. Chances are, as the remarks in Section 1.8 indicate, that we will be unable to solve (1) explicitly. This leads us to ask the following questions.

1. How are we to know that the initial-value problem (1) actually has a solution if we can't exhibit it?
2. How do we know that there is only one solution $y(t)$ of (1)? Perhaps there are two, three, or even infinitely many solutions.
3. Why bother asking the first two questions? After all, what's the use of determining whether (1) has a unique solution if we won't be able to explicitly exhibit it?

The answer to the third question lies in the observation that it is never necessary, in applications, to find the solution $y(t)$ of (1) to more than a finite number of decimal places. Usually, it is more than sufficient to find $y(t)$ to four decimal places, and this can be done quite easily with the aid of a digital computer. In fact, it is possible to compute $y(t)$ to eight, and even sixteen, decimal places. Thus, the knowledge that (1) has a unique solution $y(t)$ is our hunting license to go looking for it.

To resolve the first question, we must establish the existence of a function $y(t)$ whose value at $t=t_0$ is y_0, and whose derivative at any time t equals $f(t,y(t))$. In order to accomplish this, we must find a theorem which enables us to establish the existence of a function having certain properties, without our having to exhibit this function explicitly. If we search through the Calculus, we find that we encounter such a situation exactly once, and this is in connection with the theory of limits. As we show in Appendix B, it is often possible to prove that a sequence of functions $y_n(t)$ has a limit $y(t)$, without our having to exhibit $y(t)$. For example, we can prove that the sequence of functions

$$y_n(t)=\frac{\sin \pi t}{1^2}+\frac{\sin 2\pi t}{2^2}+\ldots+\frac{\sin n\pi t}{n^2}$$

has a limit $y(t)$ even though we cannot exhibit $y(t)$ explicitly. This suggests the following algorithm for proving the existence of a solution $y(t)$ of (1).

(a) Construct a sequence of functions $y_n(t)$ which come closer and closer to solving (1).
(b) Show that the sequence of functions $y_n(t)$ has a limit $y(t)$ on a suitable interval $t_0 \leqslant t \leqslant t_0+\alpha$.
(c) Prove that $y(t)$ is a solution of (1) on this interval.

We now show how to implement this algorithm.

(a) *Construction of the approximating sequence* $y_n(t)$

The problem of finding a sequence of functions that come closer and closer to satisfying a certain equation is one that arises quite often in mathematics. Experience has shown that it is often easiest to resolve this problem when our equation can be written in the special form

$$y(t)=L(t,y(t)), \tag{2}$$

where L may depend explicitly on y, and on integrals of functions of y. For example, we may wish to find a function $y(t)$ satisfying

$$y(t)=1+\sin\left[t+y(t)\right],$$

or

$$y(t)=1+y^2(t)+\int_0^t y^3(s)\,ds.$$

In these two cases, $L(t,y(t))$ is an abbreviation for

$$1+\sin[t+y(t)]$$

and

$$1+y^2(t)+\int_0^t y^3(s)\,ds,$$

respectively.

The key to understanding what is special about Equation (2) is to view $L(t,y(t))$ as a "machine" that takes in one function and gives back another one. For example, let

$$L(t,y(t))=1+y^2(t)+\int_0^t y^3(s)\,ds.$$

If we plug the function $y(t)=t$ into this machine, (that is, if we compute $1+t^2+\int_0^t s^3\,ds$) then the machine returns to us the function $1+t^2+t^4/4$. If we plug the function $y(t)=\cos t$ into this machine, then it returns to us the function

$$1+\cos^2 t+\int_0^t \cos^3 s\,ds=1+\cos^2 t+\sin t-\frac{\sin^3 t}{3}.$$

According to this viewpoint, we can characterize all solutions $y(t)$ of (2) as those functions $y(t)$ which the machine L leaves unchanged. In other words, if we plug a function $y(t)$ into the machine L, and the machine returns to us this same function, then $y(t)$ is a solution of (2).

We can put the initial-value problem (1) into the special form (2) by integrating both sides of the differential equation $y'=f(t,y)$ with respect to t. Specifically, if $y(t)$ satisfies (1), then

$$\int_{t_0}^t \frac{dy(s)}{ds}\,ds=\int_{t_0}^t f(s,y(s))\,ds$$

so that

$$y(t)=y_0+\int_{t_0}^t f(s,y(s))\,ds. \tag{3}$$

Conversely, if $y(t)$ is continuous and satisfies (3), then $dy/dt=f(t,y(t))$. Moreover, $y(t_0)$ is obviously y_0. Therefore, $y(t)$ is a solution of (1) if, and only if, it is a continuous solution of (3).

Equation (3) is called an integral equation, and it is in the special form (2) if we set

$$L(t,y(t))=y_0+\int_{t_0}^t f(s,y(s))\,ds.$$

This suggests the following scheme for constructing a sequence of "approximate solutions" $y_n(t)$ of (3). Let us start by guessing a solution $y_0(t)$ of

(3). The simplest possible guess is $y_0(t)=y_0$. To check whether $y_0(t)$ is a solution of (3), we compute

$$y_1(t)=y_0+\int_{t_0}^{t} f(s,y_0(s))\,ds.$$

If $y_1(t)=y_0$, then $y(t)=y_0$ is indeed a solution of (3). If not, then we try $y_1(t)$ as our next guess. To check whether $y_1(t)$ is a solution of (3), we compute

$$y_2(t)=y_0+\int_{t_0}^{t} f(s,y_1(s))\,ds,$$

and so on. In this manner, we define a sequence of functions $y_1(t)$, $y_2(t),\ldots$, where

$$y_{n+1}(t)=y_0+\int_{t_0}^{t} f(s,y_n(s))\,ds. \tag{4}$$

These functions $y_n(t)$ are called successive approximations, or Picard iterates, after the French mathematician Picard who first discovered them. Remarkably, these Picard iterates always converge, on a suitable interval, to a solution $y(t)$ of (3).

Example 1. Compute the Picard iterates for the initial-value problem

$$y'=y, \qquad y(0)=1,$$

and show that they converge to the solution $y(t)=e^t$.

Solution. The integral equation corresponding to this initial-value problem is

$$y(t)=1+\int_0^t y(s)\,ds.$$

Hence, $y_0(t)=1$

$$y_1(t)=1+\int_0^t 1\,ds=1+t$$

$$y_2(t)=1+\int_0^t y_1(s)\,ds=1+\int_0^t (1+s)\,ds=1+t+\frac{t^2}{2!}$$

and, in general,

$$y_n(t)=1+\int_0^t y_{n-1}(s)\,ds=1+\int_0^t\left[1+s+\ldots+\frac{s^{n-1}}{(n-1)!}\right]ds$$

$$=1+t+\frac{t^2}{2!}+\ldots+\frac{t^n}{n!}.$$

Since $e^t=1+t+t^2/2!+\ldots$, we see that the Picard iterates $y_n(t)$ converge to the solution $y(t)$ of this initial-value problem.

Example 2. Compute the Picard iterates $y_1(t), y_2(t)$ for the initial-value problem $y' = 1 + y^3$, $y(1) = 1$.

Solution. The integral equation corresponding to this initial-value problem is

$$y(t) = 1 + \int_1^t \left[1 + y^3(s)\right] ds.$$

Hence, $y_0(t) = 1$

$$y_1(t) = 1 + \int_1^t (1+1)\,ds = 1 + 2(t-1)$$

and

$$y_2(t) = 1 + \int_1^t \left\{1 + \left[1 + 2(s-1)\right]^3\right\} ds$$
$$= 1 + 2(t-1) + 3(t-1)^2 + 4(t-1)^3 + 2(t-1)^4.$$

Notice that it is already quite cumbersome to compute $y_3(t)$.

(b) *Convergence of the Picard iterates*

As was mentioned in Section 1.4, the solutions of nonlinear differential equations may not exist for all time t. Therefore, we cannot expect the Picard iterates $y_n(t)$ of (3) to converge for all t. To provide us with a clue, or estimate, of where the Picard iterates converge, we try to find an interval in which all the $y_n(t)$ are uniformly bounded (that is, $|y_n(t)| \leqslant K$ for some fixed constant K). Equivalently, we seek a rectangle R which contains the graphs of all the Picard iterates $y_n(t)$. Lemma 1 shows us how to find such a rectangle.

Lemma 1. *Choose any two positive numbers a and b, and let R be the rectangle: $t_0 \leqslant t \leqslant t_0 + a$, $|y - y_0| \leqslant b$. Compute*

$$M = \max_{(t,y) \text{ in } R} |f(t,y)|, \quad \textit{and set} \quad \alpha = \min\left(a, \frac{b}{M}\right).$$

Then,

$$|y_n(t) - y_0| \leqslant M(t - t_0) \tag{5}$$

for $t_0 \leqslant t \leqslant t_0 + \alpha$.

Lemma 1 states that the graph of $y_n(t)$ is sandwiched between the lines $y = y_0 + M(t - t_0)$ and $y = y_0 - M(t - t_0)$, for $t_0 \leqslant t \leqslant t_0 + \alpha$. These lines leave the rectangle R at $t = t_0 + a$ if $a \leqslant b/M$, and at $t = t_0 + b/M$ if $b/M < a$ (see Figures 1a and 1b). In either case, therefore, the graph of $y_n(t)$ is contained in R for $t_0 \leqslant t \leqslant t_0 + \alpha$.

PROOF OF LEMMA 1. We establish (5) by induction on n. Observe first that (5) is obviously true for $n = 0$, since $y_0(t) = y_0$. Next, we must show that (5) is true for $n = j + 1$ if it is true for $n = j$. But this follows immediately, for if

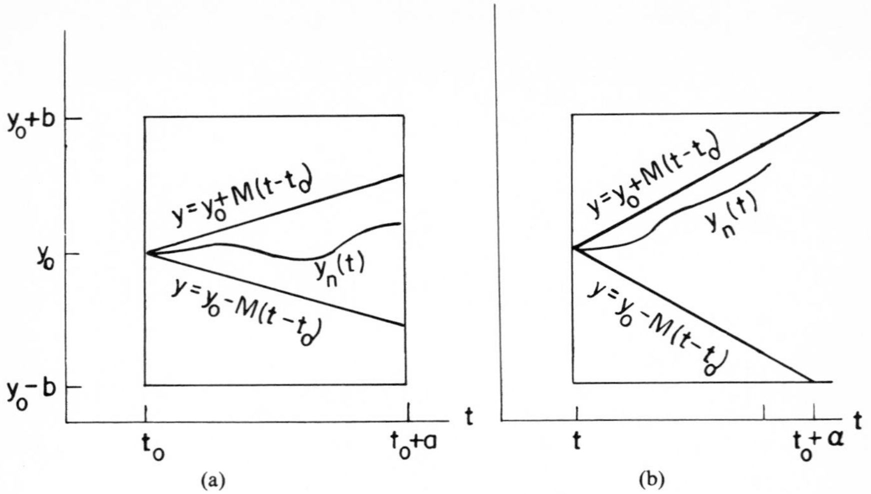

Figure 1. (a) $\alpha=a$; (b) $\alpha=b/M$

$|y_j(t)-y_0| \leqslant M(t-t_0)$, then

$$|y_{j+1}(t)-y_0| = \left| \int_{t_0}^{t} f(s,y_j(s))\,ds \right|$$

$$\leqslant \int_{t_0}^{t} |f(s,y_j(s))|\,ds \leqslant M(t-t_0)$$

for $t_0 \leqslant t \leqslant t_0+\alpha$. Consequently, (5) is true for all n, by induction. □

We now show that the Picard iterates $y_n(t)$ of (3) converge for each t in the interval $t_0 \leqslant t \leqslant t_0+\alpha$, if $\partial f/\partial y$ exists and is continuous. Our first step is to reduce the problem of showing that the sequence of functions $y_n(t)$ converges to the much simpler problem of proving that an infinite series converges. This is accomplished by writing $y_n(t)$ in the form

$$y_n(t) = y_0(t) + [y_1(t)-y_0(t)] + \ldots + [y_n(t)-y_{n-1}(t)].$$

Clearly, the sequence $y_n(t)$ converges if, and only if, the infinite series

$$[y_1(t)-y_0(t)] + [y_2(t)-y_1(t)] + \ldots + [y_n(t)-y_{n-1}(t)] + \ldots \tag{6}$$

converges. To prove that the infinite series (6) converges, it suffices to show that

$$\sum_{n=1}^{\infty} |y_n(t)-y_{n-1}(t)| < \infty. \tag{7}$$

This is accomplished in the following manner. Observe that

$$\begin{aligned}|y_n(t)-y_{n-1}(t)| &= \left|\int_{t_0}^{t}\left[f(s,y_{n-1}(s))-f(s,y_{n-2}(s))\right]ds\right|\\ &\leqslant \int_{t_0}^{t}|f(s,y_{n-1}(s))-f(s,y_{n-2}(s))|\,ds\\ &= \int_{t_0}^{t}\left|\frac{\partial f(s,\xi(s))}{\partial y}\right||y_{n-1}(s)-y_{n-2}(s)|\,ds,\end{aligned}$$

where $\xi(s)$ lies between $y_{n-1}(s)$ and $y_{n-2}(s)$. (Recall that $f(x_1)-f(x_2)=f'(\xi)(x_1-x_2)$, where ξ is some number between x_1 and x_2.) It follows immediately from Lemma 1 that the points $(s,\xi(s))$ all lie in the rectangle R for $s<t_0+\alpha$. Consequently,

$$|y_n(t)-y_{n-1}(t)|\leqslant L\int_{t_0}^{t}|y_{n-1}(s)-y_{n-2}(s)|\,ds, \qquad t_0\leqslant t\leqslant t_0+\alpha, \tag{8}$$

where

$$L=\max_{(t,y)\text{ in }R}\left|\frac{\partial f(t,y)}{\partial y}\right|. \tag{9}$$

Equation (9) defines the constant L. Setting $n=2$ in (8) gives

$$\begin{aligned}|y_2(t)-y_1(t)| &\leqslant L\int_{t_0}^{t}|y_1(s)-y_0|\,ds\leqslant L\int_{t_0}^{t}M(s-t_0)\,ds\\ &= \frac{LM(t-t_0)^2}{2}.\end{aligned}$$

This, in turn, implies that

$$\begin{aligned}|y_3(t)-y_2(t)| &\leqslant L\int_{t_0}^{t}|y_2(s)-y_1(s)|\,ds\leqslant ML^2\int_{t_0}^{t}\frac{(s-t_0)^2}{2}\,ds\\ &= \frac{ML^2(t-t_0)^3}{3!}.\end{aligned}$$

Proceeding inductively, we see that

$$|y_n(t)-y_{n-1}(t)|\leqslant\frac{ML^{n-1}(t-t_0)^n}{n!}, \quad \text{for } t_0\leqslant t\leqslant t_0+\alpha. \tag{10}$$

Therefore, for $t_0 \leqslant t \leqslant t_0 + \alpha$,

$$|y_1(t) - y_0(t)| + |y_2(t) - y_1(t)| + \dots$$

$$\leqslant M(t - t_0) + \frac{ML(t-t_0)^2}{2!} + \frac{ML^2(t-t_0)^3}{3!} + \dots$$

$$\leqslant M\alpha + \frac{ML\alpha^2}{2!} + \frac{ML^2\alpha^3}{3!} + \dots$$

$$= \frac{M}{L}\left[\alpha L + \frac{(\alpha L)^2}{2!} + \frac{(\alpha L)^3}{3!} + \dots\right]$$

$$= \frac{M}{L}(e^{\alpha L} - 1).$$

This quantity, obviously, is less than infinity. Consequently, the Picard iterates $y_n(t)$ converge for each t in the interval $t_0 \leqslant t \leqslant t_0 + \alpha$. (A similar argument shows that $y_n(t)$ converges for each t in the interval $t_0 - \beta \leqslant t \leqslant t_0$, where $\beta = \min(a, b/N)$, and N is the maximum value of $|f(t,y)|$ for (t,y) in the rectangle $t_0 - a \leqslant t \leqslant t_0, |y - y_0| \leqslant b$.) We will denote the limit of the sequence $y_n(t)$ by $y(t)$. □

(c) *Proof that $y(t)$ satisfies the initial-value problem* (1)

We will show that $y(t)$ satisfies the integral equation

$$y(t) = y_0 + \int_{t_0}^{t} f(s, y(s))\, ds \tag{11}$$

and that $y(t)$ is continuous. To this end, recall that the Picard iterates $y_n(t)$ are defined recursively through the equation

$$y_{n+1}(t) = y_0 + \int_{t_0}^{t} f(s, y_n(s))\, ds. \tag{12}$$

Taking limits of both sides of (12) gives

$$y(t) = y_0 + \lim_{n\to\infty} \int_{t_0}^{t} f(s, y_n(s))\, ds. \tag{13}$$

To show that the right-hand side of (13) equals

$$y_0 + \int_{t_0}^{t} f(s, y(s))\, ds,$$

(that is, to justify passing the limit through the integral sign) we must show that

$$\left| \int_{t_0}^{t} f(s, y(s))\, ds - \int_{t_0}^{t} f(s, y_n(s))\, ds \right|$$

approaches zero as n approaches infinity. This is accomplished in the following manner. Observe first that the graph of $y(t)$ lies in the rectangle R for $t \leqslant t_0 + \alpha$, since it is the limit of functions $y_n(t)$ whose graphs lie in R.

Hence

$$\left|\int_{t_0}^{t} f(s,y(s))\,ds - \int_{t_0}^{t} f(s,y_n(s))\,ds\right|$$
$$\leq \int_{t_0}^{t} |f(s,y(s)) - f(s,y_n(s))|\,ds \leq L\int_{t_0}^{t} |y(s)-y_n(s)|\,ds$$

where L is defined by Equation (9). Next, observe that

$$y(s)-y_n(s) = \sum_{j=n+1}^{\infty}\left[y_j(s)-y_{j-1}(s)\right]$$

since

$$y(s) = y_0 + \sum_{j=1}^{\infty}\left[y_j(s)-y_{j-1}(s)\right]$$

and

$$y_n(s) = y_0 + \sum_{j=1}^{n}\left[y_j(s)-y_{j-1}(s)\right].$$

Consequently, from (10),

$$|y(s)-y_n(s)| \leq M\sum_{j=n+1}^{\infty} L^{j-1}\frac{(s-t_0)^j}{j!}$$
$$\leq M\sum_{j=n+1}^{\infty}\frac{L^{j-1}\alpha^j}{j!} = \frac{M}{L}\sum_{j=n+1}^{\infty}\frac{(\alpha L)^j}{j!}, \tag{14}$$

and

$$\left|\int_{t_0}^{t} f(s,y(s))\,ds - \int_{t_0}^{t} f(s,y_n(s))\,ds\right| \leq M\sum_{j=n+1}^{\infty}\frac{(\alpha L)^j}{j!}\int_{t_0}^{t} ds$$
$$\leq M\alpha\sum_{j=n+1}^{\infty}\frac{(\alpha L)^j}{j!}.$$

This summation approaches zero as n approaches infinity, since it is the tail end of the convergent Taylor series expansion of $e^{\alpha L}$. Hence,

$$\lim_{n\to\infty}\int_{t_0}^{t} f(s,y_n(s))\,ds = \int_{t_0}^{t} f(s,y(s))\,ds,$$

and $y(t)$ satisfies (11).

To show that $y(t)$ is continuous, we must show that for every $\varepsilon > 0$ we can find $\delta > 0$ such that

$$|y(t+h)-y(t)| < \varepsilon \quad \text{if } |h| < \delta.$$

Now, we cannot compare $y(t+h)$ with $y(t)$ directly, since we do not know $y(t)$ explicitly. To overcome this difficulty, we choose a large integer N and observe that

$$\begin{aligned} y(t+h)-y(t)= &\left[y(t+h)-y_N(t+h)\right] \\ &+\left[y_N(t+h)-y_N(t)\right]+\left[y_N(t)-y(t)\right]. \end{aligned}$$

Specifically, we choose N so large that

$$\frac{M}{L}\sum_{j=N+1}^{\infty}\frac{(\alpha L)^j}{j!}<\frac{\varepsilon}{3}.$$

Then, from (14),

$$|y(t+h)-y_N(t+h)|<\frac{\varepsilon}{3} \quad\text{and}\quad |y_N(t)-y(t)|<\frac{\varepsilon}{3},$$

for $t<t_0+\alpha$, and h sufficiently small (so that $t+h<t_0+\alpha$.) Next, observe that $y_N(t)$ is continuous, since it is obtained from N repeated integrations of continuous functions. Therefore, we can choose $\delta>0$ so small that

$$|y_N(t+h)-y_N(t)|<\frac{\varepsilon}{3} \quad \text{for } |h|<\delta.$$

Consequently,

$$\begin{aligned} |y(t+h)-y(t)| \leqslant\ & |y(t+h)-y_N(t+h)|+|y_N(t+h)-y_N(t)| \\ &+|y_N(t)-y(t)|<\frac{\varepsilon}{3}+\frac{\varepsilon}{3}+\frac{\varepsilon}{3}=\varepsilon \end{aligned}$$

for $|h|<\delta$. Therefore, $y(t)$ is a continuous solution of the integral equation (11), and this completes our proof that $y(t)$ satisfies (1). □

In summary, we have proven the following theorem.

Theorem 2. *Let f and $\partial f/\partial y$ be continuous in the rectangle $R: t_0 \leqslant t \leqslant t_0+a$, $|y-y_0|\leqslant b$. Compute*

$$M=\max_{(t,y)\text{ in }R}|f(t,y)|, \quad \textit{and set} \quad \alpha=\min\left(a,\frac{b}{M}\right).$$

Then, the initial-value problem $y'=f(t,y)$, $y(t_0)=y_0$ has at least one solution $y(t)$ on the interval $t_0\leqslant t\leqslant t_0+\alpha$. A similar result is true for $t<t_0$.

Remark. The number α in Theorem 2 depends specifically on our choice of a and b. Different choices of a and b lead to different values of α. Moreover, α doesn't necessarily increase when a and b increase, since an increase in a or b will generally result in an increase in M.

Finally, we turn our attention to the problem of uniqueness of solutions of (1). Consider the initial-value problem

$$\frac{dy}{dt}=(\sin 2t)y^{1/3}, \qquad y(0)=0. \tag{15}$$

One solution of (15) is $y(t)=0$. Additional solutions can be obtained if we ignore the fact that $y(0)=0$ and rewrite the differential equation in the form

$$\frac{1}{y^{1/3}}\frac{dy}{dt}=\sin 2t,$$

or

$$\frac{d}{dt}\frac{3y^{2/3}}{2}=\sin 2t.$$

Then,

$$\frac{3y^{2/3}}{2}=\frac{1-\cos t}{2}=\sin^2 t$$

and $y=\pm\sqrt{8/27}\,\sin^3 t$ are two additional solutions of (15).

Now, initial-value problems that have more than one solution are clearly unacceptable in applications. Therefore, it is important for us to find out exactly what is "wrong" with the initial-value problem (15) that it has more than one solution. If we look carefully at the right-hand side of this differential equation, we see that it does not have a partial derivative with respect to y at $y=0$. This is indeed the problem, as the following theorem shows.

Theorem 2′. *Let f and $\partial f/\partial y$ be continuous in the rectangle $R: t_0 \leqslant t \leqslant t_0+a$, $|y-y_0|\leqslant b$. Compute*

$$M=\max_{(t,y)\text{ in }R}|f(t,y)|,\quad \textit{and set}\quad \alpha=\min\left(a,\frac{b}{M}\right).$$

Then, the initial-value problem

$$y'=f(t,y),\qquad y(t_0)=y_0 \tag{16}$$

has a unique solution $y(t)$ on the interval $t_0\leqslant t\leqslant t_0+\alpha$. In other words, if $y(t)$ and $z(t)$ are two solutions of (16), then $y(t)$ must equal $z(t)$ for $t_0\leqslant t \leqslant t_0+\alpha$.

PROOF. Theorem 2 guarantees the existence of at least one solution $y(t)$ of (16). Suppose that $z(t)$ is a second solution of (16). Then,

$$y(t)=y_0+\int_{t_0}^{t} f(s,y(s))\,ds \quad\text{and}\quad z(t)=y_0+\int_{t_0}^{t} f(s,z(s))\,ds.$$

Subtracting these two equations gives

$$\begin{aligned}|y(t)-z(t)|&=\left|\int_{t_0}^{t}\left[f(s,y(s))-f(s,z(s))\right]ds\right|\\ &\leqslant\int_{t_0}^{t}|f(s,y(s))-f(s,z(s))|\,ds\\ &\leqslant L\int_{t_0}^{t}|y(s)-z(s)|\,ds\end{aligned}$$

where L is the maximum value of $|\partial f/\partial y|$ for (t,y) in R. As Lemma 2 below shows, this inequality implies that $y(t)=z(t)$. Hence, the initial-value problem (16) has a unique solution $y(t)$. □

Lemma 2. *Let $w(t)$ be a nonnegative function, with*

$$w(t)\leqslant L\int_{t_0}^{t} w(s)\,ds. \tag{17}$$

Then, $w(t)$ is identically zero.

FAKE PROOF. Differentiating both sides of (17) gives

$$\frac{dw}{dt}\leqslant Lw(t), \quad \text{or} \quad \frac{dw}{dt}-Lw(t)\leqslant 0.$$

Multiplying both sides of this inequality by the integrating factor $e^{-L(t-t_0)}$ gives

$$\frac{d}{dt}e^{-L(t-t_0)}w(t)\leqslant 0, \quad \text{so that} \quad e^{-L(t-t_0)}w(t)\leqslant w(t_0)$$

for $t\geqslant t_0$. But $w(t_0)$ must be zero if $w(t)$ is nonnegative and satisfies (17). Consequently, $e^{-L(t-t_0)}w(t)\leqslant 0$, and this implies that $w(t)$ is identically zero.

The error in this proof, of course, is that we cannot differentiate both sides of an inequality, and still expect to preserve the inequality. For example, the function $f_1(t)=2t-2$ is less than $f_2(t)=t$ on the interval $[0,1]$, but $f_1'(t)$ is greater than $f_2'(t)$ on this interval. We make this proof "kosher" by the clever trick of setting

$$U(t)=\int_{t_0}^{t} w(s)\,ds.$$

Then,

$$\frac{dU}{dt}=w(t)\leqslant L\int_{t_0}^{t} w(s)\,ds=LU(t).$$

Consequently, $e^{-L(t-t_0)}U(t)\leqslant U(t_0)=0$, for $t\geqslant t_0$, and thus $U(t)=0$. This, in turn, implies that $w(t)=0$ since

$$0\leqslant w(t)\leqslant L\int_{t_0}^{t} w(s)\,ds=LU(t)=0.$$ □

Example 3. Show that the solution $y(t)$ of the initial-value problem

$$\frac{dy}{dt}=t^2+e^{-y^2}, \qquad y(0)=0$$

exists for $0\leqslant t\leqslant \frac{1}{2}$, and in this interval, $|y(t)|\leqslant 1$.

Solution. Let R be the rectangle $0 \leqslant t \leqslant \frac{1}{2}$, $|y| \leqslant 1$. Computing

$$M = \max_{(t,y) \text{ in } R} t^2 + e^{-y^2} = 1 + \left(\tfrac{1}{2}\right)^2 = \tfrac{5}{4},$$

we see that $y(t)$ exists for

$$0 \leqslant t \leqslant \min\left(\frac{1}{2}, \frac{1}{5/4}\right) = \frac{1}{2},$$

and in this interval, $|y(t)| \leqslant 1$.

Example 4. Show that the solution $y(t)$ of the initial-value problem

$$\frac{dy}{dt} = e^{-t^2} + y^3, \qquad y(0) = 1$$

exists for $0 \leqslant t \leqslant 1/9$, and in this interval, $0 \leqslant y \leqslant 2$.
Solution. Let R be the rectangle $0 \leqslant t \leqslant \frac{1}{9}$, $0 \leqslant y \leqslant 2$. Computing

$$M = \max_{(t,y) \text{ in } R} e^{-t^2} + y^3 = 1 + 2^3 = 9,$$

we see that $y(t)$ exists for

$$0 \leqslant t \leqslant \min\left(\tfrac{1}{9}, \tfrac{1}{9}\right)$$

and in this interval, $0 \leqslant y \leqslant 2$.

Example 5. What is the largest interval of existence that Theorem 2 predicts for the solution $y(t)$ of the initial-value problem $y' = 1 + y^2$, $y(0) = 0$?
Solution. Let R be the rectangle $0 \leqslant t \leqslant a$, $|y| \leqslant b$. Computing

$$M = \max_{(t,y) \text{ in } R} 1 + y^2 = 1 + b^2,$$

we see that $y(t)$ exists for

$$0 \leqslant t \leqslant \alpha = \min\left(a, \frac{b}{1+b^2}\right).$$

Clearly, the largest α that we can achieve is the maximum value of the function $b/(1+b^2)$. This maximum value is $\frac{1}{2}$. Hence, Theorem 2 predicts that $y(t)$ exists for $0 \leqslant t \leqslant \frac{1}{2}$. The fact that $y(t) = \tan t$ exists for $0 \leqslant t \leqslant \pi/2$ points out the limitation of Theorem 2.

Example 6. Suppose that $|f(t,y)| \leqslant K$ in the strip $t_0 \leqslant t < \infty$, $-\infty < y < \infty$. Show that the solution $y(t)$ of the initial-value problem $y' = f(t,y)$, $y(t_0) = y_0$ exists for all $t \geqslant t_0$.
Solution. Let R be the rectangle $t_0 \leqslant t \leqslant t_0 + a$, $|y - y_0| \leqslant b$. The quantity

$$M = \max_{(t,y) \text{ in } R} |f(t,y)|$$

is at most K. Hence, $y(t)$ exists for

$$t_0 \leqslant t \leqslant t_0 + \min(a, b/K).$$

Now, we can make the quantity $\min(a,b/K)$ as large as desired by choosing a and b sufficiently large. Therefore $y(t)$ exists for $t \geqslant t_0$.

Exercises

1. Construct the Picard iterates for the initial-value problem $y' = 2t(y+1)$, $y(0) = 0$ and show that they converge to the solution $y(t) = e^{t^2} - 1$.

2. Compute the first two Picard iterates for the initial-value problem $y' = t^2 + y^2$, $y(0) = 1$.

3. Compute the first three Picard iterates for the initial-value problem $y' = e^t + y^2$, $y(0) = 0$.

In each of Problems 4–15, show that the solution $y(t)$ of the given initial-value problem exists on the specified interval.

4. $y' = y^2 + \cos t^2, \quad y(0) = 0; \quad 0 \leqslant t \leqslant \frac{1}{2}$

5. $y' = 1 + y + y^2 \cos t, \quad y(0) = 0; \quad 0 \leqslant t \leqslant \frac{1}{3}$

6. $y' = t + y^2, \quad y(0) = 0; \quad 0 \leqslant t \leqslant \frac{1}{2}$

7. $y' = e^{-t^2} + y^2, \quad y(0) = 0; \quad 0 \leqslant t \leqslant \frac{1}{2}$

8. $y' = e^{-t^2} + y^2, \quad y(1) = 0; \quad 0 \leqslant t \leqslant \sqrt{e}\,/2$

9. $y' = e^{-t^2} + y^2, \quad y(0) = 1; \quad 0 \leqslant t \leqslant \dfrac{\sqrt{2}}{1 + (1 + \sqrt{2}\,)^2}$

10. $y' = y + e^{-y} + e^{-t}, \quad y(0) = 0; \quad 0 \leqslant t \leqslant 1$

11. $y' = y^3 + e^{-5t}, \quad y(0) = 0.4; \quad 0 \leqslant t \leqslant \frac{3}{10}$

12. $y' = e^{(y-t)^2}, \quad y(0) = 1; \quad 0 \leqslant t \leqslant \dfrac{\sqrt{3} - 1}{2} e^{-((1+\sqrt{3}}$

13. $y' = (4y + e^{-t^2})e^{2y}, \quad y(0) = 0; \quad 0 \leqslant t \leqslant \dfrac{1}{8\sqrt{e}}$

14. $y' = e^{-t} + \ln(1 + y^2), \quad y(0) = 0; \quad 0 \leqslant t < \infty$

15. $y' = \frac{1}{4}(1 + \cos 4t)y - \frac{1}{800}(1 - \cos 4t)y^2, \quad y(0) = 100; \quad 0 \leqslant t \leqslant 1$

16. Consider the initial-value problem

$$y' = t^2 + y^2, \qquad y(0) = 0, \tag{*}$$

and let R be the rectangle $0 \leqslant t \leqslant a$, $-b \leqslant y \leqslant b$.
(a) Show that the solution $y(t)$ of (*) exists for

$$0 \leqslant t \leqslant \min\left(a, \frac{b}{a^2 + b^2}\right).$$

(b) Show that the maximum value of $b/(a^2+b^2)$, for a fixed, is $1/2a$.
(c) Show that $\alpha=\min(a,\frac{1}{2}a)$ is largest when $a=1/\sqrt{2}$.
(d) Conclude that the solution $y(t)$ of (*) exists for $0\leqslant t\leqslant 1/\sqrt{2}$.

17. Prove that $y(t)=-1$ is the only solution of the initial-value problem

$$y'=t(1+y),\qquad y(0)=-1.$$

18. Find a nontrivial solution of the initial-value problem $y'=ty^a$, $y(0)=0$, $a>1$. Does this violate Theorem 2′? Explain.

19. Find a solution of the initial-value problem $y'=t\sqrt{1-y^2}$, $y(0)=1$, other than $y(t)=1$. Does this violate Theorem 2′? Explain.

20. Here is an alternate proof of Lemma 2. Let $w(t)$ be a nonnegative function with

$$w(t)\leqslant L\int_{t_0}^{t}w(s)\,ds \tag{*}$$

on the interval $t_0\leqslant t\leqslant t_0+\alpha$. Since $w(t)$ is continuous, we can find a constant A such that $0\leqslant w(t)\leqslant A$ for $t_0\leqslant t\leqslant t_0+\alpha$.
(a) Show that $w(t)\leqslant LA(t-t_0)$.
(b) Use this estimate of $w(t)$ in (*) to obtain

$$w(t)\leqslant\frac{AL^2(t-t_0)^2}{2}.$$

(c) Proceeding inductively, show that $w(t)\leqslant AL^n(t-t_0)^n/n!$, for every integer n.
(d) Conclude that $w(t)=0$ for $t_0\leqslant t\leqslant t_0+\alpha$.

1.10 Difference equations, and how to compute the interest due on your student loans

There are many situations in applied mathematics (particularly when we derive numerical approximations of the solution $y(t)$ of the initial-value problem $dy/dt=f(t,y), y(t_0)=y_0$) where we are confronted with the following problem: How large can the numbers $E_1,\dots,E_N$ be if

$$E_{n+1}\leqslant AE_n+B,\qquad n=0,1,\dots,N-1 \tag{1}$$

for some positive constants A and B, and $E_0=0$? This is a very difficult problem since it deals with *inequalities*, rather than *equalities*. Fortunately, though, we can convert the problem of solving the inequalities (1) into the simpler problem of solving a system of equalities. This is the content of the following lemma.

Lemma 1. *Let $E_1,\dots,E_N$ satisfy the inequalities*

$$E_{n+1}\leqslant AE_n+B,\qquad E_0=0$$

for some positive constants A and B. Then, E_n is less than or equal to y_n, *where*

$$y_{n+1}=Ay_n+B, \qquad y_0=0. \tag{2}$$

PROOF. We prove Lemma 1 by induction on n. To this end, observe that Lemma 1 is obviously true for $n=0$. Next, we assume that Lemma 1 is true for $n=j$. We must show that Lemma 1 is also true for $n=j+1$. That is to say, we must prove that $E_j \leqslant y_j$ implies $E_{j+1} \leqslant y_{j+1}$. But this follows immediately, for if $E_j \leqslant y_j$ then

$$E_{j+1} \leqslant AE_j+B \leqslant Ay_j+B=y_{j+1}.$$

By induction, therefore, $E_n \leqslant y_n$, $n=0,1,\dots,N$. □

Our next task is to solve Equation (2), which is often referred to as a difference equation. We will accomplish this in two steps. First we will solve the "simple" difference equation

$$y_{n+1}=y_n+B_n, \qquad y_0=y_0. \tag{3}$$

Then we will reduce the difference equation (2) to the difference equation (3) by a clever change of variables.

Equation (3) is trivial to solve. Observe that

$$\begin{aligned} y_1-y_0 &= B_0\\ y_2-y_1 &= B_1\\ &\vdots\\ y_{n-1}-y_{n-2} &= B_{n-2}\\ y_n-y_{n-1} &= B_{n-1}. \end{aligned}$$

Adding these equations gives

$$(y_n-y_{n-1})+(y_{n-1}-y_{n-2})+\dots+(y_1-y_0)=B_0+B_1+\dots+B_{n-1}.$$

Hence,

$$y_n=y_0+B_0+\dots+B_{n-1}=y_0+\sum_{j=0}^{n-1} B_j.$$

Next, we reduce the difference equation (2) to the simpler equation (3) in the following clever manner. Let

$$z_n=\frac{y_n}{A^n}, \qquad n=0,1,\dots,N.$$

Then, $z_{n+1}=y_{n+1}/A^{n+1}$. But $y_{n+1}=Ay_n+B$. Consequently,

$$z_{n+1}=\frac{y_n}{A^n}+\frac{B}{A^{n+1}}=z_n+\frac{B}{A^{n+1}}.$$

Therefore,

$$z_n = z_0 + \sum_{j=0}^{n-1} \frac{B}{A^{j+1}} = y_0 + \frac{B}{A}\left[\frac{1-\left(\frac{1}{A}\right)^n}{1-\frac{1}{A}}\right]$$

$$= y_0 + \frac{B}{A-1}\left[1-\left(\frac{1}{A}\right)^n\right]$$

and

$$y_n = A^n z_n = A^n y_0 + \frac{B}{A-1}(A^n - 1). \tag{4}$$

Finally, returning to the inequalities (1), we see that

$$E_n \leqslant \frac{B}{A-1}(A^n - 1), \qquad n = 1, 2, \ldots, N. \tag{5}$$

While collecting material for this book, this author was approached by a colleague with the following problem. He had just received a bill from the bank for the first payment on his wife's student loan. This loan was to be repaid in 10 years in 120 equal monthly installments. According to his rough estimate, the bank was overcharging him by at least 20%. Before confronting the bank's officers, though, he wanted to compute exactly the monthly payments due on this loan.

This problem can be put in the following more general framework. Suppose that P dollars are borrowed from a bank at an annual interest rate of $R\%$. This loan is to be repaid in n years in equal monthly installments of x dollars. Find x.

Our first step in solving this problem is to compute the interest due on the loan. To this end observe that the interest I_1 owed when the first payment is due is $I_1 = (r/12)P$, where $r = R/100$. The principal outstanding during the second month of the loan is $(x - I_1)$ less than the principal outstanding during the first month. Hence, the interest I_2 owed during the second month of the loan is

$$I_2 = I_1 - \frac{r}{12}(x - I_1).$$

Similarly, the interest I_{j+1} owed during the $(j+1)$st month is

$$I_{j+1} = I_j - \frac{r}{12}(x - I_j) = \left(1 + \frac{r}{12}\right)I_j - \frac{r}{12}x, \tag{6}$$

where I_j is the interest owed during the jth month.

Equation (6) is a difference equation for the numbers

$$I_1 = \frac{r}{12}P, I_2, \ldots, I_{12n}.$$

Its solution (see Exercise 4) is

$$I_j = \frac{r}{12}P\left(1+\frac{r}{12}\right)^{j-1} + x\left[1-\left(1+\frac{r}{12}\right)^{j-1}\right].$$

Hence, the total amount of interest paid on the loan is

$$\begin{aligned} I = I_1 + I_2 + \ldots + I_{12n} &= \sum_{j=1}^{12n} I_j \\ &= \frac{r}{12}P\sum_{j=1}^{12n}\left(1+\frac{r}{12}\right)^{j-1} + 12nx - x\sum_{j=1}^{12n}\left(1+\frac{r}{12}\right)^{j-1}. \end{aligned}$$

Now,

$$\sum_{j=1}^{12n}\left(1+\frac{r}{12}\right)^{j-1} = \frac{12}{r}\left[\left(1+\frac{r}{12}\right)^{12n} - 1\right].$$

Therefore,

$$\begin{aligned} I &= P\left[\left(1+\frac{r}{12}\right)^{12n} - 1\right] + 12nx - \frac{12x}{r}\left[\left(1+\frac{r}{12}\right)^{12n} - 1\right] \\ &= 12nx - P + P\left(1+\frac{r}{12}\right)^{12n} - \frac{12x}{r}\left[\left(1+\frac{r}{12}\right)^{12n} - 1\right]. \end{aligned}$$

But, $12nx - P$ must equal I, since $12nx$ is the amount of money paid the bank and P was the principal loaned. Consequently,

$$P\left(1+\frac{r}{12}\right)^{12n} - \frac{12x}{r}\left[\left(1+\frac{r}{12}\right)^{12n} - 1\right] = 0$$

and this equation implies that

$$x = \frac{\frac{r}{12}P\left(1+\frac{r}{12}\right)^{12n}}{\left(1+\frac{r}{12}\right)^{12n} - 1}. \tag{7}$$

Epilog. Using Equation (7), this author computed x for his wife's and his colleague's wife's student loans. In both cases the bank was right—to the penny.

Exercises

1. Solve the difference equation $y_{n+1} = -7y_n + 2, y_0 = 1$.

2. Find y_{37} if $y_{n+1} = 3y_n + 1$, $y_0 = 0$, $n = 0, 1, \ldots, 36$.

3. Estimate the numbers $E_0, E_1, \ldots, E_N$ if $E_0 = 0$ and
(a) $E_{n+1} \leqslant 3E_n + 1$, $n = 0, 1, \ldots, N-1$;
(b) $E_{n+1} \leqslant 2E_n + 2$, $n = 0, 1, \ldots, N-1$.

4. (a) Show that the transformation $y_j = I_{j+1}$ transforms the difference equation

$$I_{j+1} = \left(1+\frac{r}{12}\right)I_j - \frac{r}{12}x, \qquad I_1 = \frac{r}{12}P$$

into the difference equation

$$y_{j+1} = \left(1+\frac{r}{12}\right)y_j - \frac{r}{12}x, \qquad y_0 = \frac{r}{12}P.$$

(b) Use Equation (4) to find $y_{j-1} = I_j$.

5. Solve the difference equation $y_{n+1} = a_n y_n + b_n$, $y_1 = \alpha$. *Hint*: Set $z_1 = y_1$ and $z_n = y_n / a_1 \dots a_{n-1}$ for $n \geqslant 2$. Observe that

$$z_{n+1} = \frac{y_{n+1}}{a_1 \dots a_n} = \frac{a_n y_n}{a_1 \dots a_n} + \frac{b_n}{a_1 \dots a_n}$$
$$= z_n + \frac{b_n}{a_1 \dots a_n}.$$

Hence, conclude that $z_n = z_1 + \sum_{j=1}^{n-1} b_j / a_1 \dots a_j$.

6. Solve the difference equation $y_{n+1} - n y_n = 1 - n$, $y_1 = 2$.

7. Find y_{25} if $y_1 = 1$ and $(n+1)y_{n+1} - n y_n = 2^n$, $n = 1, \dots, 24$.

8. A student borrows P dollars at an annual interest rate of $R\%$. This loan is to be repayed in n years in equal monthly installments of x dollars. Find x if
(a) $P = 4250$, $R = 3$, and $n = 5$;
(b) $P = 5000$, $R = 7$, and $n = 10$.

9. A home buyer takes out a \$30,000 mortgage at an annual interest rate of 9%. This loan is to be repaid over 20 years in 240 equal monthly installments of x dollars.
(a) Compute x.
(b) Find x if the annual interest rate is 10%.

10. The quantity supplied of some commodity in a given week is obviously an increasing function of its price the previous week, while the quantity demanded in a given week is a function of its current price. Let S_j and D_j denote, respectively, the quantities supplied and demanded in the jth week, and let P_j denote the price of the commodity in the jth week. We assume that there exist positive constants a, b, and c such that

$$S_j = aP_{j-1} \quad \text{and} \quad D_j = b - cP_j.$$

(a) Show that $P_j = b/(a+c) + (-a/c)^j (P_0 - b/(a+c))$, if supply always equals demand.
(b) Show that P_j approaches $b/(a+c)$ as j approaches infinity if $a/c < 1$.
(c) Show that $P = b/(a+c)$ represents an equilibrium situation. That is to say, if supply always equals demand, and if the price ever reaches the level $b/(a+c)$, then it will always remain at that level.

2 Second-order linear differential equations

2.1 Algebraic properties of solutions

A second-order differential equation is an equation of the form

$$\frac{d^2y}{dt^2}=f\left(t,y,\frac{dy}{dt}\right). \tag{1}$$

For example, the equation

$$\frac{d^2y}{dt^2}=\sin t+3y+\left(\frac{dy}{dt}\right)^2$$

is a second-order differential equation. A function $y=y(t)$ is a solution of (1) if $y(t)$ satisfies the differential equation; that is

$$\frac{d^2y(t)}{dt^2}=f\left(t,y(t),\frac{dy(t)}{dt}\right).$$

Thus, the function $y(t)=\cos t$ is a solution of the second-order equation $d^2y/dt^2=-y$ since $d^2(\cos t)/dt^2=-\cos t$.

Second-order differential equations arise quite often in applications. The most famous second-order differential equation is Newton's second law of motion (see Section 1.6)

$$m\frac{d^2y}{dt^2}=F\left(t,y,\frac{dy}{dt}\right)$$

which governs the motion of a particle of mass m moving under the influence of a force F. In this equation, m is the mass of the particle, $y=y(t)$ is its position at time t, dy/dt is its velocity, and F is the total force acting on the particle. As the notation suggests, the force F may depend on the position and velocity of the particle, as well as on time.

In addition to the differential equation (1), we will often impose initial conditions on $y(t)$ of the form

$$y(t_0)=y_0, \qquad y'(t_0)=y_0'. \tag{1'}$$

The differential equation (1) together with the initial conditions (1′) is referred to as an initial-value problem. For example, let $y(t)$* denote the position at time t of a particle moving under the influence of gravity. Then, $y(t)$ satisfies the initial-value problem

$$\frac{d^2y}{dt^2}=-g; \qquad y(t_0)=y_0, \quad y'(t_0)=y_0',$$

where y_0 is the initial position of the particle and y_0' is the initial velocity of the particle.

Second-order differential equations are extremely difficult to solve. This should not come as a great surprise to us after our experience with first-order equations. We will only succeed in solving the special differential equation

$$\frac{d^2y}{dt^2}+p(t)\frac{dy}{dt}+q(t)y=g(t). \tag{2}$$

Fortunately, though, many of the second-order equations that arise in applications are of this form.

The differential equation (2) is called a second-order linear differential equation. We single out this equation and call it linear because both y and dy/dt appear by themselves. For example, the differential equations

$$\frac{d^2y}{dt^2}+3t\frac{dy}{dt}+(\sin t)y=e^t$$

and

$$\frac{d^2y}{dt^2}+e^t\frac{dy}{dt}+2y=1$$

are linear, while the differential equations

$$\frac{d^2y}{dt^2}+3\frac{dy}{dt}+\sin y=t^3$$

and

$$\frac{d^2y}{dt^2}+\left(\frac{dy}{dt}\right)^2=1$$

are both nonlinear, due to the presence of the $\sin y$ and $(dy/dt)^2$ terms, respectively.

We consider first the second-order linear homogeneous equation

$$\frac{d^2y}{dt^2}+p(t)\frac{dy}{dt}+q(t)y=0 \tag{3}$$

*The positive direction of y is taken upwards.

which is obtained from (2) by setting $g(t)=0$. It is certainly not obvious at this point how to find all the solutions of (3), or how to solve the initial-value problem

$$\frac{d^2y}{dt^2}+p(t)\frac{dy}{dt}+q(t)y=0; \qquad y(t_0)=y_0, \quad y'(t_0)=y_0'. \tag{4}$$

Therefore, before trying to develop any elaborate procedures for solving (4), we should first determine whether it actually has a solution. This information is contained in the following theorem.

Theorem 1. (Existence–uniqueness Theorem). *Let the functions $p(t)$ and $q(t)$ be continuous in the open interval $\alpha<t<\beta$. Then, there exists one, and only one function $y(t)$ satisfying the differential equation* (3) *on the entire interval $\alpha<t<\beta$, and the prescribed initial conditions $y(t_0)=y_0$, $y'(t_0)=y_0'$. In particular, any solution $y=y(t)$ of* (3) *which satisfies $y(t_0)=0$ and $y'(t_0)=0$ at some time $t=t_0$ must be identically zero.*

Theorem 1 is an extremely important theorem for us. On the one hand, it is our hunting license to find the unique solution $y(t)$ of (4). And, on the other hand, we will actually use Theorem 1 to help us find all the solutions of (3).

We begin our analysis of Equation (3) with the important observation that the left-hand side

$$y''+p(t)y'+q(t)y \qquad \left(y'=\frac{dy}{dt}, y''=\frac{d^2y}{dt^2}\right)$$

of the differential equation can be viewed as defining a "function of a function": with each function y having two derivatives, we associate another function, which we'll call $L[y]$, by the relation

$$L[y](t)=y''(t)+p(t)y'(t)+q(t)y(t).$$

In mathematical terminology, L is an operator which operates on functions; that is, there is a prescribed recipe for associating with each function y a new function $L[y]$.

Example 1. Let $p(t)=0$ and $q(t)=t$. Then,

$$L[y](t)=y''(t)+ty(t).$$

If $y(t)=\cos t$, then

$$L[y](t)=(\cos t)''+t\cos t=(t-1)\cos t,$$

and if $y(t)=t^3$, then

$$L[y](t)=(t^3)''+t(t^3)=t^4+6t.$$

Thus, the operator L assigns the function $(t-1)\cos t$ to the function $\cos t$, and the function $6t+t^4$ to the function t^3.

The concept of an operator acting on functions, or a "function of a function" is analogous to that of a function of a single variable t. Recall the definition of a function f on an interval I: with each number t in I we associate a new number called $f(t)$. In an exactly analogous manner, we associate with each function y having two derivatives a new function called $L[y]$. This is an extremely sophisticated mathematical concept, because in a certain sense, we are treating a function exactly as we do a point. Admittedly, this is quite difficult to grasp. It's not surprising, therefore, that the concept of a "function of a function" was not developed till the beginning of this century, and that many of the "high powered" theorems of mathematical analysis were only proven after this concept was mastered.

We now derive several important properties of the operator L, which we will use to great advantage shortly.

Property 1. $L[cy]=cL[y]$, *for any constant* c.

PROOF.
$$\begin{aligned}L[cy](t)&=(cy)''(t)+p(t)(cy)'(t)+q(t)(cy)(t)\\&=cy''(t)+cp(t)y'(t)+cq(t)y(t)\\&=c\left[y''(t)+p(t)y'(t)+q(t)y(t)\right]\\&=cL[y](t).\end{aligned}$$ □

The meaning of Property 1 is that the operator L assigns to the function (cy) c times the function it assigns to y. For example, let

$$L[y](t)=y''(t)+6y'(t)-2y(t).$$

This operator L assigns the function

$$(t^2)''+6(t^2)'-2(t^2)=2+12t-2t^2$$

to the function t^2. Hence, L must assign the function $5(2+12t-2t^2)$ to the function $5t^2$.

Property 2. $L[y_1+y_2]=L[y_1]+L[y_2]$.

PROOF.
$$\begin{aligned}L[y_1+y_2](t)&=(y_1+y_2)''(t)+p(t)(y_1+y_2)'(t)+q(t)(y_1+y_2)(t)\\&=y_1''(t)+y_2''(t)+p(t)y_1'(t)+p(t)y_2'(t)+q(t)y_1(t)+q(t)y_2(t)\\&=\left[y_1''(t)+p(t)y_1'(t)+q(t)y_1(t)\right]+\left[y_2''(t)+p(t)y_2'(t)+q(t)y_2(t)\right]\\&=L[y_1](t)+L[y_2](t).\end{aligned}$$ □

The meaning of Property 2 is that the operator L assigns to the function y_1+y_2 the sum of the functions it assigns to y_1 and y_2. For example, let

$$L[y](t)=y''(t)+2y'(t)-y(t).$$

This operator L assigns the function

$$(\cos t)''+2(\cos t)'-\cos t=-2\cos t-2\sin t$$

to the function $\cos t$, and the function

$$(\sin t)''+2(\sin t)'-\sin t=2\cos t-2\sin t$$

to the function $\sin t$. Hence, L assigns the function

$$(-2\cos t-2\sin t)+2\cos t-2\sin t=-4\sin t$$

to the function $\sin t+\cos t$.

Definition. An operator L which assigns functions to functions and which satisfies Properties 1 and 2 is called a linear operator. All other operators are nonlinear. An example of a nonlinear operator is

$$L[y](t)=y''(t)-2t[y(t)]^4.$$

This operator assigns the function

$$\left(\frac{1}{t}\right)''-2t\left(\frac{1}{t}\right)^4=\frac{2}{t^3}-\frac{2}{t^3}=0$$

to the function $1/t$, and the function

$$\left(\frac{c}{t}\right)''-2t\left(\frac{c}{t}\right)^4=\frac{2c}{t^3}-\frac{2c^4}{t^3}=\frac{2c(1-c^3)}{t^3}$$

to the function c/t. Hence, for $c\neq 0,1$, and $y(t)=1/t$, we see that $L[cy]\neq cL[y]$.

The usefulness of Properties 1 and 2 lies in the observation that the solutions $y(t)$ of the differential equation (3) are exactly those functions y for which

$$L[y](t)=y''(t)+p(t)y'(t)+q(t)y(t)=0.$$

In other words, the solutions $y(t)$ of (3) are exactly those functions y to which the operator L assigns the zero function.* Hence, if $y(t)$ is a solution of (3) then so is $cy(t)$, since

$$L[cy](t)=cL[y](t)=0.$$

If $y_1(t)$ and $y_2(t)$ are solutions of (3), then $y_1(t)+y_2(t)$ is also a solution of (3), since

$$L[y_1+y_2](t)=L[y_1](t)+L[y_2](t)=0+0=0.$$

Combining Properties 1 and 2, we see that all linear combinations

$$c_1y_1(t)+c_2y_2(t)$$

of solutions of (3) are again solutions of (3).

*The zero function is the function whose value at any time t is zero.

The preceding argument shows that we can use our knowledge of two solutions $y_1(t)$ and $y_2(t)$ of (3) to generate infinitely many other solutions. This statement has some very interesting implications. Consider, for example, the differential equation

$$\frac{d^2y}{dt^2}+y=0. \tag{5}$$

Two solutions of (5) are $y_1(t)=\cos t$ and $y_2(t)=\sin t$. Hence,

$$y(t)=c_1\cos t+c_2\sin t \tag{6}$$

is also a solution of (5), for every choice of constants c_1 and c_2. Now, Equation (6) contains two arbitrary constants. It is natural to suspect, therefore, that this expression represents the general solution of (5); that is, every solution $y(t)$ of (5) must be of the form (6). This is indeed the case, as we now show. Let $y(t)$ be any solution of (5). By the existence–uniqueness theorem, $y(t)$ exists for all t. Let $y(0)=y_0$, $y'(0)=y_0'$, and consider the function

$$\phi(t)=y_0\cos t+y_0'\sin t.$$

This function is a solution of (5) since it is a linear combination of solutions of (5). Moreover, $\phi(0)=y_0$ and $\phi'(0)=y_0'$. Thus, $y(t)$ and $\phi(t)$ satisfy the same second-order linear homogeneous equation and the same initial conditions. Therefore, by the uniqueness part of Theorem 1, $y(t)$ must be identically equal to $\phi(t)$, so that

$$y(t)=y_0\cos t+y_0'\sin t.$$

Thus, Equation (6) is indeed the general solution of (5).

Let us return now to the general linear equation (3). Suppose, in some manner, that we manage to find two solutions $y_1(t)$ and $y_2(t)$ of (3). Then, every function

$$y(t)=c_1y_1(t)+c_2y_2(t) \tag{7}$$

is again a solution of (3). Does the expression (7) represent the general solution of (3)? That is to say, does every solution $y(t)$ of (3) have the form (7)? The following theorem answers this question.

Theorem 2. *Let $y_1(t)$ and $y_2(t)$ be two solutions of* (3) *on the interval $\alpha<t<\beta$, with*

$$y_1(t)y_2'(t)-y_1'(t)y_2(t)$$

unequal to zero in this interval. Then,

$$y(t)=c_1y_1(t)+c_2y_2(t)$$

is the general solution of (3).

Proof. Let $y(t)$ be any solution of (3). We must find constants c_1 and c_2 such that $y(t)=c_1y_1(t)+c_2y_2(t)$. To this end, pick a time t_0 in the interval

(α,β) and let y_0 and y_0' denote the values of y and y' at $t=t_0$. The constants c_1 and c_2, if they exist, must satisfy the two equations

$$c_1y_1(t_0)+c_2y_2(t_0)=y_0$$
$$c_1y_1'(t_0)+c_2y_2'(t_0)=y_0'.$$

Multiplying the first equation by $y_2'(t_0)$, the second equation by $y_2(t_0)$ and subtracting gives

$$c_1\left[y_1(t_0)y_2'(t_0)-y_1'(t_0)y_2(t_0)\right]=y_0y_2'(t_0)-y_0'y_2(t_0).$$

Similarly, multiplying the first equation by $y_1'(t_0)$, the second equation by $y_1(t_0)$ and subtracting gives

$$c_2\left[y_1'(t_0)y_2(t_0)-y_1(t_0)y_2'(t_0)\right]=y_0y_1'(t_0)-y_0'y_1(t_0).$$

Hence,

$$c_1=\frac{y_0y_2'(t_0)-y_0'y_2(t_0)}{y_1(t_0)y_2'(t_0)-y_1'(t_0)y_2(t_0)}$$

and

$$c_2=\frac{y_0'y_1(t_0)-y_0y_1'(t_0)}{y_1(t_0)y_2'(t_0)-y_1'(t_0)y_2(t_0)}$$

if $y_1(t_0)y_2'(t_0)-y_1'(t_0)y_2(t_0)\neq 0$. Now, let

$$\phi(t)=c_1y_1(t)+c_2y_2(t)$$

for this choice of constants c_1,c_2. We know that $\phi(t)$ satisfies (3), since it is a linear combination of solutions of (3). Moreover, by construction, $\phi(t_0)=y_0$ and $\phi'(t_0)=y_0'$. Thus, $y(t)$ and $\phi(t)$ satisfy the same second-order linear homogeneous equation and the same initial conditions. Therefore, by the uniqueness part of Theorem 1, $y(t)$ must be identically equal to $\phi(t)$; that is,

$$y(t)=c_1y_1(t)+c_2y_2(t),\qquad \alpha<t<\beta.$$ □

Theorem 2 is an extremely useful theorem since it reduces the problem of finding all solutions of (3), of which there are infinitely many, to the much simpler problem of finding just two solutions $y_1(t),y_2(t)$. The only condition imposed on the solutions $y_1(t)$ and $y_2(t)$ is that the quantity $y_1(t)y_2'(t)-y_1'(t)y_2(t)$ be unequal to zero for $\alpha<t<\beta$. When this is the case, we say that $y_1(t)$ and $y_2(t)$ are a *fundamental* set of solutions of (3), since all other solutions of (3) can be obtained by taking linear combinations of $y_1(t)$ and $y_2(t)$.

Definition. The quantity $y_1(t)y_2'(t)-y_1'(t)y_2(t)$ is called the *Wronskian* of y_1 and y_2, and is denoted by $W(t)=W[y_1,y_2](t)$.

Theorem 2 requires that $W[y_1,y_2](t)$ be unequal to zero at all points in the interval (α,β). In actual fact, the Wronskian of any two solutions

$y_1(t), y_2(t)$ of (3) is either identically zero, or is never zero, as we now show.

Theorem 3. *Let $p(t)$ and $q(t)$ be continuous in the interval $\alpha<t<\beta$, and let $y_1(t)$ and $y_2(t)$ be two solutions of (3). Then, $W[y_1,y_2](t)$ is either identically zero, or is never zero, on the interval $\alpha<t<\beta$.*

We prove Theorem 3 with the aid of the following lemma.

Lemma 1. *Let $y_1(t)$ and $y_2(t)$ be two solutions of the linear differential equation $y''+p(t)y'+q(t)y=0$. Then, their Wronskian*

$$W(t)=W[y_1,y_2](t)=y_1(t)y_2'(t)-y_1'(t)y_2(t)$$

satisfies the first-order differential equation

$$W'+p(t)W=0.$$

Proof. Observe that

$$\begin{aligned}W'(t)&=\frac{d}{dt}(y_1y_2'-y_1'y_2)\\&=y_1y_2''+y_1'y_2'-y_1'y_2'-y_1''y_2\\&=y_1y_2''-y_1''y_2.\end{aligned}$$

Since y_1 and y_2 are solutions of $y''+p(t)y'+q(t)y=0$, we know that

$$y_2''=-p(t)y_2'-q(t)y_2$$

and

$$y_1''=-p(t)y_1'-q(t)y_1.$$

Hence,

$$\begin{aligned}W'(t)&=y_1[-p(t)y_2'-q(t)y_2]-y_2[-p(t)y_1'-q(t)y_1]\\&=-p(t)[y_1y_2'-y_1'y_2]\\&=-p(t)W(t).\end{aligned}$$

□

We can now give a very simple proof of Theorem 3.

Proof of Theorem 3. Pick any t_0 in the interval (α,β). From Lemma 1,

$$W[y_1,y_2](t)=W[y_1,y_2](t_0)\exp\left(-\int_{t_0}^{t}p(s)\,ds\right).$$

Now, $\exp\left(-\int_{t_0}^{t}p(s)\,ds\right)$ is unequal to zero for $\alpha<t<\beta$. Therefore, $W[y_1,y_2](t)$ is either identically zero, or is never zero. □

The simplest situation where the Wronskian of two functions $y_1(t),y_2(t)$ vanishes identically is when one of the functions is identically zero. More generally, the Wronskian of two functions $y_1(t),y_2(t)$ vanishes identically if one of the functions is a constant multiple of the other. If $y_2=cy_1$, say, then

$$W[y_1,y_2](t)=y_1(cy_1)'-y_1'(cy_1)=0.$$

Conversely, suppose that the Wronskian of two *solutions* $y_1(t),y_2(t)$ of (3) vanishes identically. Then, one of these solutions must be a constant multiple of the other, as we now show.

Theorem 4. *Let $y_1(t)$ and $y_2(t)$ be two solutions of* (3) *on the interval $\alpha<t<\beta$, and suppose that $W[y_1,y_2](t_0)=0$ for some t_0 in this interval. Then, one of these solutions is a constant multiple of the other.*

PROOF #1. Suppose that $W[y_1,y_2](t_0)=0$. Then, the equations

$$c_1y_1(t_0)+c_2y_2(t_0)=0$$
$$c_1y_1'(t_0)+c_2y_2'(t_0)=0$$

have a nontrivial solution c_1,c_2; that is, a solution c_1,c_2 with $|c_1|+|c_2|\neq 0$. Let $y(t)=c_1y_1(t)+c_2y_2(t)$, for this choice of constants c_1,c_2. We know that $y(t)$ is a solution of (3), since it is a linear combination of $y_1(t)$ and $y_2(t)$. Moreover, by construction, $y(t_0)=0$ and $y'(t_0)=0$. Therefore, by Theorem 1, $y(t)$ is identically zero, so that

$$c_1y_1(t)+c_2y_2(t)=0, \qquad \alpha<t<\beta,$$

If $c_1\neq 0$, then $y_1(t)=-(c_2/c_1)y_2(t)$, and if $c_2\neq 0$, then $y_2(t)=-(c_1/c_2)y_1(t)$. In either case, one of these solutions is a constant multiple of the other. □

PROOF #2. Suppose that $W[y_1,y_2](t_0)=0$. Then, by Theorem 3, $W[y_1,y_2](t)$ is identically zero. Assume that $y_1(t)y_2(t)\neq 0$ for $\alpha<t<\beta$. Then, dividing both sides of the equation

$$y_1(t)y_2'(t)-y_1'(t)y_2(t)=0$$

by $y_1(t)y_2(t)$ gives

$$\frac{y_2'(t)}{y_2(t)}-\frac{y_1'(t)}{y_1(t)}=0.$$

This equation implies that $y_1(t)=cy_2(t)$ for some constant c.

Next, suppose that $y_1(t)y_2(t)$ is zero at some point $t=t^*$ in the interval $\alpha<t<\beta$. Without loss of generality, we may assume that $y_1(t^*)=0$, since otherwise we can relabel y_1 and y_2. In this case it is simple to show (see Exercise 19) that either $y_1(t)\equiv 0$, or $y_2(t)=[y_2'(t^*)/y_1'(t^*)]y_1(t)$. This completes the proof of Theorem 4. □

complex roots

if $\Delta = b^2 - 4ac < 0$ then the characteristic eq. has two complex roots.

$$r_1 = \frac{-b + i\sqrt{4ac-b^2}}{2a}, \quad r_2 = \frac{-b - i\sqrt{4ac-b^2}}{2a}$$

and

$$y(t) = e^{\left[\frac{-b + i\sqrt{4ac-b^2}}{2a}\right]t}$$

and b Euler

$$e^{i\beta t} = \cos\beta t + i\sin\beta t$$

hence

$$y(t) = e^{-\frac{bt}{2a}}\left[\cos\left[\left(\sqrt{4ac-b^2}\right)t/2a\right] + i\sin\left[\left(\sqrt{4ac-b^2}\right)t/2a\right]\right]$$

$$y_1 = e^{-bt/2a}\cos\beta t, \quad y_2 = e^{-bt/2a}\sin\beta t$$

$$\beta = \frac{\sqrt{4ac-b^2}}{2a}$$

$\Delta < 0$

G.S

$$y(t) = e^{-tb/2a}\left[c_1\cos\beta t + c_2\sin\beta t\right]$$

$$\beta = \frac{\sqrt{4ac-b^2}}{2a}$$

Example

the solution to

$$y'' + 2y' + 4y = 0 \qquad y(0)=1,\ y'(0)=1$$

Solution:

the characteristic eq is $r^2 + 2r + 4 = 0$

its roots are $\dfrac{-2 \pm \sqrt{4-16}}{2}$

$d < 0$, therefore it has complex roots

$$-1 \pm \frac{i\sqrt{16-4}}{2}$$

$$r_{1,2} = -1 \pm i\sqrt{3}$$

---

we use the G.S

$$y(t) = e^{(-b/2a)t}\left[c_1 \cos\beta t + c_2 \sin\beta t\right]$$

$$\beta = \frac{\sqrt{4ac-b^2}}{2}$$

$$y(t) = e^{-t}\left(c_1 \cos\sqrt{3}t + c_2 \sin\sqrt{3}t\right)$$

determine c_1, c_2

$$y(0) = \quad c_1 \cos 0 + c_2 \sin 0 \quad = 1$$

$$c_1 = 1$$

$$y'(0) = 1 = -e^{-t}\left(c_1 \cos\sqrt{3}t + c_2 \sin\sqrt{3}t\right)$$

$$+ e^{-t}\left(-\sqrt{3}c_1 \sin\sqrt{3}t + c_2\sqrt{3}\cos\sqrt{3}t\right)$$

$$1 = -1 \cdot c_1 + c_2\sqrt{3}$$

$$1 = -1 + c_2\sqrt{3} \Rightarrow c_2 = 2/\sqrt{3}$$

Definition. The functions $y_1(t)$ and $y_2(t)$ are said to be *linearly dependent* on an interval I if one of these functions is a constant multiple of the other on I. The functions $y_1(t)$ and $y_2(t)$ are said to be *linearly independent* on an interval I if they are not linearly dependent on I.

Corollary to Theorem 4. *Two solutions $y_1(t)$ and $y_2(t)$ of* (3) *are linearly independent on the interval $\alpha<t<\beta$ if, and only if, their Wronskian is unequal to zero on this interval. Thus, two solutions $y_1(t)$ and $y_2(t)$ form a fundamental set of solutions of* (3) *on the interval $\alpha<t<\beta$ if, and only if, they are linearly independent on this interval.*

Exercises

1. Let $L[y](t)=y''(t)-3ty'(t)+3y(t)$. Compute
(a) $L[e^t]$, (b) $L[\cos\sqrt{3}\,t]$, (c) $L[2e^t+4\cos\sqrt{3}\,t]$,
(d) $L[t^2]$, (e) $L[5t^2]$, (f) $L[t]$, (g) $L[t^2+3t]$.

2. Let $L[y](t)=y''(t)-6y'(t)+5y(t)$. Compute
(a) $L[e^t]$, (b) $L[e^{2t}]$, (c) $L[e^{3t}]$, (d) $L[e^{rt}]$,
(e) $L[t]$, (f) $L[t^2]$, (g) $L[t^2+2t]$.

3. Show that the operator L defined by

$$L[y](t)=\int_a^t s^2 y(s)\,ds$$

is linear; that is, $L[cy]=cL[y]$ and $L[y_1+y_2]=L[y_1]+L[y_2]$.

4. Let $L[y](t)=y''(t)+p(t)y'(t)+q(t)y(t)$, and suppose that $L[t^2]=t+1$ and $L[t]=2t+2$. Show that $y(t)=t-2t^2$ is a solution of $y''+p(t)y'+q(t)y=0$.

5. (a) Show that $y_1(t)=\sqrt{t}$ and $y_2(t)=1/t$ are solutions of the differential equation

$$2t^2y''+3ty'-y=0 \qquad (*)$$

on the interval $0<t<\infty$.
(b) Compute $W[y_1,y_2](t)$. What happens as t approaches zero?
(c) Show that $y_1(t)$ and $y_2(t)$ form a fundamental set of solutions of (*) on the interval $0<t<\infty$.
(d) Solve the initial-value problem $2t^2y''+3ty'-y=0$; $y(1)=2$, $y'(1)=1$.

6. (a) Show that $y_1(t)=e^{-t^2/2}$ and $y_2(t)=e^{-t^2/2}\int_0^t e^{s^2/2}ds$ are solutions of

$$y''+ty'+y=0 \qquad (*)$$

on the interval $-\infty<t<\infty$.
(b) Compute $W[y_1,y_2](t)$.
(c) Show that y_1 and y_2 form a fundamental set of solutions of (*) on the interval $-\infty<t<\infty$.
(d) Solve the initial-value problem $y''+ty'+y=0$; $y(0)=0$, $y'(0)=1$.

7. Compute the Wronskian of the following pairs of functions.
(a) $\sin at, \cos bt$ (b) $\sin^2 t, 1-\cos 2t$
(c) e^{at}, e^{bt} (d) e^{at}, te^{at}
(e) $t, t\ln t$ (f) $e^{at}\sin bt, e^{at}\cos bt$

8. Let $y_1(t)$ and $y_2(t)$ be solutions of (3) on the interval $-\infty<t<\infty$ with $y_1(0)=3$, $y_1'(0)=1$, $y_2(0)=1$, and $y_2'(0)=\frac{1}{3}$. Show that $y_1(t)$ and $y_2(t)$ are linearly dependent on the interval $-\infty<t<\infty$.

9. (a) Let $y_1(t)$ and $y_2(t)$ be solutions of (3) on the interval $\alpha<t<\beta$, with $y_1(t_0)=1$, $y_1'(t_0)=0$, $y_2(t_0)=0$, and $y_2'(t_0)=1$. Show that $y_1(t)$ and $y_2(t)$ form a fundamental set of solutions of (3) on the interval $\alpha<t<\beta$.
(b) Show that $y(t)=y_0y_1(t)+y_0'y_2(t)$ is the solution of (3) satisfying $y(t_0)=y_0$ and $y'(t_0)=y_0'$.

10. Show that $y(t)=t^2$ can never be a solution of (3) if the functions $p(t)$ and $q(t)$ are continuous at $t=0$.

11. Let $y_1(t)=t^2$ and $y_2(t)=t|t|$.
(a) Show that y_1 and y_2 are linearly dependent on the interval $0\leqslant t\leqslant 1$.
(b) Show that y_1 and y_2 are linearly independent on the interval $-1\leqslant t\leqslant 1$.
(c) Show that $W[y_1,y_2](t)$ is identically zero.
(d) Show that y_1 and y_2 can never be two solutions of (3) on the interval $-1<t<1$ if both p and q are continuous in this interval.

12. Suppose that y_1 and y_2 are linearly independent on an interval I. Prove that $z_1=y_1+y_2$ and $z_2=y_1-y_2$ are also linearly independent on I.

13. Let y_1 and y_2 be solutions of Bessel's equation

$$t^2y''+ty'+(t^2-n^2)y=0$$

on the interval $0<t<\infty$, with $y_1(1)=1$, $y_1'(1)=0$, $y_2(1)=0$, and $y_2'(1)=1$. Compute $W[y_1,y_2](t)$.

14. Suppose that the Wronskian of any two solutions of (3) is constant in time. Prove that $p(t)=0$.

In Problems 15–18, assume that p and q are continuous, and that the functions y_1 and y_2 are solutions of the differential equation

$$y''+p(t)y'+q(t)y=0$$

on the interval $\alpha<t<\beta$.

15. Prove that if y_1 and y_2 vanish at the same point in the interval $\alpha<t<\beta$, then they cannot form a fundamental set of solutions on this interval.

16. Prove that if y_1 and y_2 achieve a maximum or minimum at the same point in the interval $\alpha<t<\beta$, then they cannot form a fundamental set of solutions on this interval.

17. Prove that if y_1 and y_2 are a fundamental set of solutions, then they cannot have a common point of inflection in $\alpha<t<\beta$ unless p and q vanish simultaneously there.

18. Suppose that y_1 and y_2 are a fundamental set of solutions on the interval $-\infty < t < \infty$. Show that there is one and only one zero of y_1 between consecutive zeros of y_2. *Hint*: Differentiate the quantity y_2/y_1 and use Rolle's Theorem.

19. Suppose that $W[y_1,y_2](t^*)=0$, and, in addition, $y_1(t^*)=0$. Prove that either $y_1(t)\equiv 0$ or $y_2(t)=[y_2'(t^*)/y_1'(t^*)]y_1(t)$. *Hint*: If $W[y_1,y_2](t^*)=0$ and $y_1(t^*)=0$, then $y_2(t^*)y_1'(t^*)=0$.

2.2 Linear equations with constant coefficients

We consider now the homogeneous linear second-order equation with constant coefficients

$$L[y]=a\frac{d^2y}{dt^2}+b\frac{dy}{dt}+cy=0 \tag{1}$$

where a, b, and c are constants, with $a\neq 0$. Theorem 2 of Section 2.1 tells us that we need only find two independent solutions y_1 and y_2 of (1); all other solutions of (1) are then obtained by taking linear combinations of y_1 and y_2. Unfortunately, Theorem 2 doesn't tell us how to find two solutions of (1). Therefore, we will try an educated guess. To this end, observe that a function $y(t)$ is a solution of (1) if a constant times its second derivative, plus another constant times its first derivative, plus a third constant times itself is identically zero. In other words, the three terms ay'', by', and cy must cancel each other. In general, this can only occur if the three functions $y(t)$, $y'(t)$, and $y''(t)$ are of the "same type". For example, the function $y(t)=t^5$ can never be a solution of (1) since the three terms $20at^3$, $5bt^4$, and ct^5 are polynomials in t of different degree, and therefore cannot cancel each other. On the other hand, the function $y(t)=e^{rt}$, r constant, has the property that both $y'(t)$ and $y''(t)$ are multiples of $y(t)$. This suggests that we try $y(t)=e^{rt}$ as a solution of (1). Computing

$$\begin{aligned}L[e^{rt}]&=a(e^{rt})''+b(e^{rt})'+c(e^{rt})\\&=(ar^2+br+c)e^{rt},\end{aligned}$$

we see that $y(t)=e^{rt}$ is a solution of (1) if, and only if

$$ar^2+br+c=0. \tag{2}$$

Equation (2) is called the *characteristic equation* of (1). It has two roots r_1,r_2 given by the quadratic formula

$$r_1=\frac{-b+\sqrt{b^2-4ac}}{2a},\qquad r_2=\frac{-b-\sqrt{b^2-4ac}}{2a}.$$

If b^2-4ac is positive, then r_1 and r_2 are real and distinct. In this case, $y_1(t)=e^{r_1t}$ and $y_2(t)=e^{r_2t}$ are two distinct solutions of (1). These solutions are clearly linearly independent (on any interval I), since e^{r_2t} is obviously not a constant multiple of e^{r_1t} for $r_2\neq r_1$. (If the reader is unconvinced of this he can compute

$$W\left[e^{r_1t},e^{r_2t}\right]=(r_2-r_1)e^{(r_1+r_2)t},$$

and observe that W is never zero. Hence, e^{r_1t} and e^{r_2t} are linearly independent on any interval I.)

Example 1. Find the general solution of the equation

$$\frac{d^2y}{dt^2}+5\frac{dy}{dt}+4y=0. \tag{3}$$

Solution. The characteristic equation $r^2+5r+4=(r+4)(r+1)=0$ has two distinct roots $r_1=-4$ and $r_2=-1$. Thus, $y_1(t)=e^{-4t}$ and $y_2(t)=e^{-t}$ form a fundamental set of solutions of (3), and every solution $y(t)$ of (3) is of the form

$$y(t)=c_1e^{-4t}+c_2e^{-t}$$

for some choice of constants c_1,c_2.

Example 2. Find the solution $y(t)$ of the initial-value problem

$$\frac{d^2y}{dt^2}+4\frac{dy}{dt}-2y=0; \qquad y(0)=1, \quad y'(0)=2.$$

Solution. The characteristic equation $r^2+4r-2=0$ has 2 roots

$$r_1=\frac{-4+\sqrt{16+8}}{2}=-2+\sqrt{6}$$

and

$$r_2=\frac{-4-\sqrt{16+8}}{2}=-2-\sqrt{6}\ .$$

Hence, $y_1(t)=e^{r_1t}$ and $y_2(t)=e^{r_2t}$ are a fundamental set of solutions of $y''+4y'-2y=0$, so that

$$y(t)=c_1e^{(-2+\sqrt{6}\,)t}+c_2e^{(-2-\sqrt{6}\,)t}$$

for some choice of constants c_1,c_2. The constants c_1 and c_2 are determined from the initial conditions

$$c_1+c_2=1 \quad \text{and} \quad (-2+\sqrt{6}\,)c_1+(-2-\sqrt{6}\,)c_2=2.$$

From the first equation, $c_2=1-c_1$. Substituting this value of c_2 into the second equation gives

$$(-2+\sqrt{6}\,)c_1-(2+\sqrt{6}\,)(1-c_1)=2, \quad \text{or} \quad 2\sqrt{6}\,c_1=4+\sqrt{6}\ .$$

Therefore, $c_1=2/\sqrt{6}\,+\frac{1}{2}$, $c_2=1-c_1=\frac{1}{2}-2/\sqrt{6}$\,, and

$$y(t)=\left(\frac{1}{2}+\frac{2}{\sqrt{6}}\right)e^{(-2+\sqrt{6}\,)t}+\left(\frac{1}{2}-\frac{2}{\sqrt{6}}\right)e^{-(2+\sqrt{6}\,)t}.$$

Exercises

Find the general solution of each of the following equations.

1. $\dfrac{d^2y}{dt^2} - y = 0$

2. $6\dfrac{d^2y}{dt^2} - 7\dfrac{dy}{dt} + y = 0$

3. $\dfrac{d^2y}{dt^2} - 3\dfrac{dy}{dt} + y = 0$

4. $3\dfrac{d^2y}{dt^2} + 6\dfrac{dy}{dt} + 2y = 0$

Solve each of the following initial-value problems.

5. $\dfrac{d^2y}{dt^2} - 3\dfrac{dy}{dt} - 4y = 0; \quad y(0) = 1,\ y'(0) = 0$

6. $2\dfrac{d^2y}{dt^2} + \dfrac{dy}{dt} - 10y = 0; \quad y(1) = 5,\ y'(1) = 2$

7. $5\dfrac{d^2y}{dt^2} + 5\dfrac{dy}{dt} - y = 0; \quad y(0) = 0,\ y'(0) = 1$

8. $\dfrac{d^2y}{dt^2} - 6\dfrac{dy}{dt} + y = 0; \quad y(2) = 1,\ y'(2) = 1$

Remark. In doing Problems 6 and 8, observe that $e^{r(t-t_0)}$ is also a solution of the differential equation $ay'' + by' + cy = 0$ if $ar^2 + br + c = 0$. Thus, to find the solution $y(t)$ of the initial-value problem $ay'' + by' + cy = 0$; $y(t_0) = y_0, y'(t_0) = y'_0$, we would write $y(t) = c_1 e^{r_1(t-t_0)} + c_2 e^{r_2(t-t_0)}$ and solve for c_1 and c_2 from the initial conditions.

9. Let $y(t)$ be the solution of the initial-value problem

$$\frac{d^2y}{dt^2} + 5\frac{dy}{dt} + 6y = 0; \qquad y(0) = 1, \quad y'(0) = V.$$

For what values of V does $y(t)$ remain nonnegative for all $t \geqslant 0$?

10. The differential equation

$$L[y] = t^2 y'' + \alpha t y' + \beta y = 0 \tag{*}$$

is known as Euler's equation. Observe that $t^2 y''$, ty', and y are all multiples of t^r if $y = t^r$. This suggests that we try $y = t^r$ as a solution of (*). Show that $y = t^r$ is a solution of (*) if $r^2 + (\alpha - 1)r + \beta = 0$.

11. Find the general solution of the equation

$$t^2 y'' + 5ty' - 5y = 0, \qquad t > 0$$

12. Solve the initial-value problem

$$t^2 y'' - ty' - 2y = 0; \qquad y(1) = 0, \quad y'(1) = 1$$

on the interval $0 < t < \infty$.

2.2.1 Complex roots

If b^2-4ac is negative, then the characteristic equation $ar^2+br+c=0$ has complex roots

$$r_1=\frac{-b+i\sqrt{4ac-b^2}}{2a} \quad \text{and} \quad r_2=\frac{-b-i\sqrt{4ac-b^2}}{2a}.$$

We would like to say that e^{r_1t} and e^{r_2t} are solutions of the differential equation

$$a\frac{d^2y}{dt^2}+b\frac{dy}{dt}+cy=0. \tag{1}$$

However, this presents us with two serious difficulties. On the one hand, the function e^{rt} is not defined, as yet, for r complex. And on the other hand, even if we succeed in defining e^{r_1t} and e^{r_2t} as complex-valued solutions of (1), we are still faced with the problem of finding two *real-valued* solutions of (1).

We begin by resolving the second difficulty, since otherwise there's no sense tackling the first problem. Assume that $y(t)=u(t)+iv(t)$ is a complex-valued solution of (1). This means, of course, that

$$a\left[u''(t)+iv''(t)\right]+b\left[u'(t)+iv'(t)\right]+c\left[u(t)+iv(t)\right]=0. \tag{2}$$

This complex-valued solution of (1) gives rise to *two* real-valued solutions, as we now show.

Lemma 1. *Let $y(t)=u(t)+iv(t)$ be a complex-valued solution of* (1), *with a, b, and c real. Then, $y_1(t)=u(t)$ and $y_2(t)=v(t)$ are two real-valued solutions of* (1). *In other words, both the real and imaginary parts of a complex-valued solution of* (1) *are solutions of* (1). (*The imaginary part of the complex number $\alpha+i\beta$ is β. Similarly, the imaginary part of the function $u(t)+iv(t)$ is $v(t)$.*)

PROOF. From Equation (2),

$$\left[au''(t)+bu'(t)+cu(t)\right]+i\left[av''(t)+bv'(t)+cv(t)\right]=0. \tag{3}$$

Now, if a complex number is zero, then both its real and imaginary parts must be zero. Consequently,

$$au''(t)+bu'(t)+cu(t)=0 \quad \text{and} \quad av''(t)+bv'(t)+cv(t)=0,$$

and this proves Lemma 1. □

The problem of defining e^{rt} for r complex can also be resolved quite easily. Let $r=\alpha+i\beta$. By the law of exponents,

$$e^{rt}=e^{\alpha t}e^{i\beta t}. \tag{4}$$

Thus, we need only define the quantity $e^{i\beta t}$, for β real. To this end, recall that

$$e^x = 1 + x + \frac{x^2}{2!} + \frac{x^3}{3!} + \dots . \tag{5}$$

Equation (5) makes sense, formally, even for x complex. This suggests that we set

$$e^{i\beta t} = 1 + i\beta t + \frac{(i\beta t)^2}{2!} + \frac{(i\beta t)^3}{3!} + \dots .$$

Next, observe that

$$\begin{aligned} 1 + i\beta t + \frac{(i\beta t)^2}{2!} + \dots &= 1 + i\beta t - \frac{\beta^2 t^2}{2!} - \frac{i\beta^3 t^3}{3!} + \frac{\beta^4 t^4}{4!} + \frac{i\beta^5 t^5}{5!} + \dots \\ &= \left[1 - \frac{\beta^2 t^2}{2!} + \frac{\beta^4 t^4}{4!} + \dots\right] + i\left[\beta t - \frac{\beta^3 t^3}{3!} + \frac{\beta^5 t^5}{5!} + \dots\right] \\ &= \cos\beta t + i \sin\beta t. \end{aligned}$$

Hence,

$$e^{(\alpha + i\beta)t} = e^{\alpha t} e^{i\beta t} = e^{\alpha t}(\cos\beta t + i\sin\beta t). \tag{6}$$

Returning to the differential equation (1), we see that

$$\begin{aligned} y(t) &= e^{[-b + i\sqrt{4ac - b^2}\,]t/2a} \\ &= e^{-bt/2a}\left[\cos\sqrt{4ac - b^2}\; t/2a + i\sin\sqrt{4ac - b^2}\; t/2a\right] \end{aligned}$$

is a complex-valued solution of (1) if $b^2 - 4ac$ is negative. Therefore, by Lemma 1,

$$y_1(t) = e^{-bt/2a}\cos\beta t \quad \text{and} \quad y_2(t) = e^{-bt/2a}\sin\beta t, \qquad \beta = \frac{\sqrt{4ac - b^2}}{2a}$$

are two real-valued solutions of (1). These two functions are linearly independent on any interval I, since their Wronskian (see Exercise 10) is never zero. Consequently, the general solution of (1) for $b^2 - 4ac < 0$ is

$$y(t) = e^{-bt/2a}\left[c_1\cos\beta t + c_2\sin\beta t\right], \qquad \beta = \frac{\sqrt{4ac - b^2}}{2a}.$$

Remark 1. Strictly speaking, we must verify that the formula

$$\frac{d}{dt}e^{rt} = re^{rt}$$

is true even for r complex, before we can assert that $e^{r_1 t}$ and $e^{r_2 t}$ are complex-valued solutions of (1). To this end, we compute

$$\begin{aligned} \frac{d}{dt}e^{(\alpha + i\beta)t} &= \frac{d}{dt}e^{\alpha t}\left[\cos\beta t + i\sin\beta t\right] \\ &= e^{\alpha t}\left[(\alpha\cos\beta t - \beta\sin\beta t) + i(\alpha\sin\beta t + \beta\cos\beta t)\right] \end{aligned}$$

and this equals $(\alpha+i\beta)e^{(\alpha+i\beta)t}$, since

$$\begin{aligned}(\alpha+i\beta)e^{(\alpha+i\beta)t}&=(\alpha+i\beta)e^{\alpha t}[\cos\beta t+i\sin\beta t]\\&=e^{\alpha t}[(\alpha\cos\beta t-\beta\sin\beta t)+i(\alpha\sin\beta t+\beta\cos\beta t)].\end{aligned}$$

Thus, $(d/dt)e^{rt}=re^{rt}$, even for r complex.

Remark 2. At first glance, one might think that e^{r_2t} would give rise to two additional solutions of (1). This is not the case, though, since

$$\begin{aligned}e^{r_2t}&=e^{-(b/2a)t}e^{-i\beta t},\qquad \beta=\sqrt{4ac-b^2}\,/2a\\&=e^{-bt/2a}[\cos(-\beta t)+i\sin(-\beta t)]=e^{-bt/2a}[\cos\beta t-i\sin\beta t].\end{aligned}$$

Hence,

$$\operatorname{Re}\{e^{r_2t}\}=e^{-bt/2a}\cos\beta t=y_1(t)$$

and

$$\operatorname{Im}\{e^{r_2t}\}=-e^{-bt/2a}\sin\beta t=-y_2(t).$$

Example 1. Find two linearly independent real-valued solutions of the differential equation

$$4\frac{d^2y}{dt^2}+4\frac{dy}{dt}+5y=0. \tag{7}$$

Solution. The characteristic equation $4r^2+4r+5=0$ has complex roots $r_1=-\frac{1}{2}+i$ and $r_2=-\frac{1}{2}-i$. Consequently,

$$e^{r_1t}=e^{(-1/2+i)t}=e^{-t/2}\cos t+ie^{-t/2}\sin t$$

is a complex-valued solution of (7). Therefore, by Lemma 1,

$$\operatorname{Re}\{e^{r_1t}\}=e^{-t/2}\cos t\quad\text{and}\quad\operatorname{Im}\{e^{r_1t}\}=e^{-t/2}\sin t$$

are two linearly independent real-valued solutions of (7).

Example 2. Find the solution $y(t)$ of the initial-value problem

$$\frac{d^2y}{dt^2}+2\frac{dy}{dt}+4y=0;\qquad y(0)=1,\quad y'(0)=1.$$

Solution. The characteristic equation $r^2+2r+4=0$ has complex roots $r_1=-1+\sqrt{3}\,i$ and $r_2=-1-\sqrt{3}\,i$. Hence,

$$e^{r_1t}=e^{(-1+\sqrt{3}\,i)t}=e^{-t}\cos\sqrt{3}\,t+ie^{-t}\sin\sqrt{3}\,t$$

is a complex-valued solution of $y''+2y'+4y=0$. Therefore, by Lemma 1, both

$$\operatorname{Re}\{e^{r_1t}\}=e^{-t}\cos\sqrt{3}\,t\quad\text{and}\quad\operatorname{Im}\{e^{r_1t}\}=e^{-t}\sin\sqrt{3}\,t$$

are real-valued solutions. Consequently,

$$y(t)=e^{-t}[c_1\cos\sqrt{3}\,t+c_2\sin\sqrt{3}\,t]$$

for some choice of constants c_1, c_2. The constants c_1 and c_2 are determined from the initial conditions

$$1 = y(0) = c_1$$

and

$$1 = y'(0) = -c_1 + \sqrt{3}\, c_2.$$

This implies that

$$c_1 = 1, c_2 = \frac{2}{\sqrt{3}} \quad \text{and} \quad y(t) = e^{-t}\left[\cos\sqrt{3}\,t + \frac{2}{\sqrt{3}}\sin\sqrt{3}\,t\right].$$

Exercises

Find the general solution of each of the following equations.

1. $\dfrac{d^2y}{dt^2} + \dfrac{dy}{dt} + y = 0$

2. $2\dfrac{d^2y}{dt^2} + 3\dfrac{dy}{dt} + 4y = 0$

3. $\dfrac{d^2y}{dt^2} + 2\dfrac{dy}{dt} + 3y = 0$

4. $4\dfrac{d^2y}{dt^2} - \dfrac{dy}{dt} + y = 0$

Solve each of the following initial-value problems.

5. $\dfrac{d^2y}{dt^2} + \dfrac{dy}{dt} + 2y = 0; \quad y(0) = 1,\ y'(0) = -2$

6. $\dfrac{d^2y}{dt^2} + 2\dfrac{dy}{dt} + 5y = 0; \quad y(0) = 0,\ y'(0) = 2$

7. Assume that $b^2 - 4ac < 0$. Show that

$$y_1(t) = e^{(-b/2a)(t-t_0)}\cos\beta(t-t_0)$$

and

$$y_2(t) = e^{(-b/2a)(t-t_0)}\sin\beta(t-t_0), \qquad \beta = \frac{\sqrt{4ac-b^2}}{2a}$$

are solutions of (1), for any number t_0.

Solve each of the following initial-value problems.

8. $2\dfrac{d^2y}{dt^2} - \dfrac{dy}{dt} + 3y = 0; \quad y(1) = 1,\ y'(1) = 1$

9. $3\dfrac{d^2y}{dt^2} - 2\dfrac{dy}{dt} + 4y = 0; \quad y(2) = 1,\ y'(2) = -1$

10. Verify that $W[e^{\alpha t}\cos\beta t,\ e^{\alpha t}\sin\beta t] = \beta e^{2\alpha t}$.

11. Show that $e^{i\omega t}$ is a complex-valued solution of the differential equation $y'' + \omega^2 y = 0$. Find two real-valued solutions.

12. Show that $(\cos t + i\sin t)^r = \cos rt + i\sin rt$. Use this result to obtain the double angle formulas $\sin 2t = 2\sin t\cos t$ and $\cos 2t = \cos^2 t - \sin^2 t$.

13. Show that

$$(\cos t_1 + i\sin t_1)(\cos t_2 + i\sin t_2) = \cos(t_1+t_2) + i\sin(t_1+t_2).$$

Use this result to obtain the trigonometric identities

$$\cos(t_1+t_2) = \cos t_1 \cos t_2 - \sin t_1 \sin t_2,$$
$$\sin(t_1+t_2) = \sin t_1 \cos t_2 + \cos t_1 \sin t_2.$$

14. Show that any complex number $a+ib$ can be written in the form $Ae^{i\theta}$, where $A=\sqrt{a^2+b^2}$ and $\tan\theta = b/a$.

15. Defining the two possible square roots of a complex number $Ae^{i\theta}$ as $\pm\sqrt{A}\,e^{i\theta/2}$, compute the square roots of i, $1+i$, $-i$, $\sqrt{i}$.

16. Use Problem 14 to find the three cube roots of i.

17. (a) Let $r_1 = \lambda + i\mu$ be a complex root of $r^2 + (\alpha-1)r + \beta = 0$. Show that

$$t^{\lambda+i\mu} = t^{\lambda}t^{i\mu} = t^{\lambda}e^{(\ln t)i\mu} = t^{\lambda}[\cos\mu\ln t + i\sin\mu\ln t]$$

is a complex-valued solution of Euler's equation

$$t^2\frac{d^2y}{dt^2} + \alpha t\frac{dy}{dt} + \beta y = 0. \tag{*}$$

(b) Show that $t^{\lambda}\cos\mu\ln t$ and $t^{\lambda}\sin\mu\ln t$ are real-valued solutions of (*).

Find the general solution of each of the following equations.

18. $t^2\dfrac{d^2y}{dt^2} + t\dfrac{dy}{dt} + y = 0, \quad t>0$

19. $t^2\dfrac{d^2y}{dt^2} + 2t\dfrac{dy}{dt} + 2y = 0, \quad t>0$

2.2.2 *Equal roots; reduction of order*

If $b^2 = 4ac$, then the characteristic equation $ar^2 + br + c = 0$ has real equal roots $r_1 = r_2 = -b/2a$. In this case, we obtain only one solution

$$y_1(t) = e^{-bt/2a}$$

of the differential equation

$$a\frac{d^2y}{dt^2} + b\frac{dy}{dt} + cy = 0. \tag{1}$$

Our problem is to find a second solution which is independent of y_1. One approach to this problem is to try some additional guesses. A second, and much more clever approach is to try and use our knowledge of $y_1(t)$ to help us find a second independent solution. More generally, suppose that we know one solution $y = y_1(t)$ of the second-order linear equation

$$L[y] = \frac{d^2y}{dt^2} + p(t)\frac{dy}{dt} + q(t)y = 0. \tag{2}$$

Can we use this solution to help us find a second independent solution?

The answer to this question is yes. Once we find one solution $y=y_1(t)$ of (2), we can reduce the problem of finding all solutions of (2) to that of solving a first-order linear homogeneous equation. This is accomplished by defining a new dependent variable v through the substitution

$$y(t)=y_1(t)v(t).$$

Then

$$\frac{dy}{dt}=v\frac{dy_1}{dt}+y_1\frac{dv}{dt}$$

and

$$\frac{d^2y}{dt^2}=v\frac{d^2y_1}{dt^2}+2\frac{dv}{dt}\frac{dy_1}{dt}+y_1\frac{d^2v}{dt^2}.$$

Consequently.

$$\begin{aligned}L[y]&=v\frac{d^2y_1}{dt^2}+2\frac{dv}{dt}\frac{dy_1}{dt}+y_1\frac{d^2v}{dt^2}+p(t)\left[v\frac{dy_1}{dt}+y_1\frac{dv}{dt}\right]+q(t)vy_1\\&=y_1\frac{d^2v}{dt^2}+\left[2\frac{dy_1}{dt}+p(t)y_1\right]\frac{dv}{dt}+\left[\frac{d^2y_1}{dt^2}+p(t)\frac{dy_1}{dt}+q(t)y_1\right]v\\&=y_1\frac{d^2v}{dt^2}+\left[2\frac{dy_1}{dt}+p(t)y_1\right]\frac{dv}{dt},\end{aligned}$$

since $y_1(t)$ is a solution of $L[y]=0$. Hence, $y(t)=y_1(t)v(t)$ is a solution of (2) if v satisfies the differential equation

$$y_1\frac{d^2v}{dt^2}+\left[2\frac{dy_1}{dt}+p(t)y_1\right]\frac{dv}{dt}=0. \tag{3}$$

Now, observe that Equation (3) is really a first-order linear equation for dv/dt. Its solution is

$$\begin{aligned}\frac{dv}{dt}&=c\exp\left[-\int\left[2\frac{y_1'(t)}{y_1(t)}+p(t)\right]dt\right]\\&=c\exp\left(-\int p(t)\,dt\right)\exp\left[-2\int\frac{y_1'(t)}{y_1(t)}dt\right]\\&=\frac{c\exp\left(-\int p(t)dt\right)}{y_1^2(t)}.\end{aligned} \tag{4}$$

Since we only need one solution $v(t)$ of (3), we set $c=1$ in (4). Integrating this equation with respect to t, and setting the constant of integration equal to zero, we obtain that $v(t) = \int u(t)\,dt$, where

$$u(t) = \frac{\exp\left(-\int p(t)\,dt\right)}{y_1^2(t)}. \tag{5}$$

Hence,

$$y_2(t) = v(t)y_1(t) = y_1(t)\int u(t)\,dt \tag{6}$$

is a second solution of (2). This solution is independent of y_1, for if $y_2(t)$ were a constant multiple of $y_1(t)$ then $v(t)$ would be constant, and consequently, its derivative would vanish identically. However, from (4)

$$\frac{dv}{dt} = \frac{\exp\left(-\int p(t)\,dt\right)}{y_1^2(t)},$$

and this quantity is never zero.

Remark 1. In writing $v(t) = \int u(t)\,dt$, we set the constant of integration equal to zero. Choosing a nonzero constant of integration would only add a constant multiple of $y_1(t)$ to $y_2(t)$. Similarly, the effect of choosing a constant c other than one in Equation (4) would be to multiply $y_2(t)$ by c.

Remark 2. The method we have just presented for solving Equation (2) is known as the method of *reduction of order*, since the substitution $y(t) = y_1(t)v(t)$ reduces the problem of solving the second-order equation (2) to that of solving a first-order equation.

Application to the case of equal roots: In the case of equal roots, we found $y_1(t) = e^{-bt/2a}$ as one solution of the equation

$$a\frac{d^2y}{dt^2} + b\frac{dy}{dt} + cy = 0. \tag{7}$$

We can find a second solution from Equations (5) and (6). It is important to realize though, that Equations (5) and (6) were derived under the assumption that our differential equation was written in the form

$$\frac{d^2y}{dt^2} + p(t)\frac{dy}{dt} + q(t)y = 0;$$

that is, the coefficient of y'' was one. In our equation, the coefficient of y'' is a. Hence, we must divide Equation (7) by a to obtain the equivalent equation

$$\frac{d^2y}{dt^2} + \frac{b}{a}\frac{dy}{dt} + \frac{c}{a}y = 0.$$

Now, we can insert $p(t)=b/a$ into (5) to obtain that

$$u(t)=\frac{\exp\left(-\int\frac{b}{a}dt\right)}{\left[e^{-bt/2a}\right]^2}=\frac{e^{-bt/a}}{e^{-bt/a}}=1.$$

Hence,

$$y_2(t)=y_1(t)\int dt=ty_1(t)$$

is a second solution of (7). The functions $y_1(t)$ and $y_2(t)$ are clearly linearly independent on the interval $-\infty<t<\infty$. Therefore, the general solution of (7) in the case of equal roots is

$$y(t)=c_1e^{-bt/2a}+c_2te^{-bt/2a}=\left[c_1+c_2t\right]e^{-bt/2a}$$

Example 1. Find the solution $y(t)$ of the initial-value problem

$$\frac{d^2y}{dt^2}+4\frac{dy}{dt}+4y=0; \qquad y(0)=1, \quad y'(0)=3.$$

Solution. The characteristic equation $r^2+4r+4=(r+2)^2=0$ has two equal roots $r_1=r_2=-2$. Hence,

$$y(t)=c_1e^{-2t}+c_2te^{-2t}$$

for some choice of constants c_1, c_2. The constants c_1 and c_2 are determined from the initial conditions

$$1=y(0)=c_1$$

and

$$3=y'(0)=-2c_1+c_2.$$

This implies that $c_1=1$ and $c_2=5$, so that $y(t)=(1+5t)e^{-2t}$.

Example 2. Find the solution $y(t)$ of the initial-value problem

$$(1-t^2)\frac{d^2y}{dt^2}+2t\frac{dy}{dt}-2y=0; \qquad y(0)=3, \quad y'(0)=-4$$

on the interval $-1<t<1$.

Solution. Clearly, $y_1(t)=t$ is one solution of the differential equation

$$(1-t^2)\frac{d^2y}{dt^2}+2t\frac{dy}{dt}-2y=0. \tag{8}$$

We will use the method of reduction of order to find a second solution $y_2(t)$ of (8). To this end, divide both sides of (8) by $1-t^2$ to obtain the equivalent equation

$$\frac{d^2y}{dt^2}+\frac{2t}{1-t^2}\frac{dy}{dt}-\frac{2}{1-t^2}y=0.$$

Then, from (5)

$$u(t)=\frac{\exp\left(-\int\frac{2t}{1-t^2}\,dt\right)}{y_1^2(t)}=\frac{e^{\ln(1-t^2)}}{t^2}=\frac{1-t^2}{t^2},$$

and

$$y_2(t)=t\int\frac{1-t^2}{t^2}\,dt=-t\left(\frac{1}{t}+t\right)=-(1+t^2)$$

is a second solution of (8). Therefore,

$$y(t)=c_1t-c_2(1+t^2)$$

for some choice of constants c_1, c_2. (Notice that all solutions of (9) are continuous at $t=\pm 1$ even though the differential equation is not defined at these points. Thus, it does not necessarily follow that the solutions of a differential equation are discontinuous at a point where the differential equation is not defined—but this is often the case.) The constants c_1 and c_2 are determined from the initial conditions

$$3=y(0)=-c_2 \quad\text{and}\quad -4=y'(0)=c_1.$$

Hence, $y(t)=-4t+3(1+t^2)$.

Exercises

Find the general solution of each of the following equations

1. $\dfrac{d^2y}{dt^2}-6\dfrac{dy}{dt}+9y=0$

2. $4\dfrac{d^2y}{dt^2}-12\dfrac{dy}{dt}+9y=0$

Solve each of the following initial-value problems.

3. $9\dfrac{d^2y}{dt^2}+6\dfrac{dy}{dt}+y=0;\quad y(0)=1,\ y'(0)=0$

4. $4\dfrac{d^2y}{dt^2}-4\dfrac{dy}{dt}+y=0;\quad y(0)=0,\ y'(0)=3$

5. Suppose $b^2=4ac$. Show that

$$y_1(t)=e^{-b(t-t_0)/2a}\quad\text{and}\quad y_2(t)=(t-t_0)e^{-b(t-t_0)/2a}$$

are solutions of (1) for every choice of t_0.

Solve the following initial-value problems.

6. $\dfrac{d^2y}{dt^2}+2\dfrac{dy}{dt}+y=0;\quad y(2)=1,\ y'(2)=-1$

7. $9\dfrac{d^2y}{dt^2}-12\dfrac{dy}{dt}+4y=0;\quad y(\pi)=0,\ y'(\pi)=2$

8. Let a, b and c be positive numbers. Prove that every solution of the differential equation $ay''+by'+cy=0$ approaches zero as t approaches infinity.

9. Here is an alternate and very elegant way of finding a second solution $y_2(t)$ of (1).

(a) Assume that $b^2=4ac$. Show that

$$L[e^{rt}]=a(e^{rt})''+b(e^{rt})'+ce^{rt}=a(r-r_1)^2e^{rt}$$

for $r_1=-b/2a$.

(b) Show that

$$(\partial/\partial r)L[e^{rt}]=L[(\partial/\partial r)e^{rt}]=L[te^{rt}]=2a(r-r_1)e^{rt}+at(r-r_1)^2e^{rt}.$$

(c) Conclude from (a) and (b) that $L[te^{r_1t}]=0$. Hence, $y_2(t)=te^{r_1t}$ is a second solution of (1).

Use the method of reduction of order to find the general solution of the following differential equations.

10. $\frac{d^2y}{dt^2}-\frac{2(t+1)}{(t^2+2t-1)}\frac{dy}{dt}+\frac{2}{(t^2+2t-1)}y=0 \quad (y_1(t)=t+1)$

11. $\frac{d^2y}{dt^2}-4t\frac{dy}{dt}+(4t^2-2)y=0 \quad (y_1(t)=e^{t^2})$

12. $(1-t^2)\frac{d^2y}{dt^2}-2t\frac{dy}{dt}+2y=0 \quad (y_1(t)=t)$

13. $(1+t^2)\frac{d^2y}{dt^2}-2t\frac{dy}{dt}+2y=0 \quad (y_1(t)=t)$

14. $(1-t^2)\frac{d^2y}{dt^2}-2t\frac{dy}{dt}+6y=0 \quad (y_1(t)=3t^2-1)$

15. $(2t+1)\frac{d^2y}{dt^2}-4(t+1)\frac{dy}{dt}+4y=0 \quad (y_1(t)=t+1)$

16. $t^2\frac{d^2y}{dt^2}+t\frac{dy}{dt}+\left(t^2-\frac{1}{4}\right)y=0 \quad \left(y_1(t)=\frac{\sin t}{\sqrt{t}}\right)$

17. Given that the equation

$$t\frac{d^2y}{dt^2}-(1+3t)\frac{dy}{dt}+3y=0$$

has a solution of the form e^{ct}, for some constant c, find the general solution.

18. (a) Show that t^r is a solution of Euler's equation

$$t^2y''+\alpha ty'+\beta y=0, \quad t>0$$

if $r^2+(\alpha-1)r+\beta=0$.

(b) Suppose that $(\alpha-1)^2=4\beta$. Using the method of reduction of order, show that $(\ln t)t^{(1-\alpha)/2}$ is a second solution of Euler's equation.

Find the general solution of each of the following equations.

19. $t^2\frac{d^2y}{dt^2}+3t\frac{dy}{dt}+y=0$

20. $t^2\frac{d^2y}{dt^2}-t\frac{dy}{dt}+y=0$

2.3 The nonhomogeneous equation

We turn our attention now to the nonhomogeneous equation

$$L[y]=\frac{d^2y}{dt^2}+p(t)\frac{dy}{dt}+q(t)y=g(t) \tag{1}$$

where the functions $p(t)$, $q(t)$ and $g(t)$ are continuous on an open interval $\alpha<t<\beta$. An important clue as to the nature of all solutions of (1) is provided by the first-order linear equation

$$\frac{dy}{dt}-2ty=-t. \tag{2}$$

The general solution of this equation is

$$y(t)=ce^{t^2}+\tfrac{1}{2}.$$

Now, observe that this solution is the sum of two terms: the first term, ce^{t^2}, is the general solution of the homogeneous equation

$$\frac{dy}{dt}-2ty=0 \tag{3}$$

while the second term, $\frac{1}{2}$, is a solution of the nonhomogeneous equation (2). In other words, every solution $y(t)$ of (2) is the sum of a particular solution, $\psi(t)=\frac{1}{2}$, with a solution ce^{t^2} of the homogeneous equation. A similar situation prevails in the case of second-order equations, as we now show.

Theorem 5. *Let $y_1(t)$ and $y_2(t)$ be two linearly independent solutions of the homogeneous equation*

$$L[y]=\frac{d^2y}{dt^2}+p(t)\frac{dy}{dt}+q(t)y=0 \tag{4}$$

and let $\psi(t)$ be any particular solution of the nonhomogeneous equation (1). *Then, every solution $y(t)$ of* (1) *must be of the form*

$$y(t)=c_1y_1(t)+c_2y_2(t)+\psi(t)$$

for some choice of constants c_1, c_2.

The proof of Theorem 5 relies heavily on the following lemma.

Lemma 1. *The difference of any two solutions of the nonhomogeneous equation* (1) *is a solution of the homogeneous equation* (4).

PROOF. Let $\psi_1(t)$ and $\psi_2(t)$ be two solutions of (1). By the linearity of L,

$$L[\psi_1-\psi_2](t)=L[\psi_1](t)-L[\psi_2](t)=g(t)-g(t)=0.$$

Hence, $\psi_1(t)-\psi_2(t)$ is a solution of the homogeneous equation (4). □

We can now give a very simple proof of Theorem 5.

PROOF OF THEOREM 5. Let $y(t)$ be any solution of (1). By Lemma 1, the function $\phi(t)=y(t)-\psi(t)$ is a solution of the homogeneous equation (4). But every solution $\phi(t)$ of the homogeneous equation (4) is of the form $\phi(t)=c_1y_1(t)+c_2y_2(t)$, for some choice of constants c_1, c_2. Therefore,

$$y(t)=\phi(t)+\psi(t)=c_1y_1(t)+c_2y_2(t)+\psi(t). \qquad \square$$

Remark. Theorem 5 is an extremely useful theorem since it reduces the problem of finding all solutions of (1) to the much simpler problem of finding just two solutions of the homogeneous equation (4), and one solution of the nonhomogeneous equation (1).

Example 1. Find the general solution of the equation

$$\frac{d^2y}{dt^2}+y=t. \tag{5}$$

Solution. The functions $y_1(t)=\cos t$ and $y_2(t)=\sin t$ are two linearly independent solutions of the homogeneous equation $y''+y=0$. Moreover, $\psi(t)=t$ is obviously a particular solution of (5). Therefore, by Theorem 5, every solution $y(t)$ of (5) must be of the form

$$y(t)=c_1\cos t+c_2\sin t+t.$$

Example 2. Three solutions of a certain second-order nonhomogeneous linear equation are

$$\psi_1(t)=t, \qquad \psi_2(t)=t+e^t, \quad \text{and} \quad \psi_3(t)=1+t+e^t.$$

Find the general solution of this equation.
Solution. By Lemma 1, the functions

$$\psi_2(t)-\psi_1(t)=e^t \quad \text{and} \quad \psi_3(t)-\psi_2(t)=1$$

are solutions of the corresponding homogeneous equation. Moreover, these functions are obviously linearly independent. Therefore, by Theorem 5, every solution $y(t)$ of this equation must be of the form

$$y(t)=c_1e^t+c_2+t.$$

EXERCISES

1. Three solutions of a certain second-order nonhomogeneous linear equation are

$$\psi_1(t)=t^2, \psi_2(t)=t^2+e^{2t}$$

and

$$\psi_3(t)=1+t^2+2e^{2t}.$$

Find the general solution of this equation.

2. Three solutions of a certain second-order linear nonhomogeneous equation are

$$\psi_1(t)=1+e^{t^2},\ \psi_2(t)=1+te^{t^2}$$

and

$$\psi_3(t)=(t+1)e^{t^2}+1$$

Find the general solution of this equation.

3. Three solutions of a second-order linear equation $L[y]=g(t)$ are

$$\psi_1(t)=3e^t+e^{t^2},\ \psi_2(t)=7e^t+e^{t^2}$$

and

$$\psi_3(t)=5e^t+e^{-t^3}+e^{t^2}.$$

Find the solution of the initial-value problem

$$L[y]=g;\qquad y(0)=1,\quad y'(0)=2.$$

4. Let a, b and c be positive constants. Show that the difference of any two solutions of the equation

$$ay''+by'+cy=g(t)$$

approaches zero as t approaches infinity.

5. Let $\psi(t)$ be a solution of the nonhomogeneous equation (1), and let $\phi(t)$ be a solution of the homogeneous equation (4). Show that $\psi(t)+\phi(t)$ is again a solution of (1).

2.4 The method of variation of parameters

In this section we describe a very general method for finding a particular solution $\psi(t)$ of the nonhomogeneous equation

$$L[y]=\frac{d^2y}{dt^2}+p(t)\frac{dy}{dt}+q(t)y=g(t), \tag{1}$$

once the solutions of the homogeneous equation

$$L[y]=\frac{d^2y}{dt^2}+p(t)\frac{dy}{dt}+q(t)y=0 \tag{2}$$

are known. The basic principle of this method is to use our knowledge of the solutions of the homogeneous equation to help us find a solution of the nonhomogeneous equation.

Let $y_1(t)$ and $y_2(t)$ be two linearly independent solutions of the homogeneous equation (2). We will try and find a particular solution $\psi(t)$ of the nonhomogeneous equation (1) of the form

$$\boxed{\psi(t)=u_1(t)y_1(t)+u_2(t)y_2(t);} \tag{3}$$

that is, we will try and find functions $u_1(t)$ and $u_2(t)$ so that the linear combination $u_1(t)y_1(t)+u_2(t)y_2(t)$ is a solution of (1). At first glance, this would appear to be a dumb thing to do, since we are replacing the problem of finding one unknown function $\psi(t)$ by the seemingly harder problem of finding two unknown functions $u_1(t)$ and $u_2(t)$. However, by playing our cards right, we will be able to find $u_1(t)$ and $u_2(t)$ as the solutions of two very simple first-order equations. We accomplish this in the following manner. Observe that the differential equation (1) imposes only one condition on the two unknown functions $u_1(t)$ and $u_2(t)$. Therefore, we have a certain "freedom" in choosing $u_1(t)$ and $u_2(t)$. Our goal is to impose an additional condition on $u_1(t)$ and $u_2(t)$ which will make the expression $L[u_1y_1+u_2y_2]$ as simple as possible. Computing

$$\begin{aligned}\frac{d}{dt}\psi(t)&=\frac{d}{dt}[u_1y_1+u_2y_2]\\&=[u_1y_1'+u_2y_2']+[u_1'y_1+u_2'y_2]\end{aligned}$$

we see that $d^2\psi/dt^2$, and consequently $L[\psi]$, will contain no second-order derivatives of u_1 and u_2 if

$$y_1(t)u_1'(t)+y_2(t)u_2'(t)=0. \tag{4}$$

This suggests that we impose the condition (4) on the functions $u_1(t)$ and $u_2(t)$. In this case, then,

$$\begin{aligned}L[\psi]&=[u_1y_1'+u_2y_2']'+p(t)[u_1y_1'+u_2y_2']+q(t)[u_1y_1+u_2y_2]\\&=u_1'y_1'+u_2'y_2'+u_1[y_1''+p(t)y_1'+q(t)y_1]+u_2[y_2''+p(t)y_2'+q(t)y_2]\\&=u_1'y_1'+u_2'y_2'\end{aligned}$$

since both $y_1(t)$ and $y_2(t)$ are solutions of the homogeneous equation $L[y]=0$. Consequently, $\psi(t)=u_1y_1+u_2y_2$ is a solution of the nonhomogeneous equation (1) if $u_1(t)$ and $u_2(t)$ satisfy the two equations

$$\begin{aligned}y_1(t)u_1'(t)+y_2(t)u_2'(t)&=0\\y_1'(t)u_1'(t)+y_2'(t)u_2'(t)&=g(t).\end{aligned}$$

Multiplying the first equation by $y_2'(t)$, the second equation by $y_2(t)$, and subtracting gives

$$[y_1(t)y_2'(t)-y_1'(t)y_2(t)]u_1'(t)=-g(t)y_2(t),$$

while multiplying the first equation by $y_1'(t)$, the second equation by $y_1(t)$, and subtracting gives

$$[y_1(t)y_2'(t)-y_1'(t)y_2(t)]u_2'(t)=g(t)y_1(t).$$

Hence,

$$u_1'(t)=-\frac{g(t)y_2(t)}{W[y_1,y_2](t)}\quad\text{and}\quad u_2'(t)=\frac{g(t)y_1(t)}{W[y_1,y_2](t)}. \tag{5}$$

Finally, we obtain $u_1(t)$ and $u_2(t)$ by integrating the right-hand sides of (5).

Remark. The general solution of the homogeneous equation (2) is

$$y(t)=c_1y_1(t)+c_2y_2(t).$$

By letting c_1 and c_2 vary with time, we obtain a solution of the nonhomogeneous equation. Hence, this method is known as the method of variation of parameters.

Example 1.
(a) Find a particular solution $\psi(t)$ of the equation

$$\frac{d^2y}{dt^2}+y=\tan t \tag{6}$$

on the interval $-\pi/2<t<\pi/2$.
(b) Find the solution $y(t)$ of (6) which satisfies the initial conditions $y(0)=1$, $y'(0)=1$.
Solution.
(a) The functions $y_1(t)=\cos t$ and $y_2(t)=\sin t$ are two linearly independent solutions of the homogeneous equation $y''+y=0$ with

$$W[y_1,y_2](t)=y_1y_2'-y_1'y_2=(\cos t)\cos t-(-\sin t)\sin t=1.$$

Thus, from (5),

$$u_1'(t)=-\tan t\sin t \quad\text{and}\quad u_2'(t)=\tan t\cos t. \tag{7}$$

Integrating the first equation of (7) gives

$$\begin{aligned}u_1(t)&=-\int\tan t\sin t\,dt=-\int\frac{\sin^2 t}{\cos t}\,dt\\&=\int\frac{\cos^2 t-1}{\cos t}\,dt=\sin t-\ln|\sec t+\tan t|.\\&=\sin t-\ln(\sec t+\tan t),\ -\frac{\pi}{2}<t<\frac{\pi}{2}\end{aligned}$$

while integrating the second equation of (7) gives

$$u_2(t)=\int\tan t\cos t\,dt=\int\sin t\,dt=-\cos t.$$

Consequently,

$$\begin{aligned}\psi(t)&=\cos t\big[\sin t-\ln(\sec t+\tan t)\big]+\sin t(-\cos t)\\&=-\cos t\ln(\sec t+\tan t)\end{aligned}$$

is a particular solution of (6) on the interval $-\pi/2<t<\pi/2$.
(b) By Theorem 5 of Section 2.3,

$$y(t)=c_1\cos t+c_2\sin t-\cos t\ln(\sec t+\tan t)$$

for some choice of constants c_1, c_2. The constants c_1 and c_2 are determined from the initial conditions

$$1=y(0)=c_1 \quad\text{and}\quad 1=y'(0)=c_2-1.$$

Hence, $c_1=1$, $c_2=2$ and

$$y(t)=\cos t+2\sin t-\cos t\ln(\sec t+\tan t).$$

Remark. Equation (5) determines $u_1(t)$ and $u_2(t)$ up to two constants of integration. We usually take these constants to be zero, since the effect of choosing nonzero constants is to add a solution of the homogeneous equation to $\psi(t)$.

Exercises

Find the general solution of each of the following equations.

1. $\dfrac{d^2y}{dt^2}+y=\sec t, \quad -\dfrac{\pi}{2}<t<\dfrac{\pi}{2}$

2. $\dfrac{d^2y}{dt^2}-4\dfrac{dy}{dt}+4y=te^{2t}$

3. $2\dfrac{d^2y}{dt^2}-3\dfrac{dy}{dt}+y=(t^2+1)e^t$

4. $\dfrac{d^2y}{dt^2}-3\dfrac{dy}{dt}+2y=te^{3t}+1$

Solve each of the following initial-value problems.

5. $3y''+4y'+y=(\sin t)e^{-t}; \quad y(0)=1,\ y'(0)=0$

6. $y''+4y'+4y=t^{5/2}e^{-2t}; \quad y(0)=y'(0)=0$

7. $y''-3y'+2y=\sqrt{t+1}\,; \quad y(0)=y'(0)=0$

8. $y''-y=f(t); \quad y(0)=y'(0)=0$

Warning. It must be remembered, while doing Problems 3 and 5, that Equation (5) was derived under the assumption that the coefficient of y'' was one.

9. Find two linearly independent solutions of $t^2y''-2y=0$ of the form $y(t)=t^r$. Using these solutions, find the general solution of $t^2y''-2y=t^2$.

10. One solution of the equation

$$y''+p(t)y'+q(t)y=0 \qquad (*)$$

is $(1+t)^2$, and the Wronskian of any two solutions of (*) is constant. Find the general solution of

$$y''+p(t)y'+q(t)y=1+t.$$

11. Find the general solution of $y''+(1/4t^2)y=f\cos t$, $t>0$, given that $y_1(t)=\sqrt{t}$ is a solution of the homogeneous equation.

12. Find the general solution of the equation

$$\frac{d^2y}{dt^2}-\frac{2t}{1+t^2}\frac{dy}{dt}+\frac{2}{1+t^2}y=1+t^2.$$

13. Show that $\sec t+\tan t$ is positive for $-\pi/2<t<\pi/2$.

2.5 The method of judicious guessing

A serious disadvantage of the method of variation of parameters is that the integrations required are often quite difficult. In certain cases, it is usually much simpler to guess a particular solution. In this section we will establish a systematic method for guessing solutions of the equation

$$a\frac{d^2y}{dt^2}+b\frac{dy}{dt}+cy=g(t) \tag{1}$$

where a, b and c are constants, and $g(t)$ has one of several special forms.

Consider first the differential equation

$$L[y]=a\frac{d^2y}{dt^2}+b\frac{dy}{dt}+cy=a_0+a_1t+\ldots+a_nt^n. \tag{2}$$

We seek a function $\psi(t)$ such that the three functions $a\psi''$, $b\psi'$ and $c\psi$ add up to a given polynomial of degree n. The obvious choice for $\psi(t)$ is a polynomial of degree n. Thus, we set

$$\psi(t)=A_0+A_1t+\ldots+A_nt^n \tag{3}$$

and compute

$$\begin{aligned}L[\psi](t)&=a\psi''(t)+b\psi'(t)+c\psi(t)\\&=a\left[2A_2+\ldots+n(n-1)A_nt^{n-2}\right]+b\left[A_1+\ldots+nA_nt^{n-1}\right]\\&\quad+c\left[A_0+A_1t+\ldots+A_nt^n\right]\\&=cA_nt^n+(cA_{n-1}+nbA_n)t^{n-1}+\ldots+(cA_0+bA_1+2aA_2).\end{aligned}$$

Equating coefficients of like powers of t in the equation

$$L[\psi](t)=a_0+a_1t+\ldots+a_nt^n$$

gives

$$cA_n=a_n,\ cA_{n-1}+nbA_n=a_{n-1},\ldots,cA_0+bA_1+2aA_2=a_0. \tag{4}$$

The first equation determines $A_n=a_n/c$, for $c\neq0$, and the remaining equations then determine $A_{n-1},\ldots,A_0$ successively. Thus, Equation (1) has a particular solution $\psi(t)$ of the form (3), for $c\neq0$.

We run into trouble when $c=0$, since then the first equation of (4) has no solution A_n. This difficulty is to be expected though, for if $c=0$, then $L[\psi]=a\psi''+b\psi'$ is a polynomial of degree $n-1$, while the right hand side of (2) is a polynomial of degree n. To guarantee that $a\psi''+b\psi'$ is a polynomial of degree n, we must take ψ as a polynomial of degree $n+1$. Thus, we set

$$\psi(t)=t\left[A_0+A_1t+\ldots+A_nt^n\right]. \tag{5}$$

We have omitted the constant term in (5) since $y=$constant is a solution of the homogeneous equation $ay''+by'=0$, and thus can be subtracted from $\psi(t)$. The coefficients $A_0, A_1,\ldots,A_n$ are determined uniquely (see Exercise 19) from the equation

$$a\psi''+b\psi'=a_0+a_1t+\ldots+a_nt^n$$

if $b\neq 0$.

Finally, the case $b=c=0$ is trivial to handle since the differential equation (2) can then be integrated immediately to yield a particular solution $\psi(t)$ of the form

$$\psi(t)=\frac{1}{a}\left[\frac{a_0t^2}{1\cdot 2}+\frac{a_1t^3}{2\cdot 3}+\ldots+\frac{a_nt^{n+2}}{(n+1)(n+2)}\right].$$

Summary The differential equation (2) has a solution $\psi(t)$ of the form

$$\psi(t)=\begin{cases}A_0+A_1t+\ldots+A_nt^n, & c\neq 0\\ t(A_0+A_1t+\ldots+A_nt^n), & c=0,\ b\neq 0.\\ t^2(A_0+A_1t+\ldots+A_nt^n), & c=b=0\end{cases}$$

Example 1. Find a particular solution $\psi(t)$ of the equation

$$L[y]=\frac{d^2y}{dt^2}+\frac{dy}{dt}+y=t^2. \tag{6}$$

Solution. We set $\psi(t)=A_0+A_1t+A_2t^2$ and compute

$$\begin{aligned}L[\psi](t)&=\psi''(t)+\psi'(t)+\psi(t)\\&=2A_2+(A_1+2A_2t)+A_0+A_1t+A_2t^2\\&=(A_0+A_1+2A_2)+(A_1+2A_2)t+A_2t^2.\end{aligned}$$

Equating coefficients of like powers of t in the equation $L[\psi](t)=t^2$ gives

$$A_2=1, \qquad A_1+2A_2=0$$

and

$$A_0+A_1+2A_2=0.$$

The first equation tells us that $A_2=1$, the second equation then tells us that $A_1=-2$, and the third equation then tells us that $A_0=0$. Hence,

$$\psi(t)=-2t+t^2$$

is a particular solution of (6).

Let us now re-do this problem using the method of variation of parameters. It is easily verified that

$$y_1(t)=e^{-t/2}\cos\sqrt{3}\,t/2 \quad\text{and}\quad y_2(t)=e^{-t/2}\sin\sqrt{3}\,t/2$$

are two solutions of the homogeneous equation $L[y]=0$. Hence,

$$\psi(t)=u_1(t)e^{-t/2}\cos\sqrt{3}\,t/2+u_2(t)e^{-t/2}\sin\sqrt{3}\,t/2$$

is a particular solution of (6), where

$$u_1(t)=\int\frac{-t^2e^{-t/2}\sin\sqrt{3}\,t/2}{W[y_1,y_2](t)}dt=\frac{-2}{\sqrt{3}}\int t^2e^{t/2}\sin\sqrt{3}\,t/2\,dt$$

and

$$u_2(t)=\int\frac{t^2e^{-t/2}\cos\sqrt{3}\,t/2}{W[y_1,y_2](t)}dt=\frac{2}{\sqrt{3}}\int t^2e^{t/2}\cos\sqrt{3}\,t/2\,dt.$$

These integrations are extremely difficult to perform. Thus, the method of guessing is certainly preferrable, in this problem at least, to the method of variation of parameters.

Consider now the differential equation

$$L[y]=a\frac{d^2y}{dt^2}+b\frac{dy}{dt}+cy=(a_0+a_1t+\ldots+a_nt^n)e^{\alpha t}. \tag{7}$$

We would like to remove the factor $e^{\alpha t}$ from the right-hand side of (7), so as to reduce this equation to Equation (2). This is accomplished by setting $y(t)=e^{\alpha t}v(t)$. Then,

$$y'=e^{\alpha t}(v'+\alpha v) \quad\text{and}\quad y''=e^{\alpha t}(v''+2\alpha v'+\alpha^2 v)$$

so that

$$L[y]=e^{\alpha t}\left[av''+(2a\alpha+b)v'+(a\alpha^2+b\alpha+c)v\right].$$

Consequently, $y(t)=e^{\alpha t}v(t)$ is a solution of (7) if, and only if,

$$a\frac{d^2v}{dt^2}+(2a\alpha+b)\frac{dv}{dt}+(a\alpha^2+b\alpha+c)v=a_0+a_1t+\ldots+a_nt^n. \tag{8}$$

In finding a particular solution $v(t)$ of (8), we must distinguish as to whether (i) $a\alpha^2+b\alpha+c\neq 0$; (ii) $a\alpha^2+b\alpha+c=0$, but $2a\alpha+b\neq 0$; and (iii) both $a\alpha^2+b\alpha+c$ and $2a\alpha+b=0$. The first case means that α is not a root of the characteristic equation

$$ar^2+br+c=0. \tag{9}$$

In other words, $e^{\alpha t}$ is not a solution of the homogeneous equation $L[y]=0$. The second condition means that α is a single root of the characteristic equation (9). This implies that $e^{\alpha t}$ is a solution of the homogeneous equation, but $te^{\alpha t}$ is not. Finally, the third condition means that α is a double root of the characteristic equation (9), so that both $e^{\alpha t}$ and $te^{\alpha t}$ are solutions of the homogeneous equation. Hence, Equation (7) has a particular solution $\psi(t)$ of the form (i) $\psi(t)=(A_0+\ldots+A_nt^n)e^{\alpha t}$, if $e^{\alpha t}$ is not a solution of the homogeneous equation; (ii) $\psi(t)=t(A_0+\ldots+A_nt^n)e^{\alpha t}$, if $e^{\alpha t}$ is a solution of the homogeneous equation but $te^{\alpha t}$ is not; and (iii) $\psi(t)=t^2(A_0+\ldots+A_nt^n)e^{\alpha t}$ if both $e^{\alpha t}$ and $te^{\alpha t}$ are solutions of the homogeneous equation.

Remark. There are two ways of computing a particular solution $\psi(t)$ of (7). Either we make the substitution $y=e^{\alpha t}v$ and find $v(t)$ from (8), or we guess a solution $\psi(t)$ of the form $e^{\alpha t}$ times a suitable polynomial in t. If α is a double root of the characteristic equation (9), or if $n\geqslant 2$, then it is advisable to set $y=e^{\alpha t}v$ and then find $v(t)$ from (8). Otherwise, we guess $\psi(t)$ directly.

Example 2. Find the general solution of the equation

$$\frac{d^2y}{dt^2}-4\frac{dy}{dt}+4y=(1+t+\ldots+t^{27})e^{2t}. \tag{10}$$

Solution. The characteristic equation $r^2-4r+4=0$ has equal roots $r_1=r_2=2$. Hence, $y_1(t)=e^{2t}$ and $y_2(t)=te^{2t}$ are solutions of the homogeneous equation $y''-4y'+4y=0$. To find a particular solution $\psi(t)$ of (10), we set $y=e^{2t}v$. Then, of necessity,

$$\frac{d^2v}{dt^2}=1+t+t^2+\ldots+t^{27}.$$

Integrating this equation twice, and setting the constants of integration equal to zero gives

$$v(t)=\frac{t^2}{1\cdot 2}+\frac{t^3}{2\cdot 3}+\ldots+\frac{t^{29}}{28\cdot 29}.$$

Hence, the general solution of (10) is

$$\begin{aligned}y(t)&=c_1e^{2t}+c_2te^{2t}+e^{2t}\left[\frac{t^2}{1\cdot 2}+\ldots+\frac{t^{29}}{28\cdot 29}\right]\\&=e^{2t}\left[c_1+c_2t+\frac{t^2}{1\cdot 2}+\ldots+\frac{t^{29}}{28\cdot 29}\right].\end{aligned}$$

It would be sheer madness (and a terrible waste of paper) to plug the expression

$$\psi(t)=t^2(A_0+A_1t+\ldots+A_{27}t^{27})e^{2t}$$

into (10) and then solve for the coefficients $A_0, A_1,\ldots,A_{27}$.

Example 3. Find a particular solution $\psi(t)$ of the equation

$$L[y]=\frac{d^2y}{dt^2}-3\frac{dy}{dt}+2y=(1+t)e^{3t}.$$

Solution. In this case, e^{3t} is not a solution of the homogeneous equation $y''-3y'+2y=0$. Thus, we set $\psi(t)=(A_0+A_1t)e^{3t}$. Computing

$$\begin{aligned}L[\psi](t)&=\psi''-3\psi'+2\psi\\&=e^{3t}\left[(9A_0+6A_1+9A_1t)-3(3A_0+A_1+3A_1t)+2(A_0+A_1t)\right]\\&=e^{3t}\left[(2A_0+3A_1)+2A_1t\right]\end{aligned}$$

and cancelling off the factor e^{3t} from both sides of the equation

$$L[\psi](t)=(1+t)e^{3t},$$

gives

$$2A_1t+(2A_0+3A_1)=1+t.$$

This implies that $2A_1=1$ and $2A_0+3A_1=1$. Hence, $A_1=\frac{1}{2}$, $A_0=-\frac{1}{4}$ and $\psi(t)=(-\frac{1}{4}+t/2)e^{3t}$.

Finally, we consider the differential equation

$$L[y]=a\frac{d^2y}{dt^2}+b\frac{dy}{dt}+cy=(a_0+a_1t+\ldots+a_nt^n)\times\begin{cases}\cos\omega t\\ \sin\omega t\end{cases}. \tag{11}$$

We can reduce the problem of finding a particular solution $\psi(t)$ of (11) to the simpler problem of finding a particular solution of (7) with the aid of the following simple but extremely useful lemma.

Lemma 1. *Let $y(t)=u(t)+iv(t)$ be a complex-valued solution of the equation*

$$L[y]=a\frac{d^2y}{dt^2}+b\frac{dy}{dt}+cy=g(t)=g_1(t)+ig_2(t) \tag{12}$$

where a, b and c are real. This means, of course, that

$$a\left[u''(t)+iv''(t)\right]+b\left[u'(t)+iv'(t)\right]+c\left[u(t)+iv(t)\right]=g_1(t)+ig_2(t). \tag{13}$$

Then, $L[u](t)=g_1(t)$ and $L[v](t)=g_2(t)$.

PROOF. Equating real and imaginary parts in (13) gives

$$au''(t)+bu'(t)+cu(t)=g_1(t)$$

and

$$av''(t)+bv'(t)+cv(t)=g_2(t).$$ □

Now, let $\phi(t)=u(t)+iv(t)$ be a particular solution of the equation

$$a\frac{d^2y}{dt^2}+b\frac{dy}{dt}+cy=(a_0+\ldots+a_nt^n)e^{i\omega t}. \tag{14}$$

The real part of the right-hand side of (14) is $(a_0+\ldots+a_nt^n)\cos\omega t$, while the imaginary part is $(a_0+\ldots+a_nt^n)\sin\omega t$. Hence, by Lemma 1

$$u(t)=\operatorname{Re}\{\phi(t)\}$$

is a solution of

$$ay''+by'+cy=(a_0+\ldots+a_nt^n)\cos\omega t$$

while

$$v(t)=\operatorname{Im}\{\phi(t)\}$$

is a solution of

$$ay''+by'+cy=(a_0+\ldots+a_nt^n)\sin\omega t.$$

Example 4. Find a particular solution $\psi(t)$ of the equation

$$L[y]=\frac{d^2y}{dt^2}+4y=\sin 2t. \tag{15}$$

Solution. We will find $\psi(t)$ as the imaginary part of a complex-valued solution $\phi(t)$ of the equation

$$L[y]=\frac{d^2y}{dt^2}+4y=e^{2it}. \tag{16}$$

To this end, observe that the characteristic equation $r^2+4=0$ has complex roots $r=\pm 2i$. Therefore, Equation (16) has a particular solution $\phi(t)$ of the form $\phi(t)=A_0te^{2it}$. Computing

$$\phi'(t)=A_0(1+2it)e^{2it} \quad\text{and}\quad \phi''(t)=A_0(4i-4t)e^{2it}$$

we see that

$$L[\phi](t)=\phi''(t)+4\phi(t)=4iA_0e^{2it}.$$

Hence, $A_0=1/4i=-i/4$ and

$$\phi(t)=-\frac{it}{4}e^{2it}=-\frac{it}{4}(\cos 2t+i\sin 2t)=\frac{t}{4}\sin 2t-i\frac{t}{4}\cos 2t.$$

Therefore, $\psi(t)=\operatorname{Im}\{\phi(t)\}=-(t/4)\cos 2t$ is a particular solution of (15).

Example 5. Find a particular solution $\psi(t)$ of the equation

$$\frac{d^2y}{dt^2}+4y=\cos 2t. \tag{17}$$

Solution. From Example 4, $\phi(t)=(t/4)\sin 2t-i(t/4)\cos 2t$ is a complex-valued solution of (16). Therefore,

$$\psi(t)=\operatorname{Re}\{\phi(t)\}=\frac{t}{4}\sin 2t$$

is a particular solution of (17).

Example 6. Find a particular solution $\psi(t)$ of the equation

$$L[y]=\frac{d^2y}{dt^2}+2\frac{dy}{dt}+y=te^t\cos t. \tag{18}$$

Solution. Observe that $te^t\cos t$ is the real part of $te^{(1+i)t}$. Therefore, we can find $\psi(t)$ as the real part of a complex-valued solution $\phi(t)$ of the equation

$$L[y]=\frac{d^2y}{dt^2}+2\frac{dy}{dt}+y=te^{(1+i)t}. \tag{19}$$

To this end, observe that $1+i$ is not a root of the characteristic equation $r^2+2r+1=0$. Therefore, Equation (19) has a particular solution $\phi(t)$ of the form $\phi(t)=(A_0+A_1t)e^{(1+i)t}$. Computing $L[\phi]=\phi''+2\phi'+\phi$, and using the identity

$$(1+i)^2+2(1+i)+1=\left[(1+i)+1\right]^2=(2+i)^2$$

we see that

$$\left[(2+i)^2A_1t+(2+i)^2A_0+2(2+i)A_1\right]=t.$$

Equating coefficients of like powers of t in this equation gives

$$(2+i)^2A_1=1$$

and

$$(2+i)A_0+2A_1=0.$$

This implies that $A_1=1/(2+i)^2$ and $A_0=-2/(2+i)^3$, so that

$$\phi(t)=\left[\frac{-2}{(2+i)^3}+\frac{t}{(2+i)^2}\right]e^{(1+i)t}.$$

After a little algebra, we find that

$$\phi(t)=\frac{e^t}{125}\left\{\left[(15t-4)\cos t+(20t-22)\sin t\right]\right.$$
$$\left.+i\left[(22-20t)\cos t+(15t-4)\sin t\right]\right\}.$$

Hence,

$$\psi(t)=\operatorname{Re}\{\phi(t)\}=\frac{e^t}{125}\left[(15t-4)\cos t+(20t-22)\sin t\right].$$

Remark. The method of judicious guessing also applies to the equation

$$L[y]=a\frac{d^2y}{dt^2}+b\frac{dy}{dt}+cy=\sum_{j=1}^{n}p_j(t)e^{\alpha_j t} \tag{20}$$

where the $p_j(t), j=1,\ldots,n$ are polynomials in t. Let $\psi_j(t)$ be a particular solution of the equation

$$L[y]=p_j(t)e^{\alpha_j t}, \qquad j=1,\dots,n.$$

Then, $\psi(t)=\sum_{j=1}^{n}\psi_j(t)$ is a solution of (20) since

$$L[\psi]=L\left[\sum_{j=1}^{n}\psi_j\right]=\sum_{j=1}^{n}L[\psi_j]=\sum_{j=1}^{n}p_j(t)e^{\alpha_j t}.$$

Thus, to find a particular solution of the equation

$$y''+y'+y=e^t+t\sin t$$

we find particular solutions $\psi_1(t)$ and $\psi_2(t)$ of the equations

$$y''+y'+y=e^t \quad \text{and} \quad y''+y'+y=t\sin t$$

respectively, and then add these two solutions together.

Exercises

Find a particular solution of each of the following equations.

1. $y''+3y=t^3-1$

2. $y''+4y'+4y=te^{\alpha t}$

3. $y''-y=t^2e^t$

4. $y''+y'+y=1+t+t^2$

5. $y''+2y'+y=e^{-t}$

6. $y''+5y'+4y=t^2e^{7t}$

7. $y''+4y=t\sin 2t$

8. $y''-6y'+9y=(3t^7-5t^4)e^{3t}$

9. $y''-2y'+5y=2\cos^2 t$

10. $y''-2y'+5y=2(\cos^2 t)e^t$

11. $y''+y'-6y=\sin t+te^{2t}$

12. $y''+y'+4y=t^2+(2t+3)(1+\cos t)$

13. $y''-3y'+2y=e^t+e^{2t}$

14. $y''+2y'=1+t^2+e^{-2t}$

15. $y''+y=\cos t\cos 2t$

16. $y''+y=\cos t\cos 2t\cos 3t$.

17. (a) Show that $\cos^3\omega t=\frac{1}{4}\mathrm{Re}\{e^{3i\omega t}+3e^{i\omega t}\}$.
Hint: $\cos\omega t=(e^{i\omega t}+e^{-i\omega t})/2$.
(b) Find a particular solution of the equation

$$10y''+0.2y'+1000y=5+20\cos^3 10t$$

18. (a) Let $L[y]=y''-2r_1y'+r_1^2y$. Show that

$$L[e^{r_1t}v(t)]=e^{r_1t}v''(t).$$

(b) Find the general solution of the equation

$$y''-6y'+9y=t^{3/2}e^{3t}.$$

19. Let $\psi(t)=t(A_0+\dots+A_nt^n)$, and assume that $b\neq 0$. Show that the equation $a\psi''+b\psi'=a_0+\dots+a_nt^n$ determines $A_0,\dots,A_n$ uniquely.

2.6 Mechanical vibrations

Consider the case where a small object of mass m is attached to an elastic spring of length l, which is suspended from a rigid horizontal support (see Figure 1). (An elastic spring has the property that if it is stretched or compressed a distance Δl which is small compared to its natural length l, then it will exert a restoring force of magnitude $k\Delta l$. The constant k is called the spring-constant, and is a measure of the stiffness of the spring.) In addition, the mass and spring may be immersed in a medium, such as oil, which impedes the motion of an object through it. Engineers usually refer to such systems as spring-mass-dashpot systems, or as seismic instruments, since they are similar, in principle, to a seismograph which is used to detect motions of the earth's surface.

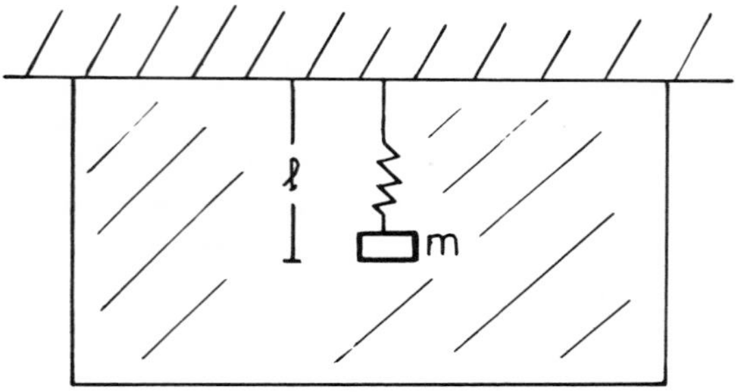

Figure 1

Spring-mass-dashpot systems have many diverse applications. For example, the shock absorbers in our automobiles are simple spring-mass-dashpot systems. Also, most heavy gun emplacements are attached to such systems so as to minimize the "recoil" effect of the gun. The usefulness of these devices will become apparent after we set up and solve the differential equation of motion of the mass m.

In calculating the motion of the mass m, it will be convenient for us to measure distances from the equilibrium position of the mass, rather than the horizontal support. The equilibrium position of the mass is that point where the mass will hang at rest if no external forces act upon it. In equilibrium, the weight mg of the mass is exactly balanced by the restoring force of the spring. Thus, in its equilibrium position, the spring has been stretched a distance Δl, where $k\Delta l = mg$. We let $y=0$ denote this equilibrium position, and we take the downward direction as positive. Let $y(t)$ denote the position of the mass at time t. To find $y(t)$, we must compute the total force acting on the mass m. This force is the sum of four separate forces W, R, D and F.

(i) The force $W = mg$ is the weight of the mass pulling it downward. This force is positive, since the downward direction is the positive y direction.

(ii) The force R is the restoring force of the spring, and it is proportional to the elongation, or compression, $\Delta l + y$ of the spring. It always acts to restore the spring to its natural length. If $\Delta l + y > 0$, then R is negative, so that $R = -k(\Delta l + y)$, and if $\Delta l + y < 0$, then R is positive, so that $R = -k(\Delta l + y)$. In either case,

$$R = -k(\Delta l + y).$$

(iii) The force D is the damping, or drag force, which the medium exerts on the mass m. (Most media, such as oil and air, tend to resist the motion of an object through it.) This force always acts in the direction opposite the direction of motion, and is usually directly proportional to the magnitude of the velocity dy/dt. If the velocity is positive; that is, the mass is moving in the downward direction, then $D = -c\,dy/dt$, and if the velocity is negative, then $D = -c\,dy/dt$. In either case,

$$D = -c\,dy/dt.$$

(iv) The force F is the external force applied to the mass. This force is directed upward or downward, depending as to whether F is positive or negative. In general, this external force will depend explicitly on time.

From Newton's second law of motion (see Section 1.6)

$$\begin{aligned} m\frac{d^2y}{dt^2} &= W + R + D + F \\ &= mg - k(\Delta l + y) - c\frac{dy}{dt} + F(t) \\ &= -ky - c\frac{dy}{dt} + F(t), \end{aligned}$$

since $mg = k\,\Delta l$. Hence, the position $y(t)$ of the mass satisfies the second-order linear differential equation

$$m\frac{d^2y}{dt^2} + c\frac{dy}{dt} + ky = F(t) \tag{1}$$

where m, c and k are nonnegative constants. We adopt here the English system of units, so that F is measured in pounds, y is measured in feet, and t is measured in seconds. In this case, the units of k are lb/ft, the units of c are lb·s/ft, and the units of m are slugs (lb·s^2/ft).

(a) *Free vibrations*:

We consider first the simplest case of free undamped motion. In this case, Equation (1) reduces to

$$m\frac{d^2y}{dt^2} + ky = 0 \quad \text{or} \quad \frac{d^2y}{dt^2} + \omega_0^2 y = 0 \tag{2}$$

where $\omega_0^2 = k/m$. The general solution of (2) is

$$y(t) = a\cos\omega_0 t + b\sin\omega_0 t. \tag{3}$$

In order to analyze the solution (3), it is convenient to rewrite it as a single cosine function. This is accomplished by means of the following lemma.

Lemma 1. *Any function $y(t)$ of the form* (3) *can be written in the simpler form*

$$y(t) = R\cos(\omega_0 t - \delta) \tag{4}$$

where $R = \sqrt{a^2+b^2}$ and $\delta = \tan^{-1} b/a$.

Proof. We will verify that the two expressions (3) and (4) are equal. To this end, compute

$$R\cos(\omega_0 t - \delta) = R\cos\omega_0 t\cos\delta + R\sin\omega_0 t\sin\delta$$

and observe from Figure 2 that $R\cos\delta = a$ and $R\sin\delta = b$. Hence,

$$R\cos(\omega_0 t - \delta) = a\cos\omega_0 t + b\sin\omega_0 t.$$ □

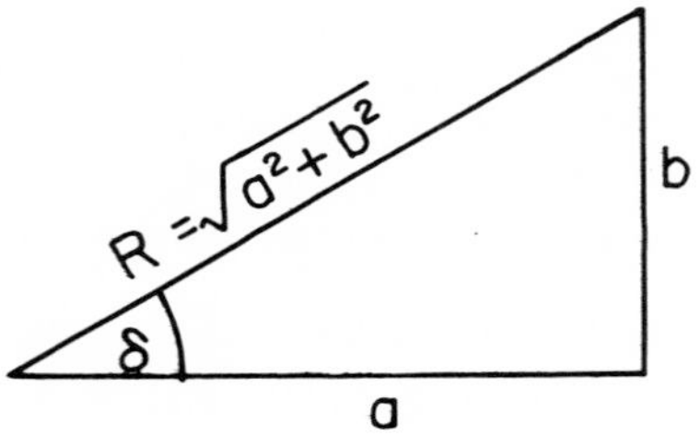

Figure 2

In Figure 3 we have graphed the function $y = R\cos(\omega_0 t - \delta)$. Notice that $y(t)$ always lies between $-R$ and $+R$, and that the motion of the mass is periodic—it repeats itself over every time interval of length $2\pi/\omega_0$. This type of motion is called simple harmonic motion; R is called the amplitude of the motion, δ the phase angle of the motion, $T_0 = 2\pi/\omega_0$ the natural period of the motion, and $\omega_0 = \sqrt{k/m}$ the natural frequency of the system.

(b) *Damped free vibrations*:

If we now include the effect of damping, then the differential equation governing the motion of the mass is

$$m\frac{d^2y}{dt^2} + c\frac{dy}{dt} + ky = 0. \tag{5}$$

The roots of the characteristic equation $mr^2 + cr + k = 0$ are

$$r_1 = \frac{-c + \sqrt{c^2 - 4km}}{2m} \quad \text{and} \quad r_2 = \frac{-c - \sqrt{c^2 - 4km}}{2m}.$$

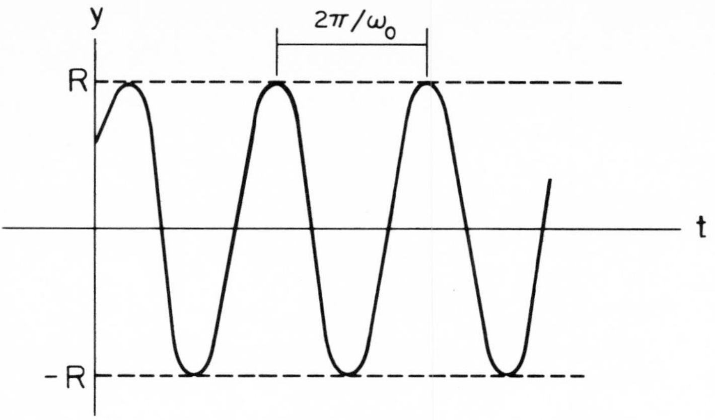

Figure 3. Graph of $y(t)=R\cos(\omega_0 t-\delta)$

Thus, there are three cases to consider, depending as to whether c^2-4km is positive, negative or zero.

(i) $c^2-4km>0$. In this case both r_1 and r_2 are negative, and every solution $y(t)$ of (5) has the form

$$y(t)=ae^{r_1 t}+be^{r_2 t}.$$

(ii) $c^2-4km=0$. In this case, every solution $y(t)$ of (5) is of the form

$$y(t)=(a+bt)e^{-ct/2m}.$$

(iii) $c^2-4km<0$. In this case, every solution $y(t)$ of (5) is of the form

$$y(t)=e^{-ct/2m}\left[a\cos\mu t+b\sin\mu t\right],\qquad \mu=\frac{\sqrt{4km-c^2}}{2m}.$$

The first two cases are referred to as overdamped and critically damped, respectively. They represent motions in which the originally displaced mass creeps back to its equilibrium position. Depending on the initial conditions, it may be possible to overshoot the equilibrium position once, but no more than once (see Exercises 2-3). The third case, which is referred to as an underdamped motion, occurs quite often in mechanical systems and represents a damped vibration. To see this, we use Lemma 1 to rewrite the function

$$y(t)=e^{-ct/2m}\left[a\cos\mu t+b\sin\mu t\right]$$

in the form

$$y(t)=Re^{-ct/2m}\cos(\mu t-\delta).$$

The displacement y oscillates between the curves $y=\pm Re^{-ct/2m}$, and thus represents a cosine curve with decreasing amplitude, as shown in Figure 4.

Now, observe that the motion of the mass always dies out eventually if there is damping in the system. In other words, any initial disturbance of

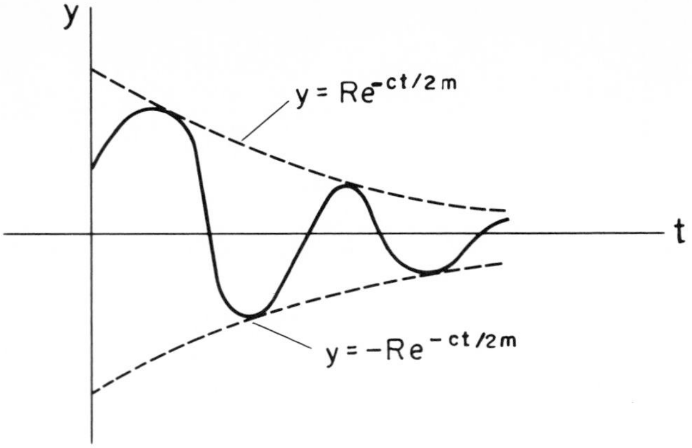

Figure 4. Graph of $Re^{-ct/2m}\cos(\mu t-\delta)$

the system is dissipated by the damping present in the system. This is one reason why spring-mass-dashpot systems are so useful in mechanical systems: they can be used to damp out any undesirable disturbances. For example, the shock transmitted to an automobile by a bump in the road is dissipated by the shock absorbers in the car, and the momentum from the recoil of a gun barrel is dissipated by a spring-mass-dashpot system attached to the gun.

(c) *Damped forced vibrations*:

If we now introduce an external force $F(t)=F_0\cos\omega t$, then the differential equation governing the motion of the mass is

$$m\frac{d^2y}{dt^2}+c\frac{dy}{dt}+ky=F_0\cos\omega t. \tag{6}$$

Using the method of judicious guessing, we can find a particular solution $\psi(t)$ of (6) of the form

$$\begin{aligned}\psi(t)&=\frac{F_0}{(k-m\omega^2)^2+c^2\omega^2}\left[(k-m\omega^2)\cos\omega t+c\omega\sin\omega t\right]\\&=\frac{F_0}{(k-m\omega^2)^2+c^2\omega^2}\left[(k-m\omega^2)^2+c^2\omega^2\right]^{1/2}\cos(\omega t-\delta)\\&=\frac{F_0\cos(\omega t-\delta)}{\left[(k-m\omega^2)^2+c^2\omega^2\right]^{1/2}}\end{aligned} \tag{7}$$

where $\tan\delta=c/(k-m\omega^2)$. Hence, every solution $y(t)$ of (6) must be of the form

$$y(t)=\phi(t)+\psi(t)=\phi(t)+\frac{F_0\cos(\omega t-\delta)}{\left[(k-m\omega^2)^2+c^2\omega^2\right]^{1/2}} \tag{8}$$

where $\phi(t)$ is a solution of the homogeneous equation

$$m\frac{d^2y}{dt^2}+c\frac{dy}{dt}+ky=0. \tag{9}$$

We have already seen though, that every solution $y=\phi(t)$ of (9) approaches zero as t approaches infinity. Thus, for large t, the equation $y(t)=\psi(t)$ describes very accurately the position of the mass m, regardless of its initial position and velocity. For this reason, $\psi(t)$ is called the steady state part of the solution (8), while $\phi(t)$ is called the transient part of the solution.

(d) *Forced free vibrations*:

We now remove the damping from our system and consider the case of forced free vibrations where the forcing term is periodic and has the form $F(t)=F_0\cos\omega t$. In this case, the differential equation governing the motion of the mass m is

$$\frac{d^2y}{dt^2}+\omega_0^2y=\frac{F_0}{m}\cos\omega t, \qquad \omega_0^2=k/m. \tag{10}$$

The case $\omega\neq\omega_0$ is uninteresting; every solution $y(t)$ of (10) has the form

$$y(t)=c_1\cos\omega_0t+c_2\sin\omega_0t+\frac{F_0}{m(\omega_0^2-\omega^2)}\cos\omega t,$$

and thus is the sum of two periodic functions of different periods. The interesting case is when $\omega=\omega_0$; that is, when the frequency ω of the external force equals the natural frequency of the system. This case is called the *resonance* case, and the differential equation of motion for the mass m is

$$\frac{d^2y}{dt^2}+\omega_0^2y=\frac{F_0}{m}\cos\omega_0t. \tag{11}$$

We will find a particular solution $\psi(t)$ of (11) as the real part of a complex-valued solution $\phi(t)$ of the equation

$$\frac{d^2y}{dt^2}+\omega_0^2y=\frac{F_0}{m}e^{i\omega_0t}. \tag{12}$$

Since $e^{i\omega_0t}$ is a solution of the homogeneous equation $y''+\omega_0^2y=0$, we know that (12) has a particular solution $\phi(t)=Ate^{i\omega_0t}$, for some constant A. Computing

$$\phi''+\omega_0^2\phi=2i\omega_0Ae^{i\omega_0t}$$

we see that

$$A = \frac{1}{2i\omega_0}\frac{F_0}{m} = \frac{-iF_0}{2m\omega_0}.$$

Hence,

$$\begin{aligned}\phi(t) &= \frac{-iF_0 t}{2m\omega_0}(\cos\omega_0 t + i\sin\omega_0 t)\\ &= \frac{F_0 t}{2m\omega_0}\sin\omega_0 t - i\frac{F_0 t}{2m\omega_0}\cos\omega_0 t\end{aligned}$$

is a particular solution of (12), and

$$\psi(t) = \operatorname{Re}\{\phi(t)\} = \frac{F_0 t}{2m\omega_0}\sin\omega_0 t$$

is a particular solution of (11). Consequently, every solution $y(t)$ of (11) is of the form

$$y(t) = c_1\cos\omega_0 t + c_2\sin\omega_0 t + \frac{F_0 t}{2m\omega_0}\sin\omega_0 t \tag{13}$$

for some choice of constants c_1, c_2.

Now, the sum of the first two terms in (13) is a periodic function of time. The third term, though, represents an oscillation with increasing amplitude, as shown in Figure 5. Thus, the forcing term $F_0\cos\omega t$, if it is in resonance with the natural frequency of the system, will always cause unbounded oscillations. Such a phenomenon was responsible for the collapse of the Tacoma Bridge, (see Section 2.6.1) and many other mechanical catastrophes.

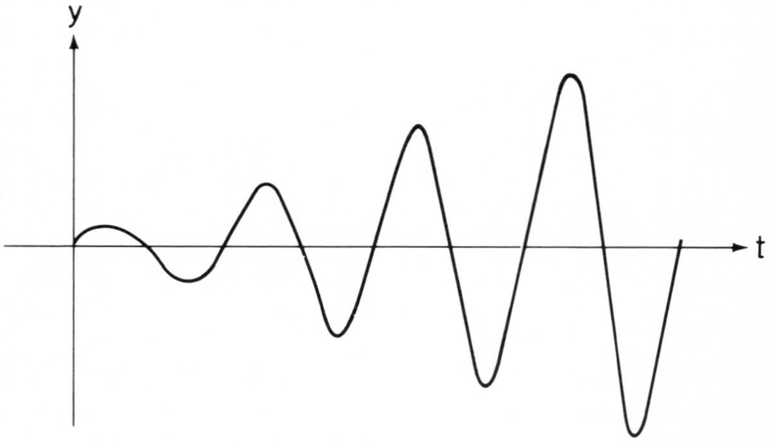

Figure 5. Graph of $f(t) = At\sin\omega_0 t$

Exercises

1. It is found experimentally that a $2\frac{1}{8}$ lb weight stretches a spring 6 in. If the weight is pulled down an additional 3 in and released, find the amplitude, period and frequency of the motion, neglecting air resistance. (The mass m of an object in terms of its weight, W, is $m = W/g = W/32.2$.)

2. Let $y(t) = Ae^{r_1 t} + Be^{r_2 t}$, with $|A| + |B| \neq 0$.
(a) Show that $y(t)$ is zero at most once.
(b) Show that $y'(t)$ is zero at most once.

3. Let $y(t) = (A + Bt)e^{rt}$, with $|A| + |B| \neq 0$.
(a) Show that $y(t)$ is zero at most once.
(b) Show that $y'(t)$ is zero at most once.

4. A small object of mass 1 slug is attached to a spring with spring-constant 2 lb/ft. This spring mass sytem is immersed in a viscous medium with damping constant 3 lb·s/ft. At time $t = 0$, the mass is lowered 6 in below its equilibrium position, and released. Show that the mass will creep back to its equilibrium position as t approaches infinity.

5. A small object of mass 1 slug is attached to a spring with spring-constant 1 lb/ft, and is immersed in a viscous medium with damping constant 2 lb·s/ft. At time $t = 0$, the mass is lowered 3 in and given an initial velocity of 1 ft/s in the upward direction. Show that the mass will overshoot its equilibrium position once, and then creep back to equilibrium.

6. A small object of mass 4 slugs is attached to an elastic spring with spring-constant 64 lb/ft, and is acted upon by an external force $F(t) = A\cos^3 \omega t$. Find all values of ω at which resonance occurs.

7. The gun of a U.S. M60 tank is attached to a spring-mass-dashpot system with spring constant $100\alpha^2$ and damping constant 200α. The mass of the gun is 100 slugs. Assume the displacement $y(t)$ of the gun from its rest position after being fired at time $t = 0$ satisfies the initial-value problem

$$100y'' + 200\alpha y' + 100\alpha^2 y = 0; \qquad y(0) = 0, \quad y'(0) = 100 \text{ ft/s}.$$

It is desired that one second later, the quantity $y^2 + (y')^2$ be less than 0.01. How large must α be to guarantee that this is so? (The spring-mass-dashpot mechanism in the M60 tanks supplied by the U.S. to Israel are critically damped, for this situation is preferrable in desert warfare where one has to fire again as quickly as possible.)

8. A spring-mass-dashpot system has the property that the spring-constant k is 9 times its mass m, and the damping constant c is 6 times its mass. At time $t = 0$, the mass, which is hanging at rest, is acted upon by an external force $F(t) = 3\sin 3t$ lb. The spring will break if it is stretched an additional 5 ft from its equilibrium position. Show that the spring will not break if $m \geqslant 1/15$ slugs.

9. A spring-mass-dashpot system with $m = 1$, $k = 2$, and $c = 2$ (in their respective units) hangs in equilibrium. At time $t = 0$, an external force $F(t) = \pi - t$ lb acts for a time interval π. Find the position of the mass at any time $t > \pi$.

10. A small object weighing $2\frac{1}{8}$ lb stretches a spring 6 in. With the weight on the spring at rest in the equilibrium position at time $t=0$, an external force $F(t) = \frac{1}{2}t$ lb is applied until time $t=t_1=7\pi/16$ seconds, at which time it is removed. Assuming no damping, find the frequency and amplitude of the resulting oscillation. (Use $g=32$ ft/s^2 in this problem.)

11. A 32.2 lb weight is attached to a spring with spring-constant $k=4$ lb/ft, and hangs in equilibrium. An external force $F(t)=(1+t+\sin 2t)$ lb is applied to the mass, beginning at time $t=0$. If the spring is stretched a length $(1/2+\pi/4)$ ft or more from its equilibrium position, then it will break. Assuming no damping present, find the time at which the spring breaks.

12. A small object of mass 1 slug is attached to a spring with spring constant $k=$ 1 lb/ft. This spring mass system is then immersed in a viscous medium with damping constant c. An external force $F(t)=(3-\cos t)$ lb is applied to the system. Determine the minimum positive value of c so that the magnitude of the steady state solution does not exceed 5 ft.

13. Determine a particular solution $\psi(t)$ of $my''+cy'+ky=F_0\cos\omega t$, of the form $\psi(t)=A\cos(\omega t-\phi)$. Show that the amplitude A is a maximum when $\omega^2=\omega_0^2-\frac{1}{2}(c/m)^2$. This value of ω is called the *resonant frequency* of the system. What happens when $\omega_0^2<\frac{1}{2}(c/m)^2$?

2.6.1 The Tacoma Bridge disaster

On July 1, 1940, the Tacoma Narrows Bridge at Puget Sound in the state of Washington was completed and opened to traffic. From the day of its opening the bridge began undergoing vertical oscillations, and it soon was nicknamed "Galloping Gertie." Strange as it may seem, traffic on the bridge increased tremendously as a result of its novel behavior. People came from hundreds of miles in their cars to enjoy the curious thrill of riding over a galloping, rolling bridge. For four months, the bridge did a thriving business. As each day passed, the authorities in charge became more and more confident of the safety of the bridge—so much so, in fact, that they were planning to cancel the insurance policy on the bridge.

Starting at about 7:00 on the morning of November 7, 1940, the bridge began undulating persistently for three hours. Segments of the span were heaving periodically up and down as much as three feet. At about 10:00 a.m., something seemed to snap and the bridge began oscillating wildly. At one moment, one edge of the roadway was twenty-eight feet higher than the other; the next moment it was twenty-eight feet lower than the other edge. At 10:30 a.m. the bridge began cracking, and finally, at 11:10 a.m. the entire bridge came crashing down. Fortunately, only one car was on the bridge at the time of its failure. It belonged to a newspaper reporter who had to abandon the car and its sole remaining occupant, a pet dog, when the bridge began its violent twisting motion. The reporter reached safety, torn and bleeding, by crawling on hands and knees, desperately clutching the curb of the bridge. His dog went down with the car and the span—the only life lost in the disaster.

There were many humorous and ironic incidents associated with the collapse of the Tacoma Bridge. When the bridge began heaving violently, the authorities notified Professor F. B. Farquharson of the University of Washington. Professor Farquharson had conducted numerous tests on a simulated model of the bridge and had assured everyone of its stability. The professor was the last man on the bridge. Even when the span was tilting more than twenty-eight feet up and down, he was making scientific observations with little or no anticipation of the imminent collapse of the bridge. When the motion increased in violence, he made his way to safety by scientifically following the yellow line in the middle of the roadway. The professor was one of the most surprised men when the span crashed into the water.

One of the insurance policies covering the bridge had been written by a local travel agent who had pocketed the premium and had neglected to report the policy, in the amount of $800,000, to his company. When he later received his prison sentence, he ironically pointed out that his embezzlement would never have been discovered if the bridge had only remained up for another week, at which time the bridge officials had planned to cancel all of the policies.

A large sign near the bridge approach advertised a local bank with the slogan "as safe as the Tacoma Bridge." Immediately following the collapse of the bridge, several representatives of the bank rushed out to remove the billboard.

After the collapse of the Tacoma Bridge, the governor of the state of Washington made an emotional speech, in which he declared "We are going to build the exact same bridge, exactly as before." Upon hearing this, the noted engineer Von Karman sent a telegram to the governor stating "If you build the exact same bridge exactly as before, it will fall into the exact same river exactly as before."

The collapse of the Tacoma Bridge was due to an aerodynamical phenomenon known as *stall flutter*. This can be explained very briefly in the following manner. If there is an obstacle in a stream of air, or liquid, then a "vortex street" is formed behind the obstacle, with the vortices flowing off at a definite periodicity, which depends on the shape and dimension of the structure as well as on the velocity of the stream (see Figure 1). As a result of the vortices separating alternately from either side of the obstacle, it is acted upon by a periodic force perpendicular to the direction of the stream, and of magnitude $F_0 \cos \omega t$. The coefficient F_0 depends on the shape of the structure. The poorer the streamlining of the structure; the larger the coefficient F_0, and hence the amplitude of the force. For example, flow around an airplane wing at small angles of attack is very smooth, so that the vortex street is not well defined and the coefficient F_0 is very

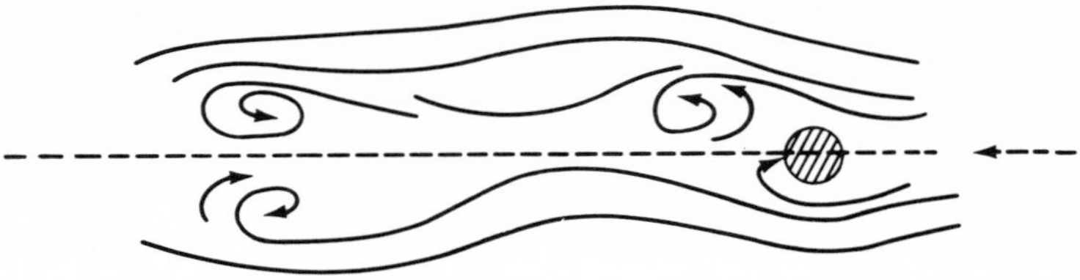

Figure 1

small. The poorly streamlined structure of a suspension bridge is another matter, and it is natural to expect that a force of large amplitude will be set up. Thus, a structure suspended in an air stream experiences the effect of this force and hence goes into a state of forced vibrations. The amount of danger from this type of motion depends on how close the natural frequency of the structure (remember that bridges are made of steel, a highly elastic material) is to the frequency of the driving force. If the two frequencies are the same, resonance occurs, and the oscillations will be destructive if the system does not have a sufficient amount of damping. It has now been established that oscillations of this type were responsible for the collapse of the Tacoma Bridge. In addition, resonances produced by the separation of vortices have been observed in steel factory chimneys, and in the periscopes of submarines.

The phenomenon of resonance was also responsible for the collapse of the Broughton suspension bridge near Manchester, England in 1831. This occurred when a column of soldiers marched in cadence over the bridge, thereby setting up a periodic force of rather large amplitude. The frequency of this force was equal to the natural frequency of the bridge. Thus, very large oscillations were induced, and the bridge collapsed. It is for this reason that soldiers are ordered to break cadence when crossing a bridge.

2.6.2 Electrical networks

We now briefly study a simple series circuit, as shown in Figure 1 below. The symbol E represents a source of electromotive force. This may be a battery or a generator which produces a potential difference (or voltage), that causes a current I to flow through the circuit when the switch S is closed. The symbol R represents a resistance to the flow of current such as that produced by a lightbulb or toaster. When current flows through a coil of wire L, a magnetic field is produced which opposes any change in the current through the coil. The change in voltage produced by the coil is proportional to the rate of change of the current, and the constant of proportionality is called the inductance L of the coil. A capacitor, or condenser, indicated by C, usually consists of two metal plates separated by a material through which very little current can flow. A capacitor has the effect of reversing the flow of current as one plate or the other becomes charged.

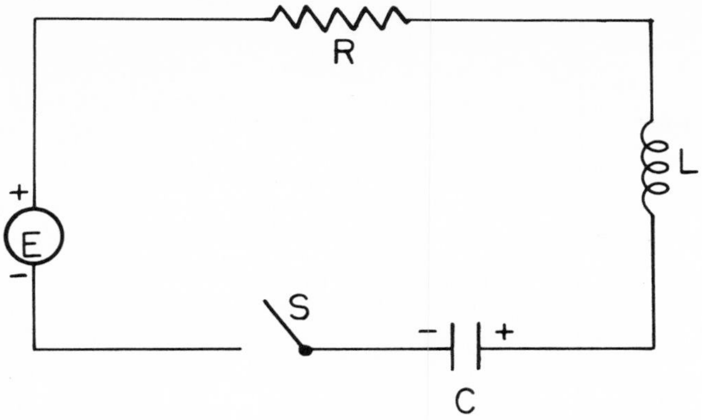

Figure 1. A simple series circuit

Let $Q(t)$ be the charge on the capacitor at time t. To derive a differential equation which is satisfied by $Q(t)$ we use the following.

Kirchoff's second law: In a closed circuit, the impressed voltage equals the sum of the voltage drops in the rest of the circuit.

Now,

(i) The voltage drop across a resistance of R ohms equals RI (Ohm's law).
(ii) The voltage drop across an inductance of L henrys equals $L(dI/dt)$.
(iii) The voltage drop across a capacitance of C farads equals Q/C.

Hence,

$$E(t) = L\frac{dI}{dt} + RI + \frac{Q}{C},$$

and since $I(t) = dQ(t)/dt$, we see that

$$L\frac{d^2Q}{dt^2} + R\frac{dQ}{dt} + \frac{Q}{C} = E(t). \tag{1}$$

Notice the resemblance of Equation (1) to the equation of a vibrating mass. Among the similarities with mechanical vibrations, electrical circuits also have the property of resonance. Unlike mechanical systems, though, resonance is put to good use in electrical systems. For example, the tuning knob of a radio is used to vary the capacitance in the tuning circuit. In this manner, the resonant frequency (see Exercise 13, Section 2.6) is changed until it agrees with the frequency of one of the incoming radio signals. The amplitude of the current produced by this signal will be much greater than that of all other signals. In this way, the tuning circuit picks out the desired station.

EXERCISES

1. Suppose that a simple series circuit has no resistance and no impressed voltage. Show that the charge Q on the capacitor is periodic in time, with frequency $\omega_0 = \sqrt{1/LC}$. The quantity $\sqrt{1/LC}$ is called the natural frequency of the circuit.

2. Suppose that a simple series circuit consisting of an inductor, a resistor and a capacitor is open, and that there is an initial charge $Q_0 = 10^{-8}$ coulombs on the capacitor. Find the charge on the capacitor and the current flowing in the circuit after the switch is closed for each of the following cases.
(a) $L = 0.5$ henrys, $C = 10^{-5}$ farads, $R = 1000$ ohms
(b) $L = 1$ henry, $C = 10^{-4}$ farads, $R = 200$ ohms
(c) $L = 2$ henrys, $C = 10^{-6}$ farads, $R = 2000$ ohms

3. A simple series circuit has an inductor of 1 henry, a capacitor of 10^{-6} farads, and a resistor of 1000 ohms. The initial charge on the capacitor is zero. If a 12 volt battery is connected to the circuit, and the circuit is closed at $t = 0$, find the charge on the capacitor 1 second later, and the steady state charge.

4. A capacitor of 10^{-3} farads is in series with an electromotive force of 12 volts and an inductor of 1 henry. At $t = 0$, both Q and I are zero.
(a) Find the natural frequency and period of the electrical oscillations.
(b) Find the maximum charge on the capacitor, and the maximum current flowing in the circuit.

5. Show that if there is no resistance in a circuit, and the impressed voltage is of the form $E_0 \sin \omega t$, then the charge on the capacitor will become unbounded as $t \to \infty$ if $\omega = \sqrt{1/LC}$. This is the phenomenon of resonance.

6. Consider the differential equation

$$L\ddot{Q} + R\dot{Q} + \frac{Q}{C} = E_0 \cos \omega t. \tag{i}$$

We find a particular solution $\psi(t)$ of (i) as the real part of a particular solution $\phi(t)$ of

$$L\ddot{Q} + R\dot{Q} + \frac{Q}{C} = E_0 e^{i\omega t}. \tag{ii}$$

(a) Show that

$$\phi(t) = \frac{E_0}{R + i\left(\omega L - \dfrac{1}{\omega C}\right)} e^{i\omega t}.$$

(b) The quantity $Z = R + i(\omega L - 1/\omega C)$ is known as the complex impedance of the circuit. The reciprocal of Z is called the admittance, and the real and imaginary parts of $1/Z$ are called the conductance and susceptance. Determine the admittance, conductance and susceptance.

7. Consider a simple series circuit with given values of L, R and C, and an impressed voltage $E_0 \sin \omega t$. For which value of ω will the steady state current be a maximum?

2.7 A model for the detection of diabetes

Diabetes mellitus is a disease of metabolism which is characterized by too much sugar in the blood and urine. In diabetes, the body is unable to burn off all its sugars, starches, and carbohydrates because of an insufficient supply of insulin. Diabetes is usually diagnosed by means of a glucose tolerance test (GTT). In this test the patient comes to the hospital after an overnight fast and is given a large dose of glucose (sugar in the form in which it usually appears in the bloodstream). During the next three to five hours several measurements are made of the concentration of glucose in the patient's blood, and these measurements are used in the diagnosis of diabetes. A very serious difficulty associated with this method of diagnosis is that there is no universally accepted criterion for interpreting the results of a glucose tolerance test. Three physicians interpreting the results of a GTT may come up with three different diagnoses. In one case recently, in Rhode Island, one physician, after reviewing the results of a GTT, came up with a diagnosis of diabetes. A second physician declared the patient to be normal. To settle the question, the results of the GTT were sent to a specialist in Boston. After examining these results, the specialist concluded that the patient was suffering from a pituitary tumor.

In the mid 1960's Drs. Rosevear and Molnar of the Mayo Clinic and Ackerman and Gatewood of the University of Minnesota discovered a fairly reliable criterion for interpreting the results of a glucose tolerance test. Their discovery arose from a very simple model they developed for the blood glucose regulatory system. Their model is based on the following simple and fairly well known facts of elementary biology.

1. Glucose plays an important role in the metabolism of any vertebrate since it is a source of energy for all tissues and organs. For each individual there is an optimal blood glucose concentration, and any excessive deviation from this optimal concentration leads to severe pathological conditions and potentially death.

2. While blood glucose levels tend to be autoregulatory, they are also influenced and controlled by a wide variety of hormones and other metabolites. Among these are the following.

(i) *Insulin*, a hormone secreted by the β cells of the pancreas. After we eat any carbohydrates, our G.I. tract sends a signal to the pancreas to secrete more insulin. In addition, the glucose in our blood directly stimulates the β cells of the pancreas to secrete insulin. It is generally believed that insulin facilitates tissue uptake of glucose by attaching itself to the impermeable membrane walls, thus allowing glucose to pass through the membranes to the center of the cells, where most of the biological and chemical activity takes place. Without sufficient insulin, the body cannot avail itself of all the energy it needs.

(ii) *Glucagon*, a hormone secreted by the α cells of the pancreas. Any excess glucose is stored in the liver in the form of glycogen. In times of need this glycogen is converted back into glucose. The hormone glucagon increases the rate of breakdown of glycogen into glucose. Evidence collected thus far clearly indicates that hypoglycemia (low blood sugar) and fasting promote the secretion of glucagon while increased blood glucose levels suppress its secretion.

(iii) *Epinephrine* (adrenalin), a hormone secreted by the adrenal medulla. Epinephrine is part of an emergency mechanism to quickly increase the concentration of glucose in the blood in times of extreme hypoglycemia. Like glucagon, epinephrine increases the rate of breakdown of glycogen into glucose. In addition, it directly inhibits glucose uptake by muscle tissue; it acts directly on the pancreas to inhibit insulin secretion; and it aids in the conversion of lactate to glucose in the liver.

(iv) *Glucocorticoids*, hormones such as cortisol which are secreted by the adrenal cortex. Glucocorticoids play an important role in the metabolism of carbohydrates.

(v) *Thyroxin*, a hormone secreted by the thyroid gland. This hormone aids the liver in forming glucose from non-carbohydrate sources such as glycerol, lactate and amino acids.

(vi) *Growth hormone* (somatotropin), a hormone secreted by the anterior pituitary gland. Not only does growth hormone affect glucose levels in a direct manner, but it also tends to "block" insulin. It is believed that growth hormone decreases the sensitivity of muscle and adipose membrane to insulin, thereby reducing the effectiveness of insulin in promoting glucose uptake.

The aim of Ackerman et al was to construct a model which would accurately describe the blood glucose regulatory system during a glucose tolerance test, and in which one or two parameters would yield criteria for distinguishing normal individuals from mild diabetics and pre-diabetics. Their model is a very simplified one, requiring only a limited number of blood samples during a GTT. It centers attention on two concentrations, that of glucose in the blood, labelled G, and that of the net hormonal concentration, labelled H. The latter is interpreted to represent the cumulative effect of all the pertinent hormones. Those hormones such as insulin which decrease blood glucose concentrations are considered to increase H, while those hormones such as cortisol which increase blood glucose concentrations are considered to decrease H. Now there are two reasons why such a simplified model can still provide an accurate description of the blood glucose regulatory system. First, studies have shown that under normal, or close to normal conditions, the interaction of one hormone, namely insulin, with blood glucose so predominates that a simple "lumped parameter model" is quite adequate. Second, evidence indicates that normoglycemia does not depend, necessarily, on the normalcy of each kinetic mechanism of the blood glucose regulatory system. Rather, it depends on the overall performance of the blood glucose regulatory system, and this system is dominated by insulin-glucose interactions.

The basic model is described analytically by the equations

$$\frac{dG}{dt} = F_1(G,H) + J(t) \tag{1}$$

$$\frac{dH}{dt} = F_2(G,H). \tag{2}$$

The dependence of F_1 and F_2 on G and H signify that changes in G and H are determined by the values of both G and H. The function $J(t)$ is the external rate at which the blood glucose concentration is being increased. Now, we assume that G and H have achieved optimal values G_0 and H_0 by the time the fasting patient has arrived at the hospital. This implies that $F_1(G_0, H_0) = 0$ and $F_2(G_0, H_0) = 0$. Since we are interested here in the deviations of G and H from their optimal values, we make the substitution

$$g = G - G_0, \qquad h = H - H_0.$$

Then,

$$\frac{dg}{dt} = F_1(G_0 + g, H_0 + h) + J(t),$$

$$\frac{dh}{dt} = F_2(G_0 + g, H_0 + h).$$

Now, observe that

$$F_1(G_0 + g, H_0 + h) = F_1(G_0, H_0) + \frac{\partial F_1(G_0, H_0)}{\partial G} g + \frac{\partial F_1(G_0, H_0)}{\partial H} h + e_1$$

and

$$F_2(G_0 + g, H_0 + h) = F_2(G_0, H_0) + \frac{\partial F_2(G_0, H_0)}{\partial G} g + \frac{\partial F_2(G_0, H_0)}{\partial H} h + e_2$$

where e_1 and e_2 are very small compared to g and h. Hence, assuming that G and H deviate only slightly from G_0 and H_0, and therefore neglecting the terms e_1 and e_2, we see that

$$\frac{dg}{dt} = \frac{\partial F_1(G_0, H_0)}{\partial G} g + \frac{\partial F_1(G_0, H_0)}{\partial H} h + J(t) \tag{3}$$

$$\frac{dh}{dt} = \frac{\partial F_2(G_0, H_0)}{\partial G} g + \frac{\partial F_2(G_0, H_0)}{\partial H} h. \tag{4}$$

Now, there are no means, a priori, of determining the numbers

$$\frac{\partial F_1(G_0, H_0)}{\partial G}, \quad \frac{\partial F_1(G_0, H_0)}{\partial H}, \quad \frac{\partial F_2(G_0, H_0)}{\partial G} \quad \text{and} \quad \frac{\partial F_2(G_0, H_0)}{\partial H}.$$

However, we can determine their signs. Referring to Figure 1, we see that dg/dt is negative for $g > 0$ and $h = 0$, since the blood glucose concentration

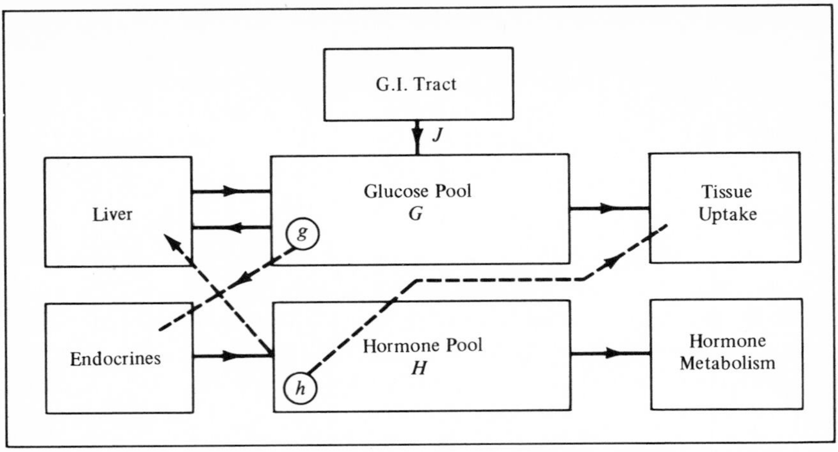

Figure 1. Simplified model of the blood glucose regulatory system

will be decreasing through tissue uptake of glucose and the storing of excess glucose in the liver in the form of glycogen. Consequently $\partial F_1(G_0, H_0)/\partial G$ must be negative. Similarly, $\partial F_1(G_0, H_0)/\partial H$ is negative since a positive value of h tends to decrease blood glucose levels by facilitating tissue uptake of glucose and by increasing the rate at which glucose is converted to glycogen. The number $\partial F_2(G_0, H_0)/\partial G$ must be positive since a positive value of g causes the endocrine glands to secrete those hormones which tend to increase H. Finally, $\partial F_2(G_0, H_0)/\partial H$ must be negative, since the concentration of hormones in the blood decreases through hormone metabolism.

Thus, we can write Equations (3) and (4) in the form

$$\frac{dg}{dt} = -m_1 g - m_2 h + J(t) \tag{5}$$

$$\frac{dh}{dt} = -m_3 h + m_4 g \tag{6}$$

where m_1, m_2, m_3, and m_4 are positive constants. Equations (5) and (6) are two first-order equations for g and h. However, since we only measure the concentration of glucose in the blood, we would like to remove the variable h. This can be accomplished as follows: Differentiating (5) with respect to t gives

$$\frac{d^2g}{dt^2} = -m_1\frac{dg}{dt} - m_2\frac{dh}{dt} + \frac{dJ}{dt}.$$

Substituting for dh/dt from (6) we obtain that

$$\frac{d^2g}{dt^2} = -m_1\frac{dg}{dt} + m_2 m_3 h - m_2 m_4 g + \frac{dJ}{dt}. \tag{7}$$

Next, observe from (5) that $m_2 h = (-dg/dt) - m_1 g + J(t)$. Consequently,

$g(t)$ satisfies the second-order linear differential equation

$$\frac{d^2g}{dt^2}+(m_1+m_3)\frac{dg}{dt}+(m_1m_3+m_2m_4)g=m_3J+\frac{dJ}{dt}.$$

We rewrite this equation in the form

$$\frac{d^2g}{dt^2}+2\alpha\frac{dg}{dt}+\omega_0^2g=S(t) \tag{8}$$

where $\alpha=(m_1+m_3)/2$, $\omega_0^2=m_1m_3+m_2m_4$, and $S(t)=m_3J+dJ/dt$.

Notice that the right-hand side of (8) is identically zero except for the very short time interval in which the glucose load is being ingested. We will learn to deal with such functions in Section 2.12. For our purposes here, let $t=0$ be the time at which the glucose load has been completely ingested. Then, for $t\geqslant 0$, $g(t)$ satisfies the second-order linear homogeneous equation

$$\frac{d^2g}{dt^2}+2\alpha\frac{dg}{dt}+\omega_0^2g=0. \tag{9}$$

This equation has positive coefficients. Hence, by the analysis in Section 2.6, (see also Exercise 8, Section 2.2.2) $g(t)$ approaches zero as t approaches infinity. Thus our model certainly conforms to reality in predicting that the blood glucose concentration tends to return eventually to its optimal concentration.

The solutions $g(t)$ of (9) are of three different types, depending as to whether $\alpha^2-\omega_0^2$ is positive, negative, or zero. These three types, of course, correspond to the overdamped, critically damped and underdamped cases discussed in Section 2.6. We will assume that $\alpha^2-\omega_0^2$ is negative; the other two cases are treated in a similar manner. If $\alpha^2-\omega_0^2<0$, then the characteristic equation of Equation (9) has complex roots. It is easily verified in this case (see Exercise 1) that every solution $g(t)$ of (9) is of the form

$$g(t)=Ae^{-\alpha t}\cos(\omega t-\delta),\qquad \omega^2=\omega_0^2-\alpha^2. \tag{10}$$

Consequently,

$$G(t)=G_0+Ae^{-\alpha t}\cos(\omega t-\delta). \tag{11}$$

Now there are five unknowns G_0, A, α, ω_0, and δ in (11). One way of determining them is as follows. The patient's blood glucose concentration before the glucose load is ingested is G_0. Hence, we can determine G_0 by measuring the patient's blood glucose concentration immediately upon his arrival at the hospital. Next, if we take four additional measurements G_1, G_2, G_3, and G_4 of the patient's blood glucose concentration at times t_1, t_2, t_3, and t_4, then we can determine A, α, ω_0, and δ from the four equations

$$G_j=G_0+Ae^{-\alpha t_j}\cos(\omega t_j-\delta),\qquad j=1,2,3,4.$$

A second, and better method of determining G_0, A, α, ω_0, and δ is to take n measurements $G_1,G_2,\dots,G_n$ of the patient's blood glucose concentration at times $t_1, t_2, \ldots, t_n$. Typically n is 6 or 7. We then find optimal values for G_0, A, α, ω_0, and δ such that the least square error

$$E = \sum_{j=1}^{n} \left[G_j - G_0 - Ae^{-\alpha t_j} \cos(\omega t_j - \delta) \right]^2$$

is minimized. The problem of minimizing E can be solved on a digital computer, and Ackerman et al (see reference at end of section) provide a complete Fortran program for determining optimal values for G_0, A, α, ω_0, and δ. This method is preferrable to the first method since Equation (11) is only an approximate formula for $G(t)$. Consequently, it is possible to find values G_0, A, α, ω_0, and δ so that Equation (11) is satisfied exactly at four points t_1, t_2, t_3, and t_4 but yields a poor fit to the data at other times. The second method usually offers a better fit to the data on the entire time interval since it involves more measurements.

In numerous experiments, Ackerman et al observed that a slight error in measuring G could produce a very large error in the value of α. Hence, any criterion for diagnosing diabetes that involves the parameter α is unreliable. However, the parameter ω_0, the natural frequency of the system, was relatively insensitive to experimental errors in measuring G. Thus, we may regard a value of ω_0 as the basic descriptor of the response to a glucose tolerance test. For discussion purposes, it is more convenient to use the corresponding natural period $T_0 = 2\pi/\omega_0$. The remarkable fact is that data from a variety of sources indicated that *a value of less than four hours for T_0 indicated normalcy, while appreciably more than four hours implied mild diabetes.*

Remark 1. The usual period between meals in our culture is about 4 hours. This suggests the interesting possibility that sociological factors may also play a role in the blood glucose regulatory system.

Remark 2. We wish to emphasize that the model described above can only be used to diagnose mild diabetes or pre-diabetes, since we have assumed throughout that the deviation g of G from its optimal value G_0 is small. Very large deviations of G from G_0 usually indicate severe diabetes or diabetes insipidus, which is a disorder of the posterior lobe of the pituitary gland.

A serious shortcoming of this simplified model is that it sometimes yields a poor fit to the data in the time period three to five hours after ingestion of the glucose load. This indicates, of course, that variables such as epinephrine and glucagon play an important role in this time period. Thus these variables should be included as separate variables in our model, rather than being lumped together with insulin. In fact, evidence indicates that levels of epinephrine may rise dramatically during the recovery phase of the GTT response, when glucose levels have been lowered below fasting

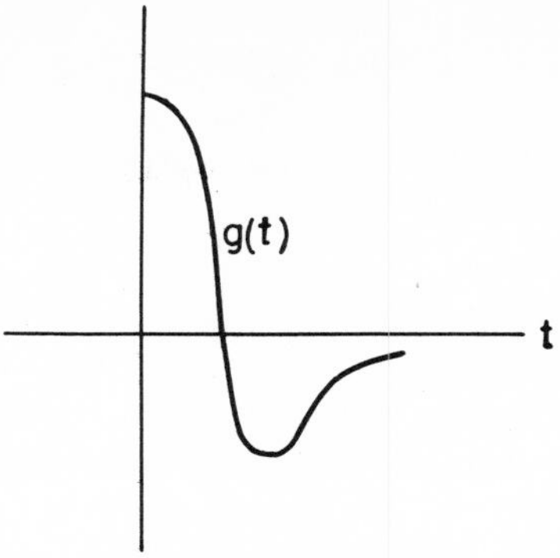

Figure 2. Graph of $g(t)$ if $\alpha^2-\omega_0^2>0$

levels. This can also be seen directly from Equation (9). If $\alpha^2-\omega_0^2>0$, then $g(t)$ may have the form described in Figure 2. Note that $g(t)$ drops very rapidly from a fairly high value to a negative one. It is quite conceivable, therefore, that the body will interpret this as an extreme emergency and thereby secrete a large amount of epinephrine.

Medical researchers have long recognized the need of including epinephrine as a separate variable in any model of the blood glucose regulatory system. However, they were stymied by the fact that there was no reliable method of measuring the concentration of epinephrine in the blood. Thus, they had to assume, for all practical purposes, the the level of epinephrine remained constant during the course of a glucose tolerance test. This author has just been informed that researchers at Rhode Island Hospital have devised an accurate method of measuring the concentration of epinephrine in the blood. Thus we will be able to develop and test more accurate models of the blood glucose regulatory system. Hopefully, this will lead to more reliable criteria for the diagnosis of diabetes.

Reference

E. Ackerman, L. Gatewood, J. Rosevear, and G. Molnar, Blood glucose regulation and diabetes, Chapter 4 in *Concepts and Models of Biomathematics*, F. Heinmets, ed., Marcel Dekker, 1969, 131–156.

Exercises

1. Derive Equation (10).

2. A patient arrives at the hospital after an overnight fast with a blood glucose concentration of 70 mg glucose/100 ml blood (mg glucose/100 ml blood = milligrams of glucose per 100 milliliters of blood). His blood glucose concentration 1 hour, 2 hours, and 3 hours after fully absorbing a large amount of glucose is 95, 65, and 75 mg glucose/100 ml blood, respectively. Show that this patient is normal. *Hint*: In the underdamped case, the time interval between two successive zeros of $G-G_0$ exceeds one half the natural period.

According to a famous diabetologist, the blood glucose concentration of a nondiabetic who has just absorbed a large amount of glucose will be at or below the fasting level in 2 hours or less. Exercises 3 and 4 compare the diagnoses of this diabetologist with those of Ackerman et al.

3. The deviation $g(t)$ of a patient's blood glucose concentration from its optimal concentration satisfies the differential equation $(d^2g/dt^2)+2\alpha(dg/dt)+\alpha^2 g=0$ immediately after he fully absorbs a large amount of glucose. The time t is measured in minutes, so that the units of α are reciprocal minutes. Show that this patient is normal according to Ackerman et al, if $\alpha>\pi/120$ (min), and that this patient is normal according to the famous diabetologist if

$$g'(0)<-\left(\tfrac{1}{120}+\alpha\right)g(0).$$

4. A patient's blood glucose concentration $G(t)$ satisfies the initial-value problem

$$\frac{d^2G}{dt^2}+\frac{1}{20\text{ (min)}}\frac{dG}{dt}+\frac{1}{2500\text{ (min)}^2}G$$

$$=\frac{1}{2500\text{ (min)}^2}75\text{ mg glucose}/100\text{ ml blood;}$$

$$G(0)=150\text{ mg glucose}/100\text{ ml blood,}$$

$$G'(0)=-\alpha G(0)/(\text{min});\qquad \alpha\geqslant\frac{1}{200}\,\frac{1-4e^{18/5}}{1-e^{18/5}}$$

immediately after he fully absorbs a large amount of glucose. This patient's optimal blood glucose concentration is 75 mg glucose/100 ml blood. Show that this patient is a diabetic according to Ackerman et al, but is normal according to the famous diabetologist.

2.8 Series solutions

We return now to the general homogeneous linear second-order equation

$$L[y]=P(t)\frac{d^2y}{dt^2}+Q(t)\frac{dy}{dt}+R(t)y=0 \tag{1}$$

with $P(t)$ unequal to zero in the interval $\alpha<t<\beta$. It was shown in Section 2.1 that every solution $y(t)$ of (1) can be written in the form $y(t)=c_1y_1(t)+c_2y_2(t)$, where $y_1(t)$ and $y_2(t)$ are any two linearly independent solutions of (1). Thus, the problem of finding all solutions of (1) is reduced to the simpler problem of finding just two solutions. In Section 2.2 we handled the special case where P, Q, and R are constants. The next simplest case is when $P(t)$, $Q(t)$, and $R(t)$ are polynomials in t. In this case, the form of the differential equation suggests that we guess a polynomial solution $y(t)$ of (1). If $y(t)$ is a polynomial in t, then the three functions $P(t)y''(t)$, $Q(t)y'(t)$, and $R(t)y(t)$ are again polynomials in t. Thus, in principle, we can determine a polynomial solution $y(t)$ of (1) by setting the sums of the coefficients of like powers of t in the expression $L[y](t)$ equal to zero. We illustrate this method with the following example.

Example 1. Find two linearly independent solutions of the equation

$$L[y]=\frac{d^2y}{dt^2}-2t\frac{dy}{dt}-2y=0. \tag{2}$$

Solution. We will try and find 2 polynomial solutions of (2). Now, it is not obvious, a priori, what the degree of any polynomial solution of (2) should be. Nor is it evident that we will be able to get away with a polynomial of finite degree. Therefore, we set

$$y(t)=a_0+a_1t+a_2t^2+\ldots=\sum_{n=0}^{\infty}a_nt^n.$$

Computing

$$\frac{dy}{dt}=a_1+2a_2t+3a_3t^2+\ldots=\sum_{n=0}^{\infty}na_nt^{n-1}$$

and

$$\frac{d^2y}{dt^2}=2a_2+6a_3t+\ldots=\sum_{n=0}^{\infty}n(n-1)a_nt^{n-2},$$

we see that $y(t)$ is a solution of (2) if

$$\begin{aligned}L[y](t)&=\sum_{n=0}^{\infty}n(n-1)a_nt^{n-2}-2t\sum_{n=0}^{\infty}na_nt^{n-1}-2\sum_{n=0}^{\infty}a_nt^n\\&=\sum_{n=0}^{\infty}n(n-1)a_nt^{n-2}-2\sum_{n=0}^{\infty}na_nt^n-2\sum_{n=0}^{\infty}a_nt^n=0.\end{aligned} \tag{3}$$

Our next step is to rewrite the first summation in (3) so that the exponent of the general term is n, instead of $n-2$. This is accomplished by increasing every n underneath the summation sign by 2, and decreasing the lower limit by 2, that is,

$$\sum_{n=0}^{\infty}n(n-1)a_nt^{n-2}=\sum_{n=-2}^{\infty}(n+2)(n+1)a_{n+2}t^n.$$

(If you don't believe this, you can verify it by writing out the first few terms in both summations. If you still don't believe this and want a formal proof, set $m=n-2$. When n is zero, m is -2 and when n is infinity, m is infinity. Therefore

$$\sum_{n=0}^{\infty}n(n-1)a_nt^{n-2}=\sum_{m=-2}^{\infty}(m+2)(m+1)a_{m+2}t^m,$$

and since m is a dummy variable, we may replace it by n.) Moreover, observe that the contribution to this sum from $n=-2$ and $n=-1$ is zero since the factor $(n+2)(n+1)$ vanishes in both these instances. Hence,

$$\sum_{n=0}^{\infty} n(n-1)a_n t^{n-2} = \sum_{n=0}^{\infty} (n+2)(n+1)a_{n+2}t^n$$

and we can rewrite (3) in the form

$$\sum_{n=0}^{\infty} (n+2)(n+1)a_{n+2}t^n - 2\sum_{n=0}^{\infty} na_n t^n - 2\sum_{n=0}^{\infty} a_n t^n = 0. \tag{4}$$

Setting the sum of the coefficients of like powers of t in (4) equal to zero gives

$$(n+2)(n+1)a_{n+2} - 2na_n - 2a_n = 0$$

so that

$$a_{n+2} = \frac{2(n+1)a_n}{(n+2)(n+1)} = \frac{2a_n}{n+2}. \tag{5}$$

Equation (5) is a recurrence formula for the coefficients $a_0, a_1, a_2, a_3, \ldots$. The coefficient a_n determines the coefficient a_{n+2}. Thus, a_0 determines a_2 through the relation $a_2 = 2a_0/2 = a_0$; a_2, in turn, determines a_4 through the relation $a_4 = 2a_2/(2+2) = a_0/2$; and so on. Similarly, a_1 determines a_3 through the relation $a_3 = 2a_1/(2+1) = 2a_1/3$; a_3, in turn, determines a_5 through the relation $a_5 = 2a_3/(3+2) = 4a_1/3\cdot 5$; and so on. Consequently, all the coefficients are determined uniquely once a_0 and a_1 are prescribed. The values of a_0 and a_1 are completely arbitrary. This is to be expected, though, for if

$$y(t) = a_0 + a_1 t + a_2 t^2 + \ldots$$

then the values of y and y' at $t=0$ are a_0 and a_1 respectively. Thus, the coefficients a_0 and a_1 must be arbitrary until specific initial conditions are imposed on y.

To find two solutions of (2), we choose two different sets of values of a_0 and a_1. The simplest possible choices are (i) $a_0 = 1, a_1 = 0$; (ii) $a_0 = 0$, $a_1 = 1$.

(i) $\qquad a_0 = 1, \qquad a_1 = 0.$

In this case, all the odd coefficients $a_1, a_3, a_5, \ldots$ are zero since $a_3 = 2a_1/3 = 0$, $a_5 = 2a_3/5 = 0$, and so on. The even coefficients are determined from the relations

$$a_2 = a_0 = 1, \qquad a_4 = \frac{2a_2}{4} = \frac{1}{2}, \qquad a_6 = \frac{2a_4}{6} = \frac{1}{2\cdot 3},$$

and so on. Proceeding inductively, we find that

$$a_{2n} = \frac{1}{2\cdot 3 \cdots n} = \frac{1}{n!}.$$

Hence,

$$y_1(t) = 1 + t^2 + \frac{t^4}{2!} + \frac{t^6}{3!} + \ldots = e^{t^2}$$

is one solution of (2).

(ii) $$a_0=0, \qquad a_1=1.$$

In this case, all the even coefficients are zero, and the odd coefficients are determined from the relations

$$a_3=\frac{2a_1}{3}=\frac{2}{3}, \qquad a_5=\frac{2a_3}{5}=\frac{2}{5}\frac{2}{3}, \qquad a_7=\frac{2a_5}{7}=\frac{2}{7}\frac{2}{5}\frac{2}{3},$$

and so on. Proceeding inductively, we find that

$$a_{2n+1}=\frac{2^n}{3\cdot 5\cdots(2n+1)}.$$

Thus,

$$y_2(t)=t+\frac{2t^3}{3}+\frac{2^2t^5}{3\cdot 5}+\ldots=\sum_{n=0}^{\infty}\frac{2^nt^{2n+1}}{3\cdot 5\cdots(2n+1)}$$

is a second solution of (2).

Now, observe that $y_1(t)$ and $y_2(t)$ are polynomials of infinite degree, even though the coefficients $P(t)=1$, $Q(t)=-2t$, and $R(t)=-2$ are polynomials of finite degree. Such polynomials are called power series. Before proceeding further, we will briefly review some of the important properties of power series.

1. An infinite series
$$y(t)=a_0+a_1(t-t_0)+a_2(t-t_0)^2+\ldots=\sum_{n=0}^{\infty}a_n(t-t_0)^n \tag{6}$$
is called a power series centered about $t=t_0$.
2. All power series have an interval (or radius) of convergence. This means that there exists a nonnegative number ρ such that the infinite series (6) converges for $|t-t_0|<\rho$, and diverges for $|t-t_0|>\rho$.
3. The power series (6) can be differentiated and integrated term by term, and the resultant series have the same interval of convergence.
4. The simplest method (if it works) for determining the interval of convergence of the power series (6) is the Cauchy ratio test. Suppose that the absolute value of a_{n+1}/a_n approaches a limit λ as n approaches infinity. Then, the power series (6) converges for $|t-t_0|<1/\lambda$, and diverges for $|t-t_0|>1/\lambda$.
5. The product of two power series $\sum_{n=0}^{\infty}a_n(t-t_0)^n$ and $\sum_{n=0}^{\infty}b_n(t-t_0)^n$ is again a power series of the form $\sum_{n=0}^{\infty}c_n(t-t_0)^n$, with $c_n=a_0b_n+a_1b_{n-1}+\ldots+a_nb_0$. The quotient
$$\frac{a_0+a_1t+a_2t^2+\ldots}{b_0+b_1t+b_2t^2+\ldots}$$
of two power series is again a power series, provided that $b_0\neq 0$.

6. Many of the functions $f(t)$ that arise in applications can be expanded in power series; that is, we can find coefficients $a_0, a_1, a_2, \ldots$ so that

$$f(t) = a_0 + a_1(t - t_0) + a_2(t - t_0)^2 + \ldots = \sum_{n=0}^{\infty} a_n (t - t_0)^n. \tag{7}$$

Such functions are said to be *analytic* at $t = t_0$, and the series (7) is called the Taylor series of f about $t = t_0$. It can easily be shown that if f admits such an expansion, then, of necessity, $a_n = f^{(n)}(t_0)/n!$, where $f^{(n)}(t) = d^n f(t)/dt^n$.

7. The interval of convergence of the Taylor series of a function $f(t)$, centered about t_0, can be determined directly through the Cauchy ratio test and other similar methods, or indirectly, through the following theorem of complex analysis.

Theorem 6. *Let the variable t assume complex values, and let z_0 be the point closest to t_0 at which "something goes wrong" with $f(t)$. Compute the distance ρ, in the complex plane, between t_0 and z_0. Then, the Taylor series of f centered about t_0 converges for $|t - t_0| < \rho$, and diverges for $|t - t_0| > \rho$.*

As an illustration of Theorem 6, consider the function $f(t) = 1/(1 + t^2)$. The Taylor series of f centered about $t = 0$ is

$$\frac{1}{1 + t^2} = 1 - t^2 + t^4 - t^6 + \ldots,$$

and this series has radius of convergence one. Although nothing is wrong with the function $(1 + t^2)^{-1}$, for t real, it goes to infinity when $t = \pm i$, and the distance of these points from the origin is one.

A second application of Theorem 6 is that the radius of convergence of the Taylor series centered about $t = 0$ of the quotient of two polynomials $a(t)$ and $b(t)$, is the magnitude of the smallest zero of $b(t)$.

At this point we make the important observation that it really wasn't necessary to assume that the functions $P(t)$, $Q(t)$, and $R(t)$ in (1) are polynomials. The method used to solve Example 1 should also be applicable to the more general differential equation

$$L[y] = P(t)\frac{d^2y}{dt^2} + Q(t)\frac{dy}{dt} + R(t)y = 0$$

where $P(t)$, $Q(t)$, and $R(t)$ are power series centered about t_0. (Of course, we would expect the algebra to be much more cumbersome in this case.) If

$$P(t) = p_0 + p_1(t - t_0) + \ldots, \qquad Q(t) = q_0 + q_1(t - t_0) + \ldots,$$
$$R(t) = r_0 + r_1(t - t_0) + \ldots$$

and $y(t) = a_0 + a_1(t - t_0) + \ldots$, then $L[y](t)$ will be the sum of three power series centered about $t = t_0$. Consequently, we should be able to find a recurrence formula for the coefficients a_n by setting the sum of the coefficients of like powers of t in the expression $L[y](t)$ equal to zero. This is the content of the following theorem, which we quote without proof.

Theorem 7. *Let the functions $Q(t)/P(t)$ and $R(t)/P(t)$ have convergent Taylor series expansions about $t=t_0$, for $|t-t_0|<\rho$. Then, every solution $y(t)$ of the differential equation*

$$P(t)\frac{d^2y}{dt^2}+Q(t)\frac{dy}{dt}+R(t)y=0 \tag{8}$$

is analytic at $t=t_0$, and the radius of convergence of its Taylor series expansion about $t=t_0$ is at least ρ. The coefficients $a_2,a_3,\ldots$, in the Taylor series expansion

$$y(t)=a_0+a_1(t-t_0)+a_2(t-t_0)^2+\ldots \tag{9}$$

are determined by plugging the series (9) into the differential equation (8) and setting the sum of the coefficients of like powers of t in this expression equal to zero.

Remark. The interval of convergence of the Taylor series expansion of any solution $y(t)$ of (8) is determined, usually, by the interval of convergence of the power series $Q(t)/P(t)$ and $R(t)/P(t)$, rather than by the interval of convergence of the power series $P(t)$, $Q(t)$, and $R(t)$. This is because the differential equation (8) must be put in the standard form

$$\frac{d^2y}{dt^2}+p(t)\frac{dy}{dt}+q(t)y=0$$

whenever we examine questions of existence and uniqueness.

Example 2.
(a) Find two linearly independent solutions of

$$L[y]=\frac{d^2y}{dt^2}+\frac{3t}{1+t^2}\frac{dy}{dt}+\frac{1}{1+t^2}y=0. \tag{10}$$

(b) Find the solution $y(t)$ of (10) which satisfies the initial conditions $y(0)=2$, $y'(0)=3$.

Solution.
(a) The *wrong* way to do this problem is to expand the functions $3t/(1+t^2)$ and $1/(1+t^2)$ in power series about $t=0$. The right way to do this problem is to multiply both sides of (10) by $1+t^2$ to obtain the equivalent equation

$$L[y]=(1+t^2)\frac{d^2y}{dt^2}+3t\frac{dy}{dt}+y=0.$$

We do the problem this way because the algebra is much less cumbersome when the coefficients of the differential equation (8) are polynomials than when they are power series. Setting $y(t)=\sum_{n=0}^{\infty}a_nt^n$, we compute

$$\begin{aligned}L[y](t)&=(1+t^2)\sum_{n=0}^{\infty}n(n-1)a_nt^{n-2}+3t\sum_{n=0}^{\infty}na_nt^{n-1}+\sum_{n=0}^{\infty}a_nt^n\\&=\sum_{n=0}^{\infty}n(n-1)a_nt^{n-2}+\sum_{n=0}^{\infty}\left[n(n-1)+3n+1\right]a_nt^n\\&=\sum_{n=0}^{\infty}(n+2)(n+1)a_{n+2}t^n+\sum_{n=0}^{\infty}(n+1)^2a_nt^n.\end{aligned}$$

Setting the sum of the coefficients of like powers of t equal to zero gives $(n+2)(n+1)a_{n+2}+(n+1)^2a_n=0$. Hence,

$$a_{n+2}=-\frac{(n+1)^2a_n}{(n+2)(n+1)}=-\frac{(n+1)a_n}{n+2}. \tag{11}$$

Equation (11) is a recurrence formula for the coefficients $a_2,a_3,\ldots$ in terms of a_0 and a_1. To find two linearly independent solutions of (10), we choose the two simplest cases (i) $a_0=1$, $a_1=0$; and (ii) $a_0=0$, $a_1=1$.

(i) $$a_0=1,\qquad a_1=0.$$

In this case, all the odd coefficients are zero since $a_3=-2a_1/3=0$, $a_5=-4a_3/5=0$, and so on. The even coefficients are determined from the relations

$$a_2=-\frac{a_0}{2}=-\frac{1}{2},\qquad a_4=-\frac{3a_2}{4}=\frac{1\cdot 3}{2\cdot 4},\qquad a_6=-\frac{5a_4}{6}=-\frac{1\cdot 3\cdot 5}{2\cdot 4\cdot 6}$$

and so on. Proceeding inductively, we find that

$$a_{2n}=(-1)^n\frac{1\cdot 3\cdots(2n-1)}{2\cdot 4\cdots 2n}=(-1)^n\frac{1\cdot 3\cdots(2n-1)}{2^nn!}.$$

Thus,

$$y_1(t)=1-\frac{t^2}{2}+\frac{1\cdot 3}{2\cdot 4}t^4+\ldots=\sum_{n=0}^{\infty}(-1)^n\frac{1\cdot 3\cdots(2n-1)}{2^nn!}t^{2n} \tag{12}$$

is one solution of (10). The ratio of the $(n+1)$st term to the nth term of $y_1(t)$ is

$$-\frac{1\cdot 3\cdots(2n-1)(2n+1)t^{2n+2}}{2^{n+1}(n+1)!}\times\frac{2^nn!}{1\cdot 3\cdots(2n-1)t^{2n}}=\frac{-(2n+1)t^2}{2(n+1)},$$

and the absolute value of this quantity approaches t^2 as n approaches infinity. Hence, by the Cauchy ratio test, the infinite series (12) converges for $|t|<1$, and diverges for $|t|>1$.

(ii) $$a_0=0,\qquad a_1=1.$$

In this case, all the even coefficients are zero, and the odd coefficients are determined from the relations

$$a_3=-\frac{2a_1}{3}=-\frac{2}{3},\qquad a_5=-\frac{4a_3}{5}=\frac{2\cdot 4}{3\cdot 5},\qquad a_7=-\frac{6a_5}{7}=-\frac{2\cdot 4\cdot 6}{3\cdot 5\cdot 7},$$

and so on. Proceeding inductively, we find that

$$a_{2n+1}=(-1)^n\frac{2\cdot 4\cdots 2n}{3\cdot 5\cdots(2n+1)}=\frac{(-1)^n 2^n n!}{3\cdot 5\cdots(2n+1)}.$$

Thus,

$$y_2(t)=t-\frac{2}{3}t^3+\frac{2\cdot 4}{3\cdot 5}t^5+\ldots=\sum_{n=0}^{\infty}\frac{(-1)^n 2^n n!}{3\cdot 5\cdots(2n+1)}t^{2n+1} \qquad (13)$$

is a second solution of (10), and it is easily verified that this solution, too, converges for $|t|<1$, and diverges for $|t|>1$. This, of course, is not very surprising, since the Taylor series expansions about $t=0$ of the functions $3t/(1+t^2)$ and $1/(1+t^2)$ only converge for $|t|<1$.

(b) The solution $y_1(t)$ satisfies the initial conditions $y(0)=1, y'(0)=0$, while $y_2(t)$ satisfies the initial conditions $y(0)=0$, $y'(0)=1$. Hence $y(t)=2y_1(t)+3y_2(t)$.

Example 3. Solve the initial-value problem

$$L[y]=\frac{d^2y}{dt^2}+t^2\frac{dy}{dt}+2ty=0;\qquad y(0)=1,\quad y'(0)=0.$$

Solution. Setting $y(t)=\sum_{n=0}^{\infty}a_n t^n$, we compute

$$\begin{aligned}L[y](t)&=\sum_{n=0}^{\infty}n(n-1)a_n t^{n-2}+t^2\sum_{n=0}^{\infty}na_n t^{n-1}+2t\sum_{n=0}^{\infty}a_n t^n\\&=\sum_{n=0}^{\infty}n(n-1)a_n t^{n-2}+\sum_{n=0}^{\infty}na_n t^{n+1}+2\sum_{n=0}^{\infty}a_n t^{n+1}\\&=\sum_{n=0}^{\infty}n(n-1)a_n t^{n-2}+\sum_{n=0}^{\infty}(n+2)a_n t^{n+1}.\end{aligned}$$

Our next step is to rewrite the first summation so that the exponent of the general term is $n+1$ instead of $n-2$. This is accomplished by increasing every n underneath the summation sign by 3, and decreasing the lower limit by 3; that is,

$$\begin{aligned}\sum_{n=0}^{\infty}n(n-1)a_n t^{n-2}&=\sum_{n=-3}^{\infty}(n+3)(n+2)a_{n+3}t^{n+1}\\&=\sum_{n=-1}^{\infty}(n+3)(n+2)a_{n+3}t^{n+1}.\end{aligned}$$

Therefore,

$$L[y](t)=\sum_{n=-1}^{\infty}(n+3)(n+2)a_{n+3}t^{n+1}+\sum_{n=0}^{\infty}(n+2)a_nt^{n+1}$$

$$=2a_2+\sum_{n=0}^{\infty}(n+3)(n+2)a_{n+3}t^{n+1}+\sum_{n=0}^{\infty}(n+2)a_nt^{n+1}.$$

Setting the sums of the coefficients of like powers of t equal to zero gives

$$2a_2=0,\quad\text{and}\quad (n+3)(n+2)a_{n+3}+(n+2)a_n=0;\qquad n=0,1,2,\ldots$$

Consequently,

$$a_2=0,\quad\text{and}\quad a_{n+3}=-\frac{a_n}{n+3};\qquad n\geqslant 0. \tag{14}$$

The recurrence formula (14) determines a_3 in terms of a_0, a_4 in terms of a_1, a_5 in terms of a_2, and so on. Since $a_2=0$, we see that $a_5,a_8,a_{11},\ldots$ are all zero, regardless of the values of a_0 and a_1. To satify the initial conditions, we set $a_0=1$ and $a_1=0$. Then, from (14), $a_4,a_7,a_{10},\ldots$ are all zero, while

$$a_3=-\frac{a_0}{3}=-\frac{1}{3},\qquad a_6=-\frac{a_3}{6}=\frac{1}{3\cdot 6},\qquad a_9=-\frac{a_6}{9}=-\frac{1}{3\cdot 6\cdot 9}$$

and so on. Proceeding inductively, we find that

$$a_{3n}=\frac{(-1)^n}{3\cdot 6\cdots 3n}=\frac{(-1)^n}{3^n1\cdot 2\cdots n}=\frac{(-1)^n}{3^nn!}.$$

Hence,

$$y(t)=1-\frac{t^3}{3}+\frac{t^6}{3\cdot 6}-\frac{t^9}{3\cdot 6\cdot 9}+\ldots=\sum_{n=0}^{\infty}\frac{(-1)^nt^{3n}}{3^nn!}.$$

By Theorem 7, this series converges for all t, since the power series t^2 and $2t$ obviously converge for all t. (We could also verify this directly using the Cauchy ratio test.)

Example 4. Solve the initial-value problem

$$L[y]=(t^2-2t)\frac{d^2y}{dt^2}+5(t-1)\frac{dy}{dt}+3y=0;\qquad y(1)=7,\ \ y'(1)=3. \tag{15}$$

Solution. Since the initial conditions are given at $t=1$, we will express the coefficients of the differential equation (15) as polynomials in $(t-1)$, and then we will find $y(t)$ as a power series centered about $t=1$. To this end, observe that

$$t^2-2t=t(t-2)=\left[(t-1)+1\right]\left[(t-1)-1\right]=(t-1)^2-1.$$

Hence, the differential equation (15) can be written in the form

$$L[y]=\left[(t-1)^2-1\right]\frac{d^2y}{dt^2}+5(t-1)\frac{dy}{dt}+3y=0.$$

Setting $y(t)=\sum_{n=0}^{\infty}a_n(t-1)^n$, we compute

$$
\begin{aligned}
L[y](t)&=\left[(t-1)^2-1\right]\sum_{n=0}^{\infty}n(n-1)a_n(t-1)^{n-2}\\
&\qquad+5(t-1)\sum_{n=0}^{\infty}na_n(t-1)^{n-1}+3\sum_{n=0}^{\infty}a_n(t-1)^n\\
&=-\sum_{n=0}^{\infty}n(n-1)a_n(t-1)^{n-2}\\
&\qquad+\sum_{n=0}^{\infty}n(n-1)a_n(t-1)^n+\sum_{n=0}^{\infty}(5n+3)a_n(t-1)^n\\
&=-\sum_{n=0}^{\infty}(n+2)(n+1)a_{n+2}(t-1)^n+\sum_{n=0}^{\infty}(n^2+4n+3)a_n(t-1)^n.
\end{aligned}
$$

Setting the sums of the coefficients of like powers of t equal to zero gives $-(n+2)(n+1)a_{n+2}+(n^2+4n+3)a_n=0$, so that

$$a_{n+2}=\frac{n^2+4n+3}{(n+2)(n+1)}a_n=\frac{n+3}{n+2}a_n,\qquad n\geqslant 0. \tag{16}$$

To satisfy the initial conditions, we set $a_0=7$ and $a_1=3$. Then, from (16),

$$
\begin{aligned}
a_2&=\frac{3}{2}a_0=\frac{3}{2}\cdot 7, & a_4&=\frac{5}{4}a_2=\frac{5\cdot 3}{4\cdot 2}\cdot 7, & a_6&=\frac{7}{6}a_4=\frac{7\cdot 5\cdot 3}{6\cdot 4\cdot 2}\cdot 7,\ldots\\
a_3&=\frac{4}{3}a_1=\frac{4}{3}\cdot 3, & a_5&=\frac{6}{5}a_3=\frac{6\cdot 4}{5\cdot 3}\cdot 3, & a_7&=\frac{8}{7}a_5=\frac{8\cdot 6\cdot 4}{7\cdot 5\cdot 3}\cdot 3,\ldots
\end{aligned}
$$

and so on. Proceeding inductively, we find that

$$a_{2n}=\frac{3\cdot 5\cdots(2n+1)}{2\cdot 4\cdots(2n)}\cdot 7\quad\text{and}\quad a_{2n+1}=\frac{4\cdot 6\cdots(2n+2)}{3\cdot 5\cdots(2n+1)}\cdot 3\qquad(\text{for } n\geqslant 1).$$

Hence,

$$
\begin{aligned}
y(t)&=7+3(t-1)+\frac{3}{2}\cdot 7(t-1)^2+\frac{4}{3}\cdot 3(t-1)^3+\ldots\\
&=7+7\sum_{n=1}^{\infty}\frac{3\cdot 5\cdots(2n+1)(t-1)^{2n}}{2^n n!}+3(t-1)+3\sum_{n=1}^{\infty}\frac{2^n(n+1)!(t-1)^{2n+1}}{3\cdot 5\cdots(2n+1)}.
\end{aligned}
$$

Example 5. Solve the initial-value problem

$$L[y]=(1-t)\frac{d^2y}{dt^2}+\frac{dy}{dt}+(1-t)y=0;\qquad y(0)=1,\quad y'(0)=1.$$

Solution. Setting $y(t)=\sum_{n=0}^{\infty}a_n t^n$, we compute

$$\begin{aligned}
L[y](t)&=(1-t)\sum_{n=0}^{\infty}n(n-1)a_n t^{n-2}\\
&\quad+\sum_{n=0}^{\infty}na_n t^{n-1}+(1-t)\sum_{n=0}^{\infty}a_n t^n\\
&=\sum_{n=0}^{\infty}n(n-1)a_n t^{n-2}-\sum_{n=0}^{\infty}n(n-1)a_n t^{n-1}\\
&\quad+\sum_{n=0}^{\infty}na_n t^{n-1}+\sum_{n=0}^{\infty}a_n t^n-\sum_{n=0}^{\infty}a_n t^{n+1}\\
&=\sum_{n=0}^{\infty}(n+2)(n+1)a_{n+2}t^n-\sum_{n=0}^{\infty}n(n-2)a_n t^{n-1}\\
&\quad+\sum_{n=0}^{\infty}a_n t^n-\sum_{n=0}^{\infty}a_n t^{n+1}\\
&=\sum_{n=0}^{\infty}(n+2)(n+1)a_{n+2}t^n-\sum_{n=0}^{\infty}(n+1)(n-1)a_{n+1}t^n\\
&\quad+\sum_{n=0}^{\infty}a_n t^n-\sum_{n=1}^{\infty}a_{n-1}t^n\\
&=2a_2+a_1+a_0\\
&\quad+\sum_{n=1}^{\infty}\{(n+2)(n+1)a_{n+2}-(n+1)(n-1)a_{n+1}+a_n-a_{n-1}\}t^n.
\end{aligned}$$

Setting the coefficients of each power of t equal to zero gives

$$a_2=-\frac{a_1+a_0}{2}\quad\text{and}\quad a_{n+2}=\frac{(n+1)(n-1)a_{n+1}-a_n+a_{n-1}}{(n+2)(n+1)},\qquad n\geqslant 1. \tag{17}$$

To satisfy the initial conditions, we set $a_0=1$ and $a_1=1$. Then, from (17),

$$a_2=-1,\qquad a_3=\frac{-a_1+a_0}{6}=0,\qquad a_4=\frac{3a_3-a_2+a_1}{12}=\frac{1}{6},$$
$$a_5=\frac{8a_4-a_3+a_2}{20}=\frac{1}{60},\qquad a_6=\frac{15a_5-a_4+a_3}{30}=\frac{1}{360}$$

and so on. Unfortunately, though, we cannot discern a general pattern for the coefficients a_n as we did in the previous examples. (This is because the coefficient a_{n+2} depends on the values of a_{n+1}, a_n, and a_{n-1}, while in our previous examples, the coefficient a_{n+2} depended on only one of its predecessors.) This is not a serious problem though, for we can find the coefficients a_n quite easily with the aid of a digital computer.

Exercises

Find the general solution of each of the following equations.

1. $y''+ty'+y=0$

2. $y''-ty=0$

3. $(2+t^2)y''-ty'-3y=0$

4. $y''-t^3y=0$

Solve each of the following initial-value problems.

5. $t(2-t)y''-6(t-1)y'-4y=0;\quad y(1)=1,\ y'(1)=0$

6. $y''+t^2y=0;\quad y(0)=2,\ y'(0)=-1$

7. $y''-t^3y=0;\quad y(0)=0,\ y'(0)=-2$

8. $y''+(t^2+2t+1)y'-(4+4t)y=0;\quad y(-1)=0,\ y'(-1)=1$

9. The equation $y''-2ty'+\lambda y=0$, λ constant, is known as the Hermite differential equation, and it appears in many areas of mathematics and physics.
(a) Find 2 linearly independent solutions of the Hermite equation.
(b) Show that the Hermite equation has a polynomial solution of degree n if $\lambda=2n$. This polynomial, when properly normalized; that is, when multiplied by a suitable constant, is known as the Hermite polynomial $H_n(t)$.

10. The equation $(1-t^2)y''-2ty'+\alpha(\alpha+1)y=0$, α constant, is known as the Legendre differential equation, and it appears in many areas of mathematics and physics.
(a) Find 2 linearly independent solutions of the Legendre equation.
(b) Show that the Legendre differential equation has a polynomial solution of degree n if $\alpha=n$.
(c) The Legendre polynomial $P_n(t)$ is defined as the polynomial solution of the Legendre equation with $\alpha=n$ which satisfies the condition $P_n(1)=1$. Find $P_0(t)$, $P_1(t)$, $P_2(t)$, and $P_3(t)$.

11. The equation $(1-t^2)y''-ty'+\alpha^2y=0$, α constant, is known as the Tchebycheff differential equation, and it appears in many areas of mathematics and physics.
(a) Find 2 linearly independent solutions of the Tchebycheff equation.
(b) Show that the Tchebycheff equation has a polynomial solution of degree n if $\alpha=n$. These polynomials, when properly normalized, are called the Tchebycheff polynomials.

12. (a) Find 2 linearly independent solutions of

$$y''+t^3y'+3t^2y=0.$$

(b) Find the first 5 terms in the Taylor series expansion about $t=0$ of the solution $y(t)$ of the initial-value problem

$$y''+t^3y'+3t^2y=e^t;\qquad y(0)=0,\quad y'(0)=0.$$

2.8.1 *Singular points; the method of Frobenius*

The differential equation

$$L[y] = P(t)\frac{d^2y}{dt^2} + Q(t)\frac{dy}{dt} + R(t)y = 0 \tag{1}$$

is said to be singular at $t = t_0$ if $P(t_0) = 0$. In this case, we can say very little about the behavior of the solutions $y(t)$ of (1) near $t = t_0$. Most probably, the solutions of (1) aren't even continuous, let alone analytic, at $t = t_0$. However, there are certain special cases where we can find solutions $y(t)$ of (1) of the form

$$y(t) = (t-t_0)^r[a_0 + a_1(t-t_0) + \dots] = \sum_{n=0}^{\infty} a_n (t-t_0)^{n+r}. \tag{2}$$

We illustrate this method, which is known as the method of Frobenius, with the following example.

Example 1. Find 2 linearly independent solutions of

$$L[y] = 2t\frac{d^2y}{dt^2} + \frac{dy}{dt} + ty = 0, \qquad 0 < t < \infty. \tag{3}$$

Solution. Let

$$y(t) = \sum_{n=0}^{\infty} a_n t^{n+r}, \qquad a_0 \neq 0.$$

Computing

$$y'(t) = \sum_{n=0}^{\infty} (n+r)a_n t^{n+r-1}$$

and

$$y''(t) = \sum_{n=0}^{\infty} (n+r)(n+r-1)a_n t^{n+r-2},$$

we see that

$$L[y] = t^r\left[2\sum_{n=0}^{\infty}(n+r)(n+r-1)a_n t^{n-1} + \sum_{n=0}^{\infty}(n+r)a_n t^{n-1} + \sum_{n=0}^{\infty}a_n t^{n+1}\right]$$

$$= t^r\left[2\sum_{n=0}^{\infty}(n+r)(n+r-1)a_n t^{n-1} + \sum_{n=0}^{\infty}(n+r)a_n t^{n-1} + \sum_{n=2}^{\infty}a_{n-2} t^{n-1}\right]$$

$$= [2r(r-1)a_0 + ra_0]t^{r-1} + [2(1+r)ra_1 + (1+r)a_1]t^r$$

$$+ \sum_{n=2}^{\infty}[2(n+r)(n+r-1)a_n + (n+r)a_n + a_{n-2}]t^{n+r-1}.$$

Setting the coefficients of each power of t equal to zero gives

(i) $2r(r-1)a_0+ra_0=r(2r-1)a_0=0$
(ii) $2(r+1)ra_1+(r+1)a_1=(r+1)(2r+1)a_1=0$
and
(iii) $2(n+r)(n+r-1)a_n+(n+r)a_n=(n+r)[2(n+r)-1]a_n=-a_{n-2},\ n\geqslant 2.$

The first equation determines r: it implies that $r=0$ or $r=\frac{1}{2}$. The second equation then forces a_1 to be zero, and the third equation determines a_n for $n\geqslant 2$.

(i) $r=0$: In this case, the recurrence formula (iii) is

$$a_n=\frac{-a_{n-2}}{n(2n-1)},\qquad n\geqslant 2.$$

Since $a_1=0$, we see that all the odd coefficients are zero. The even coefficients are determined from the relations

$$a_2=\frac{-a_0}{2\cdot 3},\qquad a_4=\frac{-a_2}{4\cdot 7}=\frac{a_0}{2\cdot 4\cdot 3\cdot 7},\qquad a_6=\frac{-a_4}{6\cdot 11}=\frac{-a_0}{2\cdot 4\cdot 6\cdot 3\cdot 7\cdot 11},$$

and so on. Setting $a_0=1$, we see that

$$y_1(t)=1-\frac{t^2}{2\cdot 3}+\frac{t^4}{2\cdot 4\cdot 3\cdot 7}+\dots=1+\sum_{n=1}^{\infty}\frac{(-1)^n t^{2n}}{2^n n!3\cdot 7\cdots(4n-1)}$$

is one solution of (3) on the interval $0<t<\infty$. (It is easily verified that this series converges for all t.)

(ii) $r=\frac{1}{2}$: In this case, the recurrence formula (iii) is

$$a_n=\frac{-a_{n-2}}{\left(n+\frac{1}{2}\right)\left[2\left(n+\frac{1}{2}\right)-1\right]}=\frac{-a_{n-2}}{n(2n+1)}.$$

Again, all the odd coefficients are zero. The even coefficients are determined from the relations

$$a_2=\frac{-a_0}{2\cdot 5},\qquad a_4=\frac{-a_2}{4\cdot 9}=\frac{a_0}{2\cdot 4\cdot 5\cdot 9},\qquad a_6=\frac{-a_4}{6\cdot 13}=\frac{-a_0}{2\cdot 4\cdot 6\cdot 5\cdot 9\cdot 13},$$

and so on. Setting $a_0=1$, we see that

$$y_2(t)=t^{1/2}\left[1-\frac{t^2}{2\cdot 5}+\frac{t^4}{2\cdot 4\cdot 5\cdot 9}+\dots\right]$$
$$=t^{1/2}\left[1+\sum_{n=1}^{\infty}\frac{(-1)^n t^{2n}}{2^n n!5\cdot 9\cdots(4n+1)}\right]$$

is a second solution of (3) on the interval $0<t<\infty$.

We return now to the differential equation

$$L[y]=P(t)\frac{d^2y}{dt^2}+Q(t)\frac{dy}{dt}+R(t)y=0, \qquad t\geqslant t_0 \tag{4}$$

and assume that $t=t_0$ is a singular point of (4). For simplicity, we set $t_0=0$. It is possible, sometimes, to solve Equation (4) by the method of Frobenius. However, we must make the very restrictive assumption that the functions

$$\frac{tQ(t)}{P(t)} \quad \text{and} \quad \frac{t^2R(t)}{P(t)}$$

do not "behave very badly" at $t=0$. Specifically, we assume that the functions $tQ(t)/P(t)$ and $t^2R(t)/P(t)$ are analytic at $t=0$. Equivalently,

$$\frac{Q(t)}{P(t)}=\frac{b_0}{t}+b_1+b_2t+\ldots, \quad \text{and} \quad \frac{R(t)}{P(t)}=\frac{c_0}{t^2}+\frac{c_1}{t}+c_2+\ldots.$$

In this case we say that $t=0$ is a regular singular point of (4). Then, we can always find at least one solution $y(t)$ of (4) of the form

$$y(t)=t^r(a_0+a_1t+a_2t^2+\ldots)=\sum_{n=0}^{\infty}a_nt^{n+r} \tag{5}$$

for some constant r (possibly complex). This is the content of the following theorem.

Theorem 8. *Consider the differential equation* (4), *where* $t=0$ *is a regular singular point. Then, the functions* $tQ(t)/P(t)$ *and* $t^2R(t)/P(t)$ *are analytic at* $t=0$ *with power series expansions*

$$\frac{tQ(t)}{P(t)}=b_0+b_1t+b_2t^2+\ldots \quad \textit{and} \quad \frac{t^2R(t)}{P(t)}=c_0+c_1t+c_2t^2+\ldots$$

which converge for $|t|<\rho$. *Let* r_1 *and* r_2 *be the two roots of the equation*

$$r(r-1)+b_0r+c_0=0,$$

with $r_1\geqslant r_2$ *if they are real. Then, Equation* (4) *has two linearly independent solutions* $y_1(t)$ *and* $y_2(t)$ *on the interval* $0<t<\rho$ *of the following form*:

(a) *If r_1-r_2 is not a positive integer, then*

$$y_1(t)=t^{r_1}\sum_{n=0}^{\infty}a_nt^n \quad and \quad y_2(t)=t^{r_2}\sum_{n=0}^{\infty}b_nt^n.$$

(b) *If $r_1=r_2$, then*

$$y_1(t)=t^{r_1}\sum_{n=0}^{\infty}a_nt^n \quad and \quad y_2(t)=y_1(t)\ln t+t^{r_1}\sum_{n=1}^{\infty}b_nt^n.$$

(c) *If $r_1-r_2=N$, a positive integer, then*

$$y_1(t)=t^{r_1}\sum_{n=0}^{\infty}a_nt^n \quad and \quad y_2(t)=ay_1(t)\ln t+t^{r_2}\sum_{n=0}^{\infty}b_nt^n.$$

The constant a may turn out to be zero.

PROOF.
(a) First, we multiply both sides of (4) by $t^2P(t)$ to obtain the equivalent equation

$$L[y]=t^2\frac{d^2y}{dt^2}+t^2\frac{Q(t)}{P(t)}\frac{dy}{dt}+t^2\frac{R(t)}{P(t)}y=0.$$

By assumption, this equation can be written in the form

$$L[y]=t^2\frac{d^2y}{dt^2}+t\left[b_0+b_1t+b_2t^2+\dots\right]\frac{dy}{dt}$$
$$+\left[c_0+c_1t+c_2t^2+\dots\right]y=0.$$

Set $y(t)=\sum_{n=0}^{\infty}a_nt^{n+r}$, with $a_0\neq 0$. Computing

$$y'(t)=\sum_{n=0}^{\infty}(n+r)a_nt^{n+r-1}$$

and

$$y''(t)=\sum_{n=0}^{\infty}(n+r)(n+r-1)a_nt^{n+r-2}$$

we see that

$$L[y](t)=t^r\left\{\sum_{n=0}^{\infty}(n+r)(n+r-1)a_nt^n+\left[\sum_{m=0}^{\infty}b_mt^m\right]\left[\sum_{n=0}^{\infty}(n+r)a_nt^n\right]\right.$$
$$\left.+\left[\sum_{m=0}^{\infty}c_mt^m\right]\left[\sum_{n=0}^{\infty}a_nt^n\right]\right\}=0.$$

Setting the sum of the coefficients of like powers of t equal to zero gives

$$r(r-1)+b_0 r+c_0=0 \tag{6}$$

and

$$\left[(n+r)(n+r-1)+b_0(n+r)+c_0\right]a_n=-\sum_{k=1}^{n}\left[b_k(n-k+r)+c_k\right]a_{n-k} \tag{7}$$

for $n\geqslant 1$. Equation (6) is called the *indicial equation* of (4). It is a quadratic equation in r, and its roots determine the two possible values of r for which there may be solutions of (4) of the form (5). Once a root r of (6) is determined, we can use Equation (7) to find the coefficients a_n, $n\geqslant 1$, *provided that*

$$(n+r)(n+r-1)+b_0(n+r)+c_0\neq 0, \qquad n\geqslant 1. \tag{8}$$

Now, let $F(r)=r(r-1)+b_0 r+c_0$. Notice that the inequality (8) can be written succinctly in the form $F(r+n)\neq 0$, $n\geqslant 1$. This implies that $r+n$ is not a root of the indicial equation, for every positive integer n. Therefore, if $r_1-r_2\neq N$, a positive integer, then (4) has two solutions $y_1(t)$ and $y_2(t)$ of the form

$$y_1(t)=t^{r_1}\sum_{n=0}^{\infty}a_n t^n \quad\text{and}\quad y_2(t)=t^{r_2}\sum_{n=0}^{\infty}b_n t^n.$$

It can be shown that these two series converge wherever the two series $tQ(t)/P(t)$ and $t^2R(t)/P(t)$ converge.
(b) Suppose that $r_1=r_2$. Then, we certainly have one solution $y_1(t)$ of (4) of the form

$$y_1(t)=t^{r_1}\sum_{n=0}^{\infty}a_n t^n.$$

It can be shown that (4) also has a solution $y_2(t)$ of the form

$$y_2(t)=y_1(t)\ln t+t^{r_1}\sum_{n=1}^{\infty}b_n t^n.$$

We omit the proof of this assertion. We wish to point out, though, that once one solution $y_1(t)$ of (4) is known, we can find a second solution by the method of reduction of order.
(c) Suppose that $r_1=r_2+N$, N a positive integer. Then, we can certainly find one solution $y_1(t)$ of (4) of the form $y_1(t)=t^{r_1}\sum_{n=0}^{\infty}a_n t^n$. It may also be possible to find a second solution $y_2(t)$ of (4) of the form $y_2(t)=t^{r_2}\sum_{n=0}^{\infty}b_n t^n$. This will depend on whether the equation

$$F(r_2+N)a_N=-\sum_{k=1}^{N}\left[b_k(N-k+r_2)+c_k\right]a_{N-k}$$

is consistent; that is, whether

$$\sum_{k=1}^{N}\left[b_k(N-k+r_2)+c_k\right]a_{N-k}=0. \tag{9}$$

If (9) is true, then we can find a second solution $y_2(t)$ of (4) of the form $y_2(t)=t^{r_2}\sum_{n=0}^{\infty}b_nt^n$. (In this case, the coefficient b_N will be arbitrary.) If (9) is not true, then we can find a second solution $y_2(t)$ of (4) of the form

$$y_2(t)=ay_1(t)\ln t+t^{r_2}\sum_{n=0}^{\infty}b_nt^n.$$

Again, we omit the proof of this assertion, but point out that a second solution $y_2(t)$ can always be obtained from $y_1(t)$ using the method of reduction of order. □

Remark. If r is a complex root of the indicial equation (6), then $y(t)=t^r\sum_{n=0}^{\infty}a_nt^n$ is a complex-valued solution of (4). (To define $t^{\alpha+i\beta}$, we use the formula

$$t^{\alpha+i\beta}=t^{\alpha}t^{i\beta}=t^{\alpha}e^{i\beta\ln t}=t^{\alpha}\left[\cos(\beta\ln t)+i\sin(\beta\ln t)\right].)$$

It is easily verified in this case, that both the real and imaginary parts of $y(t)$ are real-valued solutions of (4).

Example 2. Use the method of Frobenius to solve Bessel's equation of order zero

$$t^2\frac{d^2y}{dt^2}+t\frac{dy}{dt}+t^2y=0,\qquad t>0. \tag{10}$$

Solution. Here, $P(t)=t^2$, $Q(t)=t$ and $R(t)=t^2$, so that

$$t\frac{Q(t)}{P(t)}=\frac{t^2}{t^2}=1\quad\text{and}\quad t^2\frac{R(t)}{P(t)}=t^2\frac{t^2}{t^2}=t^2$$

are both analytic functions of t at $t=0$. Therefore, $t=0$ is a regular singular point of (10), and we can find a solution $y(t)$ of (10) of the form $y(t)=\sum_{n=0}^{\infty}a_nt^{n+r}$. Computing,

$$y'(t)=\sum_{n=0}^{\infty}(n+r)a_nt^{n+r-1}$$

and

$$y''(t) = \sum_{n=0}^{\infty} (n+r)(n+r-1)a_n t^{n+r-2},$$

we see that

$$\begin{aligned} L[y] &= t^r \left\{ \sum_{n=0}^{\infty} (n+r)(n+r-1)a_n t^n + \sum_{n=0}^{\infty} (n+r)a_n t^n + \sum_{n=0}^{\infty} a_n t^{n+2} \right\} \\ &= t^r \left\{ \sum_{n=0}^{\infty} (n+r)(n+r-1)a_n t^n + \sum_{n=0}^{\infty} (n+r)a_n t^n + \sum_{n=2}^{\infty} a_{n-2} t^n \right\}. \end{aligned}$$

Setting the sum of the coefficients of like powers of t equal to zero gives

(i) $[r(r-1)+r]a_0 = r^2 a_0 = 0$

(ii) $[(r+1)r+(r+1)]a_1 = (r+1)^2 a_1 = 0$

and

(iii) $[(r+n)(r+n-1)+r+n]a_n = (r+n)^2 a_n = -a_{n-2}, \quad n \geq 2.$

Equation (i) is the indicial equation of (10), and it has a double root $r=0$. Since $(n+1)^2 \neq 0$, Equation (ii) forces a_1 to be zero, and the recurrence formula (iii), which we now write in the form $a_n = -a_{n-2}/n^2$, determines $a_2, a_3, \ldots$ in terms of a_0. Specifically, all the odd coefficients are zero, and

$$a_2 = \frac{-a_0}{2^2}, \qquad a_4 = \frac{-a_2}{4^2} = \frac{a_0}{2^2 \cdot 4^2}, \qquad a_6 = \frac{-a_4}{6^2} = \frac{a_0}{2^2 \cdot 4^2 \cdot 6^2},$$

and so on. Setting $a_0 = 1$, we see that

$$y(t) = 1 - \frac{t^2}{2^2} + \frac{t^4}{2^2 \cdot 4^2} + \ldots = \sum_{n=0}^{\infty} \frac{(-1)^n t^{2n}}{2^2 \cdot 4^2 \cdots (2n)^2}$$

is one solution of (10). This solution is known as the **Bessel function** of order zero, and is denoted by $J_0(t)$. A second solution

$$y_2(t) = J_0(t) \int \frac{dt}{tJ_0^2(t)}$$

can be found using the method of reduction of order. This solution, when properly normalized, is known as the Neumann function of order zero, and is denoted by $Y_0(t)$. Notice that $Y_0(t)$ behaves like $\ln t$ at $t=0$, since $1/tJ_0^2(t)$ behaves like $1/t$ at $t=0$.

Exercises

In each of Problems 1–6, determine whether the specified value of t is a regular singular point of the given differential equation.

1. $t(t-2)^2 y'' + ty' + y = 0;\ t=0$

2. $t(t-2)^2 y'' + ty' + y = 0;\ t=2$

3. $(\sin t)y''+(\cos t)y'+\frac{1}{t}y=0;\ t=0$

4. $(e^t-1)y''+e^ty'+y=0;\ t=0$

5. $(1-t^2)y''+\frac{1}{\sin(t+1)}y'+y=0;\ t=-1$

6. $t^3y''+(\sin t^2)y'+ty=0;\ t=0$

Find the general solution of the following equations.

7. $t^2y''-ty'-(t^2+\frac{5}{4})y=0$

8. $t^2y''+(t-t^2)y'-y=0$

9. $ty''-(t^2+2)y'+ty=0$

10. $t^2y''+(t^2-3t)y'+3y=0$

11. $2t^2y''+3ty'-(1+t)y=0$

12. $t^2y''+(3t-t^2)y'-ty=0$

13. $2ty''+(1+t)y'-2y=0$

14. $t^2y''+ty'-(1+t)y=0$

15. $4ty''+3y'-3y=0$

16. $ty''-(4+t)y'+2y=0$

17. $t^2y''+t(t+1)y'-y=0$

18. $ty''+ty'+2y=0$

19. $ty''+(1-t^2)y'+4ty=0$

20. $ty''-y=0$

21. (a) Show that the equation

$$2(\sin t)y''+(1-t)y'-2y=0$$

has two solutions $y_1(t)$ and $y_2(t)$ of the form

$$y_1(t)=\sum_{n=0}^{\infty}a_nt^n,\qquad y_2(t)=t^{1/2}\sum_{n=0}^{\infty}b_nt^n.$$

(b) Find the first 5 terms in these infinite series expansions of $y_1(t)$ and $y_2(t)$, assuming that $a_0=b_0=1$.

22. Let $y(t)=u(t)+iv(t)$ be a complex-valued solution of (4), with $P(t)$, $Q(t)$ and $R(t)$ real. Show that both $u(t)$ and $v(t)$ are real-valued solutions of (4).

23. (a) Show that the indicial equation of

$$t^2y''+ty'+(1+t)y=0 \tag{*}$$

has complex roots $r=\pm i$.

(b) Show that (*) has 2 linearly independent solutions $y(t)$ of the form

$$y(t)=(\sin\ln t)\sum_{n=0}^{\infty}a_nt^n+(\cos\ln t)\sum_{n=0}^{\infty}b_nt^n.$$

24. The equation $t^2y''+ty'+(t^2-\nu^2)y=0$, $t>0$, is known as Bessel's equation of order ν.

(a) Find a power series solution

$$J_\nu(t)=\frac{t^\nu}{2^\nu\nu!}\sum_{n=0}^{\infty}a_nt^n,\qquad a_0=1,$$

of Bessel's equation. This function $J_\nu(t)$ is called the Bessel function of order ν.

(b) Find a second solution

$$Y_\nu(t) = J_\nu(t) \int \frac{dt}{tJ_\nu^2(t)}$$

by the method of reduction of order. The function $Y_\nu(t)$, after proper normalization, is called the Neumann function of order ν.

25. The differential equation $ty'' + (1-t)y' + \lambda y = 0$, λ constant, is called the Laguerre differential equation.
(a) Show that the indicial equation of the Laguerre equation is $r^2 = 0$.
(b) Find a solution $y(t)$ of the Laguerre equation of the form $y(t) = \sum_{n=0}^{\infty} a_n t^n$.
(c) Show that this solution reduces to a polynomial of degree n if $\lambda = n$.

26. The differential equation

$$t(1-t)y'' + [\gamma - (1+\alpha+\beta)t]y' - \alpha\beta y = 0,$$

where α, β and γ are constants, is known as the hypergeometric equation.
(a) Show that $t=0$ is a regular singular point of the hypergeometric equation, and that the roots of the indicial equation are 0 and $1-\gamma$.
(b) Show that $t=1$ is also a regular singular point of the hypergeometric equation, and that the roots of the indicial equation are now 0 and $\gamma - \alpha - \beta$.
(c) Assume that γ is not an integer. Find 2 solutions $y_1(t)$ and $y_2(t)$ of the hypergeometric equation of the form

$$y_1(t) = \sum_{n=0}^{\infty} a_n t^n, \qquad y_2(t) = t^{1-\gamma} \sum_{n=0}^{\infty} b_n t^n.$$

2.9 The method of Laplace transforms

In this section we describe a very different and extremely clever way of solving the initial-value problem

$$a\frac{d^2y}{dt^2} + b\frac{dy}{dt} + cy = f(t); \qquad y(0) = y_0, \quad y'(0) = y_0' \tag{1}$$

where a, b and c are constants. This method, which is known as the method of Laplace transforms, is especially useful in two cases which arise quite often in applications. The first case is when $f(t)$ is a discontinuous function of time. The second case is when $f(t)$ is zero except for a very short time interval in which it is very large.

To put the method of Laplace transforms into proper perspective, we consider the following hypothetical situation. Suppose that we want to multiply the numbers 3.163 and 16.38 together, but that we have forgotten completely how to multiply. We only remember how to add. Being good mathematicians, we ask ourselves the following question.

Question: Is it possible to reduce the problem of multiplying the two numbers 3.163 and 16.38 together to the simpler problem of adding two numbers together?

The answer to this question, of course, is yes, and is obtained as follows. First, we consult our logarithm tables and find that $\ln 3.163 = 1.15152094$, and $\ln 16.38 = 2.79606108$. Then, we add these two numbers together to yield 3.94758202. Finally, we consult our anti-logarithm tables and find that $3.94758202 = \ln 51.80994$. Hence, we conclude that $3.163 \times 16.38 = 51.80994$.

The key point in this analysis is that the operation of multiplication is replaced by the simpler operation of addition when we work with the logarithms of numbers, rather than with the numbers themselves. We represent this schematically in Table 1. In the method to be discussed below, the unknown function $y(t)$ will be replaced by a new function $Y(s)$, known as the Laplace transform of $y(t)$. This association will have the property that $y'(t)$ will be replaced by $sY(s) - y(0)$. Thus, the operation of differentiation with respect to t will be replaced, essentially, by the operation of multiplication with respect to s. In this manner, we will replace the initial-value problem (1) by an algebraic equation which can be solved explicitly for $Y(s)$. Once we know $Y(s)$, we can consult our "anti-Laplace transform" tables and recover $y(t)$.

Table 1

a	$\longrightarrow$	$\ln a$
b	$\longrightarrow$	$\ln b$
$a \cdot b$	$\longrightarrow$	$\ln a + \ln b$

We begin with the definition of the Laplace transform.

Definition. Let $f(t)$ be defined for $0 \leqslant t < \infty$. The Laplace transform of $f(t)$, which is denoted by $F(s)$, or $\mathcal{L}\{f(t)\}$, is given by the formula

$$F(s) = \mathcal{L}\{f(t)\} = \int_0^\infty e^{-st} f(t)\,dt \tag{2}$$

where

$$\int_0^\infty e^{-st} f(t)\,dt = \lim_{A \to \infty} \int_0^A e^{-st} f(t)\,dt.$$

Example 1. Compute the Laplace transform of the function $f(t) = 1$.

Solution. From (2),

$$\mathcal{L}\{f(t)\} = \lim_{A \to \infty} \int_0^A e^{-st}\,dt = \lim_{A \to \infty} \frac{1 - e^{-sA}}{s}$$

$$= \begin{cases} \dfrac{1}{s}, & s > 0 \\ \infty, & s \leqslant 0 \end{cases}.$$

Example 2. Compute the Laplace transform of the function e^{at}.
Solution. From (2),

$$\mathcal{L}\{e^{at}\} = \lim_{A\to\infty}\int_0^A e^{-st}e^{at}\,dt = \lim_{A\to\infty}\frac{e^{(a-s)A}-1}{a-s}$$

$$= \begin{cases} \dfrac{1}{s-a}, & s>a \\ \infty, & s\leqslant a \end{cases}.$$

Example 3. Compute the Laplace transform of the functions $\cos\omega t$ and $\sin\omega t$.
Solution. From (2),

$$\mathcal{L}\{\cos\omega t\} = \int_0^\infty e^{-st}\cos\omega t\,dt \quad\text{and}\quad \mathcal{L}\{\sin\omega t\} = \int_0^\infty e^{-st}\sin\omega t\,dt.$$

Now, observe that

$$\mathcal{L}\{\cos\omega t\} + i\mathcal{L}\{\sin\omega t\} = \int_0^\infty e^{-st}e^{i\omega t}\,dt = \lim_{A\to\infty}\int_0^A e^{(i\omega-s)t}\,dt$$

$$= \lim_{A\to\infty}\frac{e^{(i\omega-s)A}-1}{i\omega-s}$$

$$= \begin{cases} \dfrac{1}{s-i\omega} = \dfrac{s+i\omega}{s^2+\omega^2}, & s>0 \\ \text{undefined}, & s\leqslant 0 \end{cases}.$$

Equating real and imaginary parts in this equation gives

$$\mathcal{L}\{\cos\omega t\} = \frac{s}{s^2+\omega^2} \quad\text{and}\quad \mathcal{L}\{\sin\omega t\} = \frac{\omega}{s^2+\omega^2}, \qquad s>0.$$

Equation (2) associates with every function $f(t)$ a new function, which we call $F(s)$. As the notation $\mathcal{L}\{f(t)\}$ suggests, the Laplace transform is an operator acting on functions. It is also a linear operator, since

$$\mathcal{L}\{c_1f_1(t)+c_2f_2(t)\} = \int_0^\infty e^{-st}\left[c_1f_1(t)+c_2f_2(t)\right]dt$$

$$= c_1\int_0^\infty e^{-st}f_1(t)\,dt + c_2\int_0^\infty e^{-st}f_2(t)\,dt$$

$$= c_1\mathcal{L}\{f_1(t)\} + c_2\mathcal{L}\{f_2(t)\}.$$

It is to be noted, though, that whereas $f(t)$ is defined for $0\leqslant t<\infty$, its Laplace transform is usually defined in a different interval. For example, the Laplace transform of e^{2t} is only defined for $2<s<\infty$, and the Laplace transform of e^{8t} is only defined for $8<s<\infty$. This is because the integral (2) will only exist, in general, if s is sufficiently large.

One very serious difficulty with the definition (2) is that this integral may fail to exist for every value of s. This is the case, for example, if $f(t)=$

e^{t^2} (see Exercise 13). To guarantee that the Laplace transform of $f(t)$ exists at least in some interval $s > s_0$, we impose the following conditions on $f(t)$.

(i) The function $f(t)$ is piecewise continuous. This means that $f(t)$ has at most a finite number of discontinuities on any interval $0 \leqslant t \leqslant A$, and both the limit from the right and the limit from the left of f exist at every point of discontinuity. In other words, $f(t)$ has only a finite number of "jump discontinuities" in any finite interval. The graph of a typical piecewise continuous function $f(t)$ is described in Figure 1.

(ii) The function $f(t)$ is of exponential order, that is, there exist constants M and c such that

$$|f(t)| \leqslant Me^{ct}, \qquad 0 \leqslant t < \infty.$$

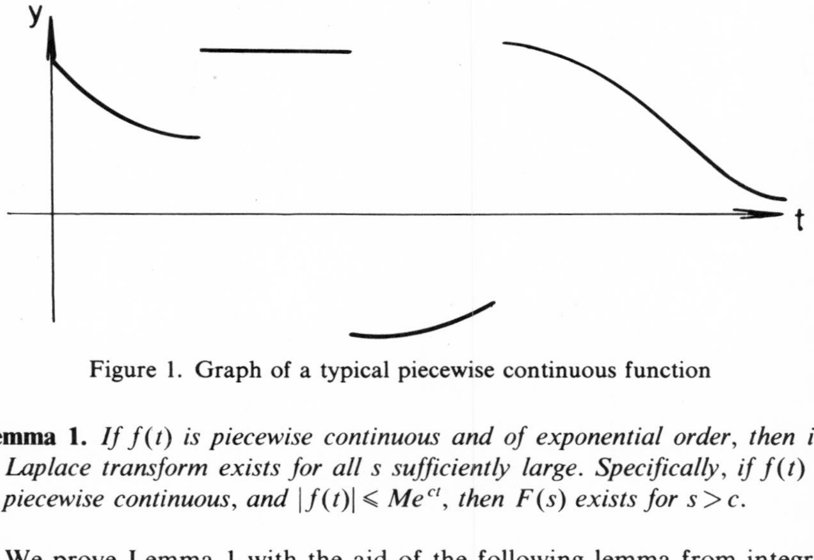

Figure 1. Graph of a typical piecewise continuous function

Lemma 1. *If $f(t)$ is piecewise continuous and of exponential order, then its Laplace transform exists for all s sufficiently large. Specifically, if $f(t)$ is piecewise continuous, and $|f(t)| \leqslant Me^{ct}$, then $F(s)$ exists for $s > c$.*

We prove Lemma 1 with the aid of the following lemma from integral calculus, which we quote without proof.

Lemma 2. *Let $g(t)$ be piecewise continuous. Then, the improper integral $\int_0^\infty g(t)\,dt$ exists if $\int_0^\infty |g(t)|\,dt$ exists. To prove that this latter integral exists, it suffices to show that there exists a constant K such that*

$$\int_0^A |g(t)|\,dt \leqslant K$$

for all A.

Remark. Notice the similarity of Lemma 2 with the theorem of infinite series (see Appendix B) which states that the infinite series $\sum a_n$ converges if $\sum |a_n|$ converges, and that $\sum |a_n|$ converges if there exists a constant K such that $|a_1|+\ldots+|a_n|\leqslant K$ for all n.

We are now in a position to prove Lemma 1.

PROOF OF LEMMA 1. Since $f(t)$ is piecewise continuous, the integral $\int_0^A e^{-st}f(t)\,dt$ exists for all A. To prove that this integral has a limit for all s sufficiently large, observe that

$$\int_0^A |e^{-st}f(t)|\,dt \leqslant M\int_0^A e^{-st}e^{ct}\,dt$$

$$=\frac{M}{c-s}\left[e^{(c-s)A}-1\right]\leqslant\frac{M}{s-c}$$

for $s>c$. Consequently, by Lemma 2, the Laplace transform of $f(t)$ exists for $s>c$. Thus, from here on, we tacitly assume that $|f(t)|\leqslant Me^{ct}$, and $s>c$. □

The real usefulness of the Laplace transform in solving differential equations lies in the fact that the Laplace transform of $f'(t)$ is very closely related to the Laplace transform of $f(t)$. This is the content of the following important lemma.

Lemma 3. *Let* $F(s)=\mathcal{L}\{f(t)\}$. *Then*

$$\mathcal{L}\{f'(t)\}=s\mathcal{L}\{f(t)\}-f(0)=sF(s)-f(0).$$

PROOF. The proof of Lemma 3 is very elementary; we just write down the formula for the Laplace transform of $f'(t)$ and integrate by parts. To wit,

$$\mathcal{L}\{f'(t)\}=\lim_{A\to\infty}\int_0^A e^{-st}f'(t)\,dt$$

$$=\lim_{A\to\infty}e^{-st}f(t)\Big|_0^A+\lim_{A\to\infty}s\int_0^A e^{-st}f(t)\,dt$$

$$=-f(0)+s\lim_{A\to\infty}\int_0^A e^{-st}f(t),dt$$

$$=-f(0)+sF(s).$$ □

Our next step is to relate the Laplace transform of $f''(t)$ to the Laplace transform of $f(t)$. This is the content of Lemma 4.

Lemma 4. *Let* $F(s)=\mathcal{L}\{f(t)\}$. *Then,*

$$\mathcal{L}\{f''(t)\}=s^2F(s)-sf(0)-f'(0).$$

PROOF. Using Lemma 3 twice, we see that

$$\begin{aligned}\mathcal{L}\{f''(t)\} &= s\mathcal{L}\{f'(t)\} - f'(0)\\ &= s[sF(s) - f(0)] - f'(0)\\ &= s^2F(s) - sf(0) - f'(0). \end{aligned}$$ □

We have now developed all the machinery necessary to reduce the problem of solving the initial-value problem

$$a\frac{d^2y}{dt^2} + b\frac{dy}{dt} + cy = f(t); \qquad y(0) = y_0, \quad y'(0) = y_0' \tag{3}$$

to that of solving an algebraic equation. Let $Y(s)$ and $F(s)$ be the Laplace transforms of $y(t)$ and $f(t)$ respectively. Taking Laplace transforms of both sides of the differential equation gives

$$\mathcal{L}\{ay''(t) + by'(t) + cy(t)\} = F(s).$$

By the linearity of the Laplace transform operator,

$$\mathcal{L}\{ay''(t) + by'(t) + cy(t)\} = a\mathcal{L}\{y''(t)\} + b\mathcal{L}\{y'(t)\} + c\mathcal{L}\{y(t)\},$$

and from Lemmas 3 and 4

$$\mathcal{L}\{y'(t)\} = sY(s) - y_0, \qquad \mathcal{L}\{y''(t)\} = s^2Y(s) - sy_0 - y_0'.$$

Hence,

$$a[s^2Y(s) - sy_0 - y_0'] + b[sY(s) - y_0] + cY(s) = F(s)$$

and this *algebraic equation* implies that

$$Y(s) = \frac{(as+b)y_0}{as^2+bs+c} + \frac{ay_0'}{as^2+bs+c} + \frac{F(s)}{as^2+bs+c}. \tag{4}$$

Equation (4) tells us the Laplace transform of the solution $y(t)$ of (3). To find $y(t)$, we must consult our anti, or inverse, Laplace transform tables. Now, just as $Y(s)$ is expressed explicitly in terms of $y(t)$; that is, $Y(s) = \int_0^\infty e^{-st}y(t)\,dt$, we can write down an explicit formula for $y(t)$. However, this formula, which is written symbolically as $y(t) = \mathcal{L}^{-1}\{Y(s)\}$, involves an integration with respect to a complex variable, and this is beyond the scope of this book. Therefore, instead of using this formula, we will derive several elegant properties of the Laplace transform operator in the next section. These properties will enable us to invert many Laplace transforms by inspection; that is, by recognizing "which functions they are the Laplace transform of".

Example 4. Solve the initial-value problem

$$\frac{d^2y}{dt^2} - 3\frac{dy}{dt} + 2y = e^{3t}; \qquad y(0) = 1, \quad y'(0) = 0.$$

Solution. Let $Y(s) = \mathcal{L}\{y(t)\}$. Taking Laplace transforms of both sides of the differential equation gives

$$s^2Y(s) - s - 3[sY(s) - 1] + 2Y(s) = \frac{1}{s-3}$$

$$\mathcal{L}\frac{d^2y}{dt} = s^2Y(s) -$$

and this implies that

$$Y(s) = \frac{1}{(s-3)(s^2-3s+2)} + \frac{s-3}{s^2-3s+2}$$

$$= \frac{1}{(s-1)(s-2)(s-3)} + \frac{s-3}{(s-1)(s-2)}. \tag{5}$$

To find $y(t)$, we expand each term on the right-hand side of (5) in partial fractions. Thus, we write

$$\frac{1}{(s-1)(s-2)(s-3)} = \frac{A}{s-1} + \frac{B}{s-2} + \frac{C}{s-3}.$$

This implies that

$$A(s-2)(s-3) + B(s-1)(s-3) + C(s-1)(s-2) = 1. \tag{6}$$

Setting $s=1$ in (6) gives $A = \frac{1}{2}$; setting $s=2$ gives $B = -1$; and setting $s=3$ gives $C = \frac{1}{2}$. Hence,

$$\frac{1}{(s-1)(s-2)(s-3)} = \frac{1}{2}\frac{1}{s-1} - \frac{1}{s-2} + \frac{1}{2}\frac{1}{s-3}.$$

Similarly, we write

$$\frac{s-3}{(s-1)(s-2)} = \frac{D}{s-1} + \frac{E}{s-2}$$

and this implies that

$$D(s-2) + E(s-1) = s-3. \tag{7}$$

Setting $s=1$ in (7) gives $D=2$, while setting $s=2$ gives $E=-1$. Hence,

$$Y(s) = \frac{1}{2}\frac{1}{s-1} - \frac{1}{s-2} + \frac{1}{2}\frac{1}{s-3} + \frac{2}{s-1} - \frac{1}{s-2}$$

$$= \frac{5}{2}\frac{1}{s-1} - \frac{2}{s-2} + \frac{1}{2}\frac{1}{s-3}.$$

Now, we recognize the first term as being the Laplace transform of $\frac{5}{2}e^t$. Similarly, we recognize the second and third terms as being the Laplace transforms of $-2e^{2t}$ and $\frac{1}{2}e^{3t}$, respectively. Therefore,

$$Y(s) = \mathcal{L}\{\tfrac{5}{2}e^t - 2e^{2t} + \tfrac{1}{2}e^{3t}\}$$

so that

$$y(t)=\tfrac{5}{2}e^{t}-2e^{2t}+\tfrac{1}{2}e^{3t}.$$

Remark. We have cheated a little bit in this problem because there are actually infinitely many functions whose Laplace transform is a given function. For example, the Laplace transform of the function

$$z(t)=\begin{cases}\tfrac{5}{2}e^{t}-2e^{2t}+\tfrac{1}{2}e^{3t}, & t\neq 1,2,\text{ and }3\\ 0, & t=1,2,3\end{cases}$$

is also $Y(s)$, since $z(t)$ differs from $y(t)$ at only three points.* However, there is only one *continuous* function $y(t)$ whose Laplace transform is a given function $Y(s)$, and it is in this sense that we write $y(t)=\mathcal{L}^{-1}\{Y(s)\}$.

We wish to emphasize that Example 4 is just by way of illustrating the method of Laplace transforms for solving initial-value problems. The best way of solving this particular initial-value problem is by the method of judicious guessing. However, even though it is longer to solve this particular initial-value problem by the method of Laplace transforms, there is still something "nice and satisfying" about this method. If we had done this problem by the method of judicious guessing, we would have first computed a particular solution $\psi(t)=\tfrac{1}{2}e^{3t}$. Then, we would have found two independent solutions e^{t} and e^{2t} of the homogeneous equation, and we would have written

$$y(t)=c_1e^{t}+c_2e^{2t}+\tfrac{1}{2}e^{3t}$$

as the general solution of the differential equation. Finally, we would have computed $c_1=\tfrac{5}{2}$ and $c_2=-2$ from the initial conditions. What is unsatisfying about this method is that we first had to find *all* the solutions of the differential equation before we could find the specific solution $y(t)$ which we were interested in. The method of Laplace transforms, on the other hand, enables us to find $y(t)$ directly, without first finding all solutions of the differential equation.

Exercises

Determine the Laplace transform of each of the following functions.

1. t

2. t^n

3. $e^{at}\cos bt$

4. $e^{at}\sin bt$

5. $\cos^2 at$

6. $\sin^2 at$

7. $\sin at\cos bt$

8. $t^2\sin t$

*If $f(t)=g(t)$ except at a finite number of points, then $\int_a^b f(t)\,dt=\int_a^b g(t)\,dt$.

9. Given that $\int_0^\infty e^{-x^2}dx = \sqrt{\pi}/2$, find $\mathcal{L}\{t^{-1/2}\}$. *Hint*: Make the change of variable $u = \sqrt{t}$ in (2).

Show that each of the following functions are of exponential order.

10. t^n **11.** $\sin at$ **12.** $e^{\sqrt{t}}$

13. Show that e^{t^2} does not possess a Laplace transform. *Hint*: Show that $e^{t^2-st} > e^t$ for $t > s+1$.

14. Suppose that $f(t)$ is of exponential order. Show that $F(s) = \mathcal{L}\{f(t)\}$ approaches 0 as $s \to \infty$.

Solve each of the following initial-value problems.

15. $y'' - 5y' + 4y = e^{2t}$; $y(0) = 1$, $y'(0) = -1$

16. $2y'' + y' - y = e^{3t}$; $y(0) = 2$, $y'(0) = 0$

Find the Laplace transform of the solution of each of the following initial-value problems.

17. $y'' + 2y' + y = e^{-t}$; $y(0) = 1$, $y'(0) = 3$

18. $y'' + y = t^2 \sin t$; $y(0) = y'(0) = 0$

19. $y'' + 3y' + 7y = \cos t$; $y(0) = 0$, $y'(0) = 2$

20. $y'' + y' + y = t^3$; $y(0) = 2$, $y'(0) = 0$

21. Prove that all solutions $y(t)$ of $ay'' + by' + cy = f(t)$ are of exponential order if $f(t)$ is of exponential order. *Hint*: Show that all solutions of the homogeneous equation are of exponential order. Obtain a particular solution using the method of variation of parameters, and show that it, too, is of exponential order.

22. Let $F(s) = \mathcal{L}\{f(t)\}$. Prove that

$$\mathcal{L}\left\{\frac{d^n f(t)}{dt^n}\right\} = s^n F(s) - s^{n-1} f(0) - \ldots - \frac{df^{(n-1)}(0)}{dt^{n-1}}.$$

Hint: Try induction.

23. Solve the initial-value problem

$$y''' - 6y'' + 11y' - 6y = e^{4t}; \qquad y(0) = y'(0) = y''(0) = 0$$

24. Solve the initial-value problem

$$y'' - 3y' + 2y = e^{-t}; \qquad y(t_0) = 1, \quad y'(t_0) = 0$$

by the method of Laplace transforms. *Hint*: Let $\phi(t) = y(t + t_0)$.

2.10 Some useful properties of Laplace transforms

In this section we derive several important properties of Laplace transforms. Using these properties, we will be able to compute the Laplace transform of most functions without performing tedious integrations, and to invert many Laplace transforms by inspection.

Property 1. If $\mathcal{L}\{f(t)\}=F(s)$, then

$$\mathcal{L}\{-tf(t)\}=\frac{d}{ds}F(s).$$

PROOF. By definition, $F(s)=\int_0^\infty e^{-st}f(t)\,dt$. Differentiating both sides of this equation with respect to s gives

$$\begin{aligned}\frac{d}{ds}F(s)&=\frac{d}{ds}\int_0^\infty e^{-st}f(t)\,dt\\&=\int_0^\infty \frac{\partial}{\partial s}(e^{-st})f(t)\,dt=\int_0^\infty -te^{-st}f(t)\,dt\\&=\mathcal{L}\{-tf(t)\}.\end{aligned}$$ □

Property 1 states that the Laplace transform of the function $-tf(t)$ is the derivative of the Laplace transform of $f(t)$. Thus, if we know the Laplace transform $F(s)$ of $f(t)$, then, we don't have to perform a tedious integration to find the Laplace transform of $tf(t)$; we need only differentiate $F(s)$ and multiply by -1.

Example 1. Compute the Laplace transform of te^t.
Solution. The Laplace transform of e^t is $1/(s-1)$. Hence, by Property 1, the Laplace transform of te^t is

$$\mathcal{L}\{te^t\}=-\frac{d}{ds}\frac{1}{s-1}=\frac{1}{(s-1)^2}.$$

Example 2. Compute the Laplace transform of t^{13}.
Solution. Using Property 1 thirteen times gives

$$\mathcal{L}\{t^{13}\}=(-1)^{13}\frac{d^{13}}{ds^{13}}\mathcal{L}\{1\}=(-1)^{13}\frac{d^{13}}{ds^{13}}\frac{1}{s}=\frac{(13)!}{s^{14}}.$$

The main usefulness of Property 1 is in inverting Laplace transforms, as the following examples illustrate.

Example 3. What function has Laplace transform $-1/(s-2)^2$?
Solution. Observe that

$$-\frac{1}{(s-2)^2}=\frac{d}{ds}\frac{1}{s-2} \quad \text{and} \quad \frac{1}{s-2}=\mathcal{L}\{e^{2t}\}.$$

Hence, by Property 1,

$$\mathcal{L}^{-1}\left\{-\frac{1}{(s-2)^2}\right\}=-te^{2t}.$$

Example 4. What function has Laplace transform $-4s/(s^2+4)^2$?
Solution. Observe that

$$-\frac{4s}{(s^2+4)^2}=\frac{d}{ds}\frac{2}{s^2+4} \quad \text{and} \quad \frac{2}{s^2+4}=\mathcal{L}\{\sin 2t\}.$$

Hence, by Property 1,

$$\mathcal{L}^{-1}\left\{-\frac{4s}{(s^2+4)^2}\right\}=-t\sin 2t.$$

Example 5. What function has Laplace transform $1/(s-4)^3$?
Solution. We recognize that

$$\frac{1}{(s-4)^3}=\frac{d^2}{ds^2}\frac{1}{2}\frac{1}{s-4}.$$

Hence, using Property 1 twice, we see that

$$\frac{1}{(s-4)^3}=\mathcal{L}\left\{\frac{1}{2}t^2e^{4t}\right\}.$$

Property 2. If $F(s)=\mathcal{L}\{f(t)\}$, then

$$\mathcal{L}\{e^{at}f(t)\}=F(s-a).$$

PROOF. By definition,

$$\mathcal{L}\{e^{at}f(t)\}=\int_0^\infty e^{-st}e^{at}f(t)\,dt=\int_0^\infty e^{(a-s)t}f(t)\,dt$$

$$=\int_0^\infty e^{-(s-a)t}f(t)\,dt\equiv F(s-a). \qquad \square$$

Property 2 states that the Laplace transform of $e^{at}f(t)$ evaluated at the point s equals the Laplace transform of $f(t)$ evaluated at the point $(s-a)$. Thus, if we know the Laplace transform $F(s)$ of $f(t)$, then we don't have to perform an integration to find the Laplace transform of $e^{at}f(t)$; we need only replace every s in $F(s)$ by $s-a$.

Example 6. Compute the Laplace transform of $e^{3t}\sin t$.
Solution. The Laplace transform of $\sin t$ is $1/(s^2+1)$. Therefore, to compute the Laplace transform of $e^{3t}\sin t$, we need only replace every s by $s-3$; that is,

$$\mathcal{L}\{e^{3t}\sin t\} = \frac{1}{(s-3)^2+1}.$$

The real usefulness of Property 2 is in inverting Laplace transforms, as the following examples illustrate.

Example 7. What function $g(t)$ has Laplace transform

$$G(s) = \frac{s-7}{25+(s-7)^2}?$$

Solution. Observe that

$$F(s) = \frac{s}{s^2+5^2} = \mathcal{L}\{\cos 5t\}$$

and that $G(s)$ is obtained from $F(s)$ by replacing every s by $s-7$. Hence, by Property 2,

$$\frac{s-7}{(s-7)^2+25} = \mathcal{L}\{e^{7t}\cos 5t\}.$$

Example 8. What function has Laplace transform $1/(s^2-4s+9)$?
Solution. One way of solving this problem is to expand $1/(s^2-4s+9)$ in partial fractions. A much better way is to complete the square of s^2-4s+9. Thus, we write

$$\frac{1}{s^2-4s+9} = \frac{1}{s^2-4s+4+(9-4)} = \frac{1}{(s-2)^2+5}.$$

Now,

$$\frac{1}{s^2+5} = \mathcal{L}\left\{\frac{1}{\sqrt{5}}\sin\sqrt{5}\,t\right\}.$$

Hence, by Property 2,

$$\frac{1}{s^2-4s+9} = \frac{1}{(s-2)^2+5} = \mathcal{L}\left\{\frac{1}{\sqrt{5}}e^{2t}\sin\sqrt{5}\,t\right\}.$$

Example 9. What function has Laplace transform $s/(s^2-4s+9)$?
Solution. Observe that

$$\frac{s}{s^2-4s+9} = \frac{s-2}{(s-2)^2+5} + \frac{2}{(s-2)^2+5}.$$

The function $s/(s^2+5)$ is the Laplace transform of $\cos\sqrt{5}\,t$. Therefore, by

Property 2,

$$\frac{s-2}{(s-2)^2+5} = \mathcal{L}\{e^{2t}\cos\sqrt{5}\,t\},$$

and

$$\frac{s}{s^2-4s+9} = \mathcal{L}\left\{e^{2t}\cos\sqrt{5}\,t + \frac{2}{\sqrt{5}}e^{2t}\sin\sqrt{5}\,t\right\}.$$

In the previous section we showed that the Laplace transform is a linear operator; that is

$$\mathcal{L}\{c_1 f_1(t) + c_2 f_2(t)\} = c_1\mathcal{L}\{f_1(t)\} + c_2\mathcal{L}\{f_2(t)\}.$$

Thus, if we know the Laplace transforms $F_1(s)$ and $F_2(s)$, of $f_1(t)$ and $f_2(t)$, then we don't have to perform any integrations to find the Laplace transform of a linear combination of $f_1(t)$ and $f_2(t)$; we need only take the same linear combination of $F_1(s)$ and $F_2(s)$. For example, two functions which appear quite often in the study of differential equations are the hyperbolic cosine and hyperbolic sine functions. These functions are defined by the equations

$$\cosh at = \frac{e^{at}+e^{-at}}{2}, \qquad \sinh at = \frac{e^{at}-e^{-at}}{2}.$$

Therefore, by the linearity of the Laplace transform,

$$\begin{aligned}\mathcal{L}\{\cosh at\} &= \frac{1}{2}\mathcal{L}\{e^{at}\} + \frac{1}{2}\mathcal{L}\{e^{-at}\}\\ &= \frac{1}{2}\left[\frac{1}{s-a} + \frac{1}{s+a}\right] = \frac{s}{s^2-a^2}\end{aligned}$$

and

$$\begin{aligned}\mathcal{L}\{\sinh at\} &= \frac{1}{2}\mathcal{L}\{e^{at}\} - \frac{1}{2}\mathcal{L}\{e^{-at}\}\\ &= \frac{1}{2}\left[\frac{1}{s-a} - \frac{1}{s+a}\right] = \frac{a}{s^2-a^2}.\end{aligned}$$

Exercises

Use Properties 1 and 2 to find the Laplace transform of each of the following functions.

1. t^n **2.** $t^n e^{at}$ **3.** $t\sin at$ **4.** $t^2\cos at$

5. $t^{5/2}$ (see Exercise 9, Section 2.9)

6. Let $F(s) = \mathcal{L}\{f(t)\}$, and suppose that $f(t)/t$ has a limit as t approaches zero. Prove that

$$\mathcal{L}\{f(t)/t\} = \int_s^\infty F(u)\,du. \qquad (*)$$

(The assumption that $f(t)/t$ has a limit as $t\to 0$ guarantees that the integral on the right-hand side of (*) exists.)

7. Use Equation (*) of Problem 6 to find the Laplace transform of each of the following functions:
(a) $\frac{\sin t}{t}$ (b) $\frac{\cos at - 1}{t}$ (c) $\frac{e^{at} - e^{bt}}{t}$

Find the inverse Laplace transform of each of the following functions. In several of these problems, it will be helpful to write the functions

$$p_1(s) = \frac{\alpha_1 s^3 + \beta_1 s^2 + \gamma_1 s + \delta_1}{(as^2 + bs + c)(ds^2 + es + f)} \quad \text{and} \quad p_2(s) = \frac{\alpha_1 s^2 + \beta_1 s + \gamma}{(as + b)(cs^2 + ds + e)}$$

in the simpler form

$$p_1(s) = \frac{As + B}{as^2 + bs + c} + \frac{Cs + D}{ds^2 + es + f} \quad \text{and} \quad p_2(s) = \frac{A}{as + b} + \frac{Cs + D}{cs^2 + ds + e}.$$

8. $\frac{s}{(s+a)^2 + b^2}$

9. $\frac{s^2 - 5}{s^3 + 4s^2 + 3s}$

10. $\frac{1}{s(s^2+4)}$

11. $\frac{s}{s^2 - 3s - 12}$

12. $\frac{1}{(s^2 + a^2)(s^2 + b^2)}$

13. $\frac{3s}{(s+1)^4}$

14. $\frac{1}{s(s+4)^2}$

15. $\frac{s}{(s+1)^2(s^2+1)}$

16. $\frac{1}{(s^2+1)^2}$

17. Let $F(s) = \mathcal{L}\{f(t)\}$. Show that

$$f(t) = -\frac{1}{t}\mathcal{L}^{-1}\{F'(s)\}.$$

Thus, if we know how to invert $F'(s)$, then we can also invert $F(s)$.

18. Use the result of Problem 17 to invert each of the following Laplace transforms
(a) $\ln\left(\frac{s+a}{s-a}\right)$ (b) $\arctan\frac{a}{s}$ (c) $\ln\left(1 - \frac{a^2}{s^2}\right)$

Solve each of the following initial-value problems by the method of Laplace transforms.

19. $y'' + y = \sin t; \quad y(0) = 1, y'(0) = 2$

20. $y'' + y = t\sin t; \quad y(0) = 1, y'(0) = 2$

21. $y'' - 2y' + y = te^t; \quad y(0) = 0, y'(0) = 0$

22. $y'' - 2y' + 7y = \sin t; \quad y(0) = 0, y'(0) = 0$

23. $y'' + y' + y = 1 + e^{-t}; \quad y(0) = 3, y'(0) = -5$

24. $y'' + y = \begin{cases} 2, & 0 \leqslant t \leqslant 3 \\ 3t - 7, & 3 < t < \infty \end{cases}; \quad y(0) = 0, y'(0) = 0$

2.11 Differential equations with discontinuous right-hand sides

In many applications, the right-hand side of the differential equation $ay''+by'+cy=f(t)$ has a jump discontinuity at one or more points. For example, a particle may be moving under the influence of a force $f_1(t)$, and suddenly, at time t_1, an additional force $f_2(t)$ is applied to the particle. Such equations are often quite tedious and cumbersome to solve, using the methods developed in Sections 2.4 and 2.5. In this section we show how to handle such problems by the method of Laplace transforms. We begin by computing the Laplace transform of several simple discontinuous functions.

The simplest example of a function with a single jump discontinuity is the function

$$H_c(t)=\begin{cases}0, & 0\leqslant t<c\\ 1, & t\geqslant c\end{cases}.$$

This function, whose graph is given in Figure 1, is often called the unit step function, or the Heaviside function. Its Laplace transform is

$$\begin{aligned}\mathcal{L}\{H_c(t)\}&=\int_0^\infty e^{-st}H_c(t)\,dt=\int_c^\infty e^{-st}\,dt\\ &=\lim_{A\to\infty}\int_c^A e^{-st}\,dt=\lim_{A\to\infty}\frac{e^{-cs}-e^{-sA}}{s}\\ &=\frac{e^{-cs}}{s},\qquad s>0.\end{aligned}$$

Next, let f be any function defined on the interval $0\leqslant t<\infty$, and let g be the function obtained from f by moving the graph of f over c units to the right, as shown in Figure 2. More precisely, $g(t)=0$ for $0\leqslant t<c$, and $g(t)=f(t-c)$ for $t\geqslant c$. For example, if $c=2$ then the value of g at $t=7$ is the value of f at $t=5$. A convenient analytical expression for $g(t)$ is

$$g(t)=H_c(t)f(t-c).$$

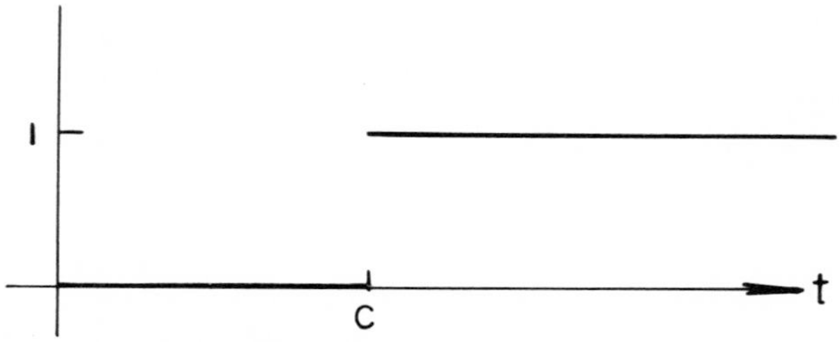

Figure 1. Graph of $H_c(t)$

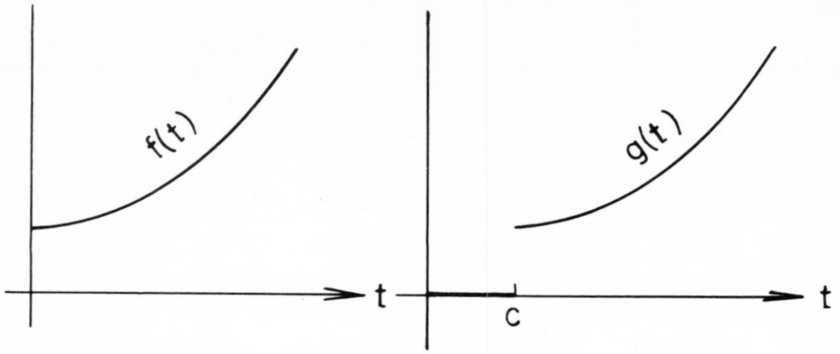

Figure 2

The factor $H_c(t)$ makes g zero for $0 \leqslant t < c$, and replacing the argument t of f by $t-c$ moves f over c units to the right. Since $g(t)$ is obtained in a simple manner from $f(t)$, we would expect that its Laplace transform can also be obtained in a simple manner from the Laplace transform of $f(t)$. This is indeed the case, as we now show.

Property 3. Let $F(s) = \mathcal{L}\{f(t)\}$. Then,

$$\mathcal{L}\{H_c(t)f(t-c)\} = e^{-cs}F(s).$$

PROOF. By definition,

$$\begin{aligned}\mathcal{L}\{H_c(t)f(t-c)\} &= \int_0^\infty e^{-st}H_c(t)f(t-c)\,dt\\ &= \int_c^\infty e^{-st}f(t-c)\,dt.\end{aligned}$$

This integral suggests the substitution

$$\xi = t-c.$$

Then,

$$\begin{aligned}\int_c^\infty e^{-st}f(t-c)\,dt &= \int_0^\infty e^{-s(\xi+c)}f(\xi)\,d\xi\\ &= e^{-cs}\int_0^\infty e^{-s\xi}f(\xi)\,d\xi\\ &= e^{-cs}F(s).\end{aligned}$$

Hence, $\mathcal{L}\{H_c(t)f(t-c)\} = e^{-cs}\mathcal{L}\{f(t)\}$. □

Example 1. What function has Laplace transform e^{-s}/s^2?
Solution. We know that $1/s^2$ is the Laplace transform of the function t.

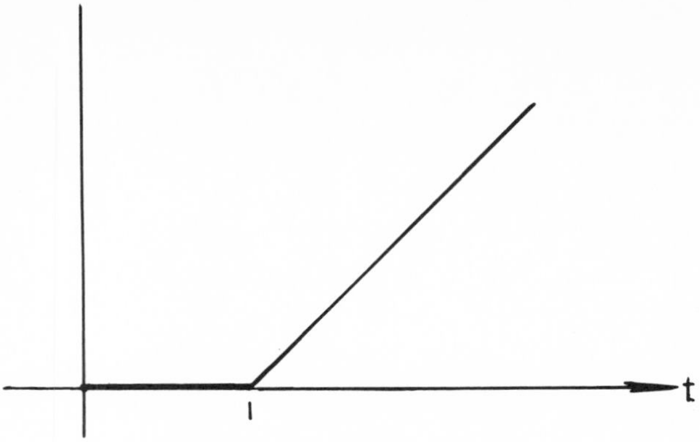

Figure 3. Graph of $H_1(t)\,(t-1)$

Hence, by Property 3

$$\frac{e^{-s}}{s^2}=\mathcal{L}\{H_1(t)(t-1)\}.$$

The graph of $H_1(t)(t-1)$ is given in Figure 3.

Example 2. What function has Laplace transform $e^{-3s}/(s^2-2s-3)$?
Solution. Observe that

$$\frac{1}{s^2-2s-3}=\frac{1}{s^2-2s+1-4}=\frac{1}{(s-1)^2-2^2}.$$

Since $1/(s^2-2^2)=\mathcal{L}\{\frac{1}{2}\sinh 2t\}$, we conclude from Property 2 that

$$\frac{1}{(s-1)^2-2^2}=\mathcal{L}\left\{\frac{1}{2}e^t\sinh 2t\right\}.$$

Consequently, from Property 3,

$$\frac{e^{-3s}}{s^2-2s-3}=\mathcal{L}\left\{\frac{1}{2}H_3(t)e^{t-3}\sinh 2(t-3)\right\}.$$

Example 3. Let $f(t)$ be the function which is t for $0\leqslant t<1$, and 0 for $t\geqslant 1$. Find the Laplace transform of f without performing any integrations.
Solution. Observe that $f(t)$ can be written in the form

$$f(t)=t\left[H_0(t)-H_1(t)\right]=t-tH_1(t).$$

Hence, from Property 1,

$$\begin{aligned}\mathcal{L}\{f(t)\}&=\mathcal{L}\{t\}-\mathcal{L}\{tH_1(t)\}\\&=\frac{1}{s^2}+\frac{d}{ds}\frac{e^{-s}}{s}=\frac{1}{s^2}-\frac{e^{-s}}{s}-\frac{e^{-s}}{s^2}.\end{aligned}$$

Example 4. Solve the initial-value problem

$$\frac{d^2y}{dt^2}-3\frac{dy}{dt}+2y=f(t)=\begin{cases}1, & 0\leqslant t<1; \quad 0, \; 1\leqslant t<2;\\ 1, & 2\leqslant t<3; \quad 0, \; 3\leqslant t<4;\\ 1, & 4\leqslant t<5; \quad 0, \; 5\leqslant t<\infty.\end{cases}\qquad y(0)=0,\ \ y'(0)=0$$

Solution. Let $Y(s)=\mathcal{L}\{y(t)\}$ and $F(s)=\mathcal{L}\{f(t)\}$. Taking Laplace transforms of both sides of the differential equation gives $(s^2-3s+2)Y(s)=F(s)$, so that

$$Y(s)=\frac{F(s)}{s^2-3s+2}=\frac{F(s)}{(s-1)(s-2)}.$$

One way of computing $F(s)$ is to write $f(t)$ in the form

$$f(t)=\left[H_0(t)-H_1(t)\right]+\left[H_2(t)-H_3(t)\right]+\left[H_4(t)-H_5(t)\right].$$

Hence, by the linearity of the Laplace transform

$$F(s)=\frac{1}{s}-\frac{e^{-s}}{s}+\frac{e^{-2s}}{s}-\frac{e^{-3s}}{s}+\frac{e^{-4s}}{s}-\frac{e^{-5s}}{s}.$$

A second way of computing $F(s)$ is to evaluate the integral

$$\int_0^\infty e^{-st}f(t)\,dt=\int_0^1 e^{-st}\,dt+\int_2^3 e^{-st}\,dt+\int_4^5 e^{-st}\,dt$$
$$=\frac{1-e^{-s}}{s}+\frac{e^{-2s}-e^{-3s}}{s}+\frac{e^{-4s}-e^{-5s}}{s}.$$

Consequently,

$$Y(s)=\frac{1-e^{-s}+e^{-2s}-e^{-3s}+e^{-4s}-e^{-5s}}{s(s-1)(s-2)}.$$

Our next step is to expand $1/s(s-1)(s-2)$ in partial fractions; i.e., we write

$$\frac{1}{s(s-1)(s-2)}=\frac{A}{s}+\frac{B}{s-1}+\frac{C}{s-2}.$$

This implies that

$$A(s-1)(s-2)+Bs(s-2)+Cs(s-1)=1. \tag{1}$$

Setting $s=0$ in (1) gives $A=\frac{1}{2}$; setting $s=1$ gives $B=-1$; and setting $s=2$ gives $C=\frac{1}{2}$. Thus,

$$\frac{1}{s(s-1)(s-2)}=\frac{1}{2}\frac{1}{s}-\frac{1}{s-1}+\frac{1}{2}\frac{1}{s-2}$$
$$=\mathcal{L}\left\{\frac{1}{2}-e^{t}+\frac{1}{2}e^{2t}\right\}.$$

Consequently, from Property 3,

$$\begin{aligned} y(t) = &\left[\tfrac{1}{2} - e^{t} + \tfrac{1}{2}e^{2t}\right] - H_1(t)\left[\tfrac{1}{2} - e^{(t-1)} + \tfrac{1}{2}e^{2(t-1)}\right] \\ &+ H_2(t)\left[\tfrac{1}{2} - e^{(t-2)} + \tfrac{1}{2}e^{2(t-2)}\right] - H_3(t)\left[\tfrac{1}{2} - e^{(t-3)} + \tfrac{1}{2}e^{2(t-3)}\right] \\ &+ H_4(t)\left[\tfrac{1}{2} - e^{(t-4)} + \tfrac{1}{2}e^{2(t-4)}\right] - H_5(t)\left[\tfrac{1}{2} - e^{(t-5)} + \tfrac{1}{2}e^{2(t-5)}\right]. \end{aligned}$$

Remark. It is easily verified that the function

$$\tfrac{1}{2} - e^{(t-n)} + \tfrac{1}{2}e^{2(t-n)}$$

and its derivative are both zero at $t = n$. Hence, both $y(t)$ and $y'(t)$ are continuous functions of time, even though $f(t)$ is discontinuous at $t = 1, 2, 3, 4,$ and 5. More generally, both the solution $y(t)$ of the initial-value problem

$$a\frac{d^2y}{dt^2} + b\frac{dy}{dt} + cy = f(t); \qquad y(t_0) = y_0, \quad y'(t_0) = y_0'$$

and its derivative $y'(t)$ are always continuous functions of time, if $f(t)$ is piecewise continuous. We will indicate the proof of this result in Section 2.12.

Exercises

Find the solution of each of the following initial-value problems.

1. $y'' + 2y' + y = 2(t-3)H_3(t); \quad y(0) = 2, y'(0) = 1$

2. $y'' + y' + y = H_\pi(t) - H_{2\pi}(t); \quad y(0) = 1, y'(0) = 0$

3. $y'' + 4y = \begin{cases} 1, & 0 \leqslant t < 4 \\ 0, & t > 4 \end{cases}; \quad y(0) = 3, y'(0) = -2$

4. $y'' + y = \begin{cases} \sin t, & 0 \leqslant t < \pi \\ \cos t, & \pi \leqslant t < \infty \end{cases}; \quad y(0) = 1, y'(0) = 0$

5. $y'' + y = \begin{cases} \cos t, & 0 \leqslant t < \pi/2 \\ 0, & \pi/2 \leqslant t < \infty \end{cases}; \quad y(0) = 3, y'(0) = -1$

6. $y'' + 2y' + y = \begin{cases} \sin 2t, & 0 \leqslant t < \pi/2 \\ 0, & \pi/2 \leqslant t < \infty \end{cases}; \quad y(0) = 1, y'(0) = 0$

7. $y'' + y' + 7y = \begin{cases} t, & 0 \leqslant t < 2 \\ 0, & 2 \leqslant t < \infty \end{cases}; \quad y(0) = 0, y'(0) = 0$

8. $y'' + y = \begin{cases} t^2, & 0 \leqslant t < 1 \\ 0, & 1 \leqslant t < \infty \end{cases}; \quad y(0) = 0, y'(0) = 0$

9. $y'' - 2y' + y = \begin{cases} 0, & 0 \leqslant t < 1 \\ t, & 1 \leqslant t < 2 \\ 0, & 2 \leqslant t < \infty \end{cases}; \quad y(0) = 0, y'(0) = 1$

10. Find the Laplace transform of $|\sin t|$. *Hint*: Observe that

$$|\sin t| = \sin t + 2\sum_{n=1}^{\infty} H_{n\pi}(t)\sin(t-n\pi).$$

11. Solve the initial-value problem of Example 4 by the method of judicious guessing. *Hint*: Find the general solution of the differential equation in each of the intervals $0<t<1$, $1<t<2$, $2<t<3$, $3<t<4$, $4<t<5$, $5<t<\infty$, and choose the arbitrary constants so that $y(t)$ and $y'(t)$ are continuous at the points $t=1, 2, 3, 4$, and 5.

2.12 The Dirac delta function

In many physical and biological applications we are often confronted with an initial-value problem

$$a\frac{d^2y}{dt^2}+b\frac{dy}{dt}+cy=f(t); \qquad y(0)=y_0,\quad y'(0)=y_0' \tag{1}$$

where we do not know $f(t)$ explicitly. Such problems usually arise when we are dealing with phenomena of an impulsive nature. In these situations, the only information we have about $f(t)$ is that it is identically zero except for a very short time interval $t_0 \leqslant t \leqslant t_1$, and that its integral over this time interval is a given number $I_0 \neq 0$. If I_0 is not very small, then $f(t)$ will be quite large in the interval $t_0 \leqslant t \leqslant t_1$. Such functions are called impulsive functions, and the graph of a typical $f(t)$ is given in Figure 1.

In the early 1930's the Nobel Prize winning physicist P. A. M. Dirac developed a very controversial method for dealing with impulsive functions. His method is based on the following argument. Let t_1 get closer and closer to t_0. Then the function $f(t)/I_0$ approaches the function which is 0 for $t \neq t_0$, and ∞ for $t=t_0$, and whose integral over any interval containing t_0 is 1. We will denote this function, which is known as the Dirac delta function, by $\delta(t-t_0)$. Of course, $\delta(t-t_0)$ is not an ordinary function. However, says Dirac, let us formally operate with $\delta(t-t_0)$ as if it really were an ordinary

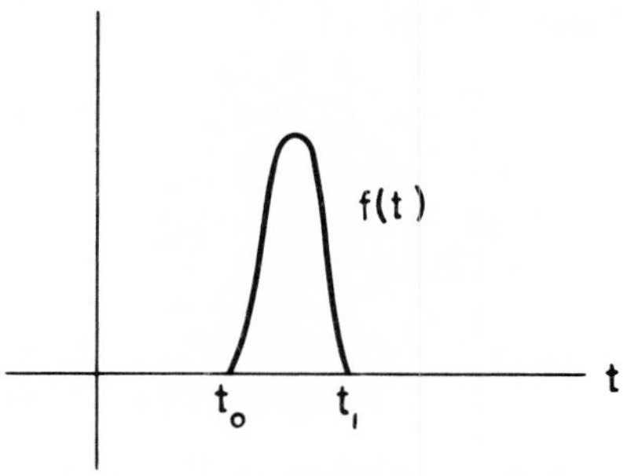

Figure 1. The graph of a typical impulsive function $f(t)$

function. Then, if we set $f(t) = I_0\delta(t-t_0)$ in (1) and impose the condition

$$\int_a^b g(t)\delta(t-t_0)\,dt = \begin{cases} g(t_0) & \text{if } a \leqslant t_0 \leqslant b \\ 0 & \text{otherwise} \end{cases} \tag{2}$$

for any continuous function $g(t)$, we will always obtain the correct solution $y(t)$.

Remark. Equation (2) is certainly a very reasonable condition to impose on $\delta(t-t_0)$. To see this, suppose that $f(t)$ is an impulsive function which is positive for $t_0 < t < t_1$, zero otherwise, and whose integral over the interval $[t_0, t_1]$ is 1. For any continuous function $g(t)$,

$$\left[\min_{t_0 \leqslant t \leqslant t_1} g(t)\right] f(t) \leqslant g(t)f(t) \leqslant \left[\max_{t_0 \leqslant t \leqslant t_1} g(t)\right] f(t).$$

Consequently,

$$\int_{t_0}^{t_1} \left[\min_{t_0 \leqslant t \leqslant t_1} g(t)\right] f(t)\,dt \leqslant \int_{t_0}^{t_1} g(t)f(t)\,dt \leqslant \int_{t_0}^{t_1} \left[\max_{t_0 \leqslant t \leqslant t_1} g(t)\right] f(t)\,dt,$$

or

$$\min_{t_0 \leqslant t \leqslant t_1} g(t) \leqslant \int_{t_0}^{t_1} g(t)f(t)\,dt \leqslant \max_{t_0 \leqslant t \leqslant t_1} g(t).$$

Thus, as $t_1 \to t_0$, $\int_{t_0}^{t_1} g(t)f(t)\,dt \to g(t_0)$.

Now, most mathematicians, of course, usually ridiculed this method. "How can you make believe that $\delta(t-t_0)$ is an ordinary function if it is obviously not," they asked. However, they never laughed too loud since Dirac and his followers always obtained the right answer. In the late 1940's, in one of the great success stories of mathematics, the French mathematician Laurent Schwartz succeeded in placing the delta function on a firm mathematical foundation. He accomplished this by enlarging the class of all functions so as to include the delta function. In this section we will first present a physical justification of the method of Dirac. Then we will illustrate how to solve the initial-value problem (1) by the method of Laplace transforms. Finally, we will indicate very briefly the "germ" of Laurent Schwartz's brilliant idea.

Physical justification of the method of Dirac. Newton's second law of motion is usually written in the form

$$\frac{d}{dt} mv(t) = f(t) \tag{3}$$

where m is the mass of the particle, v is its velocity, and $f(t)$ is the total force acting on the particle. The quantity mv is called the momentum of the particle. Integrating Equation (3) between t_0 and t_1 gives

$$mv(t_1) - mv(t_0) = \int_{t_0}^{t_1} f(t)\,dt.$$

This equation says that the change in momentum of the particle from time t_0 to time t_1 equals $\int_{t_0}^{t_1} f(t)\,dt$. Thus, the physically important quantity is the integral of the force, which is known as the impulse imparted by the force, rather than the force itself. Now, we may assume that $a>0$ in Equation (1), for otherwise we can multiply both sides of the equation by -1 to obtain $a>0$. In this case (see Section 2.6) we can view $y(t)$, for $t \leqslant t_0$, as the position at time t of a particle of mass a moving under the influence of the force $-b(dy/dt) - cy$. At time t_0 a force $f(t)$ is applied to the particle, and this force acts over an extremely short time interval $t_0 \leqslant t \leqslant t_1$. Since the time interval is extremely small, we may assume that the position of the particle does not change while the force $f(t)$ acts. Thus the sum result of the impulsive force $f(t)$ is that the velocity of the particle jumps by an amount I_0/a at time t_0. In other words, $y(t)$ satisfies the initial-value problem

$$a\frac{d^2y}{dt^2} + b\frac{dy}{dt} + cy = 0; \qquad y(0) = y_0, \quad y'(0) = y_0'$$

for $0 \leqslant t < t_0$, and

$$a\frac{d^2y}{dt^2} + b\frac{dy}{dt} + cy = 0; \qquad y(t_0) = z_0, \quad y'(t_0) = z_0' + \frac{I_0}{a} \tag{4}$$

for $t \geqslant t_0$, where z_0 and z_0' are the position and velocity of the particle just before the impulsive force acts. It is clear, therefore, that any method which correctly takes into account the momentum I_0 transferred to the particle at time t_0 by the impulsive force $f(t)$ must yield the correct answer. It is also clear that we always keep track of the momentum I_0 transferred to the particle by $f(t)$ if we replace $f(t)$ by $I_0\delta(t-t_0)$ and obey Equation (2). Hence the method of Dirac will always yield the correct answer.

Remark. We can now understand why any solution $y(t)$ of the differential equation

$$a\frac{d^2y}{dt^2} + b\frac{dy}{dt} + cy = f(t), \qquad f(t) \text{ a piecewise continuous function,}$$

is a continuous function of time even though $f(t)$ is discontinuous. To wit, since the integral of a piecewise continuous function is continuous, we see that $y'(t)$, must vary continuously with time. Consequently, $y(t)$ must also vary continuously with time.

Solution of Equation (1) *by the method of Laplace transforms*. In order to solve the initial-value problem (1) by the method of Laplace transforms, we need only know the Laplace transform of $\delta(t-t_0)$. This is obtained directly from the definition of the Laplace transform and Equation (2), for

$$\mathcal{L}\{\delta(t-t_0)\} \equiv \int_0^\infty e^{-st}\delta(t-t_0)\,dt = e^{-st_0} \qquad (\text{for } t_0 \geqslant 0).$$

Example 1. Find the solution of the initial-value problem

$$\frac{d^2y}{dt^2} - 4\frac{dy}{dt} + 4y = 3\delta(t-1) + \delta(t-2); \qquad y(0)=1, \quad y'(0)=1.$$

Solution. Let $Y(s)=\mathcal{L}\{y(t)\}$. Taking Laplace transforms of both sides of the differential equation gives

$$s^2Y - s - 1 - 4(sY-1) + 4Y = 3e^{-s} + e^{-2s}$$

or

$$(s^2-4s+4)Y(s) = s - 3 + 3e^{-s} + e^{-2s}.$$

Consequently,

$$Y(s) = \frac{s-3}{(s-2)^2} + \frac{3e^{-s}}{(s-2)^2} + \frac{e^{-2s}}{(s-2)^2}.$$

Now, $1/(s-2)^2 = \mathcal{L}\{te^{2t}\}$. Hence,

$$\frac{3e^{-s}}{(s-2)^2} + \frac{e^{-2s}}{(s-2)^2} = \mathcal{L}\{3H_1(t)(t-1)e^{2(t-1)} + H_2(t)(t-2)e^{2(t-2)}\}.$$

To invert the first term of $Y(s)$, observe that

$$\frac{s-3}{(s-2)^2} = \frac{s-2}{(s-2)^2} - \frac{1}{(s-2)^2} = \mathcal{L}\{e^{2t}\} - \mathcal{L}\{te^{2t}\}.$$

Thus, $y(t) = (1-t)e^{2t} + 3H_1(t)(t-1)e^{2(t-1)} + H_2(t)(t-2)e^{2(t-2)}$.

It is instructive to do this problem the long way, that is, to find $y(t)$ separately in each of the intervals $0 \leqslant t < 1$, $1 \leqslant t < 2$ and $2 \leqslant t < \infty$. For $0 \leqslant t < 1$, $y(t)$ satisfies the initial-value problem

$$\frac{d^2y}{dt^2} - 4\frac{dy}{dt} + 4y = 0; \qquad y(0)=1, \quad y'(0)=1.$$

The characteristic equation of this differential equation is $r^2-4r+4=0$, whose roots are $r_1=r_2=2$. Hence, any solution $y(t)$ must be of the form $y(t)=(a_1+a_2t)e^{2t}$. The constants a_1 and a_2 are determined from the initial conditions

$$1 = y(0) = a_1 \quad \text{and} \quad 1 = y'(0) = 2a_1 + a_2.$$

Hence, $a_1 = 1$, $a_2 = -1$ and $y(t) = (1-t)e^{2t}$ for $0 \leqslant t < 1$. Now $y(1)=0$ and $y'(1) = -e^2$. At time $t=1$ the derivative of $y(t)$ is suddenly increased by 3.

Consequently, for $1 \leqslant t < 2$, $y(t)$ satisfies the initial-value problem

$$\frac{d^2y}{dt^2} - 4\frac{dy}{dt} + 4y = 0; \qquad y(1) = 0, \quad y'(1) = 3 - e^2.$$

Since the initial conditions are given at $t = 1$, we write this solution in the form $y(t) = [b_1 + b_2(t-1)]e^{2(t-1)}$ (see Exercise 1). The constants b_1 and b_2 are determined from the initial conditions

$$0 = y(1) = b_1 \quad \text{and} \quad 3 - e^2 = y'(1) = 2b_1 + b_2.$$

Thus, $b_1 = 0$, $b_2 = 3 - e^2$ and $y(t) = (3 - e^2)(t-1)e^{2(t-1)}$, $1 \leqslant t < 2$. Now, $y(2) = (3 - e^2)e^2$ and $y'(2) = 3(3 - e^2)e^2$. At time $t = 2$ the derivative of $y(t)$ is suddenly increased by 1. Consequently, for $2 \leqslant t < \infty$, $y(t)$ satisfies the initial-value problem

$$\frac{d^2y}{dt^2} - 4\frac{dy}{dt} + 4y = 0; \qquad y(2) = e^2(3 - e^2), \quad y'(2) = 1 + 3e^2(3 - e^2).$$

Hence $y(t) = [c_1 + c_2(t-2)]e^{2(t-2)}$. The constants c_1 and c_2 are determined from the equations

$$e^2(3 - e^2) = c_1 \quad \text{and} \quad 1 + 3e^2(3 - e^2) = 2c_1 + c_2.$$

Thus,

$$c_1 = e^2(3 - e^2), \qquad c_2 = 1 + 3e^2(3 - e^2) - 2e^2(3 - e^2) = 1 + e^2(3 - e^2)$$

and $y(t) = [e^2(3 - e^2) + (1 + e^2(3 - e^2))(t-2)]e^{2(t-2)}$, $t \geqslant 2$. The reader should verify that this expression agrees with the expression obtained for $y(t)$ by the method of Laplace transforms.

Example 2. A particle of mass 1 is attached to a spring dashpot mechanism. The stiffness constant of the spring is 1 lb/ft and the drag force exerted by the dashpot mechanism on the particle is twice its velocity. At time $t = 0$, when the particle is at rest, an external force e^{-t} is applied to the system. At time $t = 1$, an additional force $f(t)$ of very short duration is applied to the particle. This force imparts an impulse of 3 lb·s to the particle. Find the position of the particle at any time t greater than 1.

Solution. Let $y(t)$ be the distance of the particle from its equilibrium position. Then, $y(t)$ satisfies the initial-value problem

$$\frac{d^2y}{dt^2} + 2\frac{dy}{dt} + y = e^{-t} + 3\delta(t-1); \qquad y(0) = 0, \quad y'(0) = 0.$$

Let $Y(s) = \mathcal{L}\{y(t)\}$. Taking Laplace transforms of both sides of the differential equation gives

$$(s^2 + 2s + 1)Y(s) = \frac{1}{s+1} + 3e^{-s}, \quad \text{or} \quad Y(s) = \frac{1}{(s+1)^3} + \frac{3e^{-s}}{(s+1)^2}.$$

Since

$$\frac{1}{(s+1)^3}=\mathcal{L}\left\{\frac{t^2e^{-t}}{2}\right\} \quad \text{and} \quad \frac{3e^{-s}}{(s+1)^2}=3\mathcal{L}\left\{H_1(t)(t-1)e^{-(t-1)}\right\}$$

we see that

$$y(t)=\frac{t^2e^{-t}}{2}+3H_1(t)(t-1)e^{-(t-1)}.$$

Consequently, $y(t)=\frac{1}{2}t^2e^{-t}+3(t-1)e^{-(t-1)}$ for $t>1$.

We conclude this section with a very brief description of Laurent Schwartz's method for placing the delta function on a rigorous mathematical foundation. The main step in his method is to rethink our notion of "function." In Calculus, we are taught to recognize a function by its value at each time t. A much more subtle (and much more difficult) way of recognizing a function is by what it does to other functions. More precisely, let f be a piecewise continuous function defined for $-\infty<t<\infty$. To each function ϕ which is infinitely often differentiable and which vanishes for $|t|$ sufficiently large, we assign a number $K[\phi]$ according to the formula

$$K[\phi]=\int_{-\infty}^{\infty}\phi(t)f(t)\,dt. \tag{5}$$

As the notation suggests, K is an operator acting on functions. However, it differs from the operators introduced previously in that it associates a number, rather than a function, with ϕ. For this reason, we say that $K[\phi]$ is a functional, rather than a function. Now, observe that the association $\phi\to K[\phi]$ is a linear association, since

$$\begin{aligned}K[c_1\phi_1+c_2\phi_2]&=\int_{-\infty}^{\infty}(c_1\phi_1+c_2\phi_2)(t)f(t)\,dt\\&=c_1\int_{-\infty}^{\infty}\phi_1(t)f(t)\,dt+c_2\int_{-\infty}^{\infty}\phi_2(t)f(t)\,dt\\&=c_1K[\phi_1]+c_2K[\phi_2].\end{aligned}$$

Hence every piecewise continuous function defines, through (5), a linear functional on the space of all infinitely often differentiable functions which vanish for $|t|$ sufficiently large.

Now consider the functional $K[\phi]$ defined by the relation $K[\phi]=\phi(t_0)$. K is a linear functional since

$$K[c_1\phi_1+c_2\phi_2]=c_1\phi_1(t_0)+c_2\phi_2(t_0)=c_1K[\phi_1]+c_2K[\phi_2].$$

To mimic (5), we write K symbolically in the form

$$K[\phi]=\int_{-\infty}^{\infty}\phi(t)\delta(t-t_0)\,dt. \tag{6}$$

In this sense, $\delta(t-t_0)$ is a "generalized function." It is important to realize though, that we cannot speak of the value of $\delta(t-t_0)$ at any time t. The only meaningful quantity is the expression $\int_{-\infty}^{\infty}\phi(t)\delta(t-t_0)\,dt$, and we must always assign the value $\phi(t_0)$ to this expression.

Admittedly, it is very difficult to think of a function in terms of the linear functional (5) that it induces. The advantage to this way of thinking, though, is that it is now possible to assign a derivative to every piecewise continuous function and to every "generalized function." To wit, suppose that $f(t)$ is a differentiable function. Then $f'(t)$ induces the linear functional

$$K'[\phi]=\int_{-\infty}^{\infty}\phi(t)f'(t)\,dt. \tag{7}$$

Integrating by parts and using the fact that $\phi(t)$ vanishes for $|t|$ sufficiently large, we see that

$$K'[\phi]=\int_{-\infty}^{\infty}\left[-\phi'(t)\right]f(t)\,dt=K[-\phi']. \tag{8}$$

Now, notice that the formula $K'[\phi]=K[-\phi']$ makes sense even if $f(t)$ is not differentiable. This motivates the following definition.

Definition. To every linear functional $K[\phi]$ we assign the new linear functional $K'[\phi]$ by the formula $K'[\phi]=K[-\phi']$. The linear functional $K'[\phi]$ is called the derivative of $K[\phi]$ since if $K[\phi]$ is induced by a differentiable function $f(t)$ then $K'[\phi]$ is induced by $f'(t)$.

Finally, we observe from (8) that the derivative of the delta function $\delta(t-t_0)$ is the linear functional which assigns to each function ϕ the number $-\phi'(t_0)$, for if $K[\phi]=\phi(t_0)$ then $K'[\phi]=K[-\phi']=-\phi'(t_0)$. Thus,

$$\int_{-\infty}^{\infty}\phi(t)\delta'(t-t_0)\,dt=-\phi'(t_0)$$

for all differentiable functions $\phi(t)$.

Exercises

1. Let a be a fixed constant. Show that every solution of the differential equation $(d^2y/dt^2)+2\alpha(dy/dt)+\alpha^2y=0$ can be written in the form

$$y(t)=[c_1+c_2(t-a)]e^{-\alpha(t-a)}.$$

2. Solve the initial-value problem $(d^2y/dt^2)+4(dy/dt)+5y=f(t)$; $y(0)=1$, $y'(0)=0$, where $f(t)$ is an impulsive force which acts on the extremely short time interval $1\leqslant t\leqslant 1+\tau$, and $\int_1^{1+\tau}f(t)\,dt=2$.

3. (a) Solve the initial-value problem $(d^2y/dt^2)-3(dy/dt)+2y=f(t)$; $y(0)=1$, $y'(0)=0$, where $f(t)$ is an impulsive function which acts on the extremely short time interval $2\leqslant t\leqslant 2+\tau$, and $\int_2^{2+\tau} f(t)\,dt=-1$.

(b) Solve the initial-value problem $(d^2y/dt^2)-3(dy/dt)+2y=0$; $y(0)=1$, $y'(0)=0$, on the interval $0\leqslant t\leqslant 2$. Compute $z_0=y(2)$ and $z_0'=y'(2)$. Then solve the initial-value problem

$$\frac{d^2y}{dt^2}-3\frac{dy}{dt}+2y=0;\qquad y(2)=z_0,\quad y'(2)=z_0'-1,\qquad 2\leqslant t<\infty.$$

Compare this solution with the solution of part (a).

4. A particle of mass 1 is attached to a spring dashpot mechanism. The stiffness constant of the spring is 3 lb/ft, and the drag force exerted on the particle by the dashpot mechanism is 4 times its velocity. At time $t=0$, the particle is stretched $\frac{1}{4}$ ft from its equilibrium position. At time $t=3$ seconds, an impulsive force of very short duration is applied to the system. This force imparts an impulse of 2 lb·s to the particle. Find the displacement of the particle from its equilibrium position.

In Exercises 5–7 solve the given initial-value problem.

5. $\dfrac{d^2y}{dt^2}+y=\sin t+\delta(t-\pi)$; $y(0)=0$, $y'(0)=0$

6. $\dfrac{d^2y}{dt^2}+\dfrac{dy}{dt}+y=2\delta(t-1)-\delta(t-2)$; $y(0)=1$, $y'(0)=0$

7. $\dfrac{d^2y}{dt^2}+2\dfrac{dy}{dt}+y=e^{-t}+3\delta(t-3)$; $y(0)=0$, $y'(0)=3$

8. (a) Solve the initial-value problem

$$\frac{d^2y}{dt^2}+y=\sum_{j=0}^{\infty}\delta(t-j\pi),\qquad y(0)=y'(0)=0,$$

and show that

$$y(t)=\begin{cases}\sin t, & n \text{ even}\\ 0, & n \text{ odd}\end{cases}$$

in the interval $n\pi<t<(n+1)\pi$.

(b) Solve the initial-value problem

$$\frac{d^2y}{dt^2}+y=\sum_{j=0}^{\infty}\delta(t-2j\pi),\qquad y(0)=y'(0)=0,$$

and show that $y(t)=(n+1)\sin t$ in the interval $2n\pi<t<2(n+1)\pi$.

This example indicates why soldiers are instructed to break cadence when marching across a bridge. To wit, if the soldiers are in step with the natural frequency of the steel in the bridge, then a resonance situation of the type (b) may be set up.

9. Let $f(t)$ be the function which is $\frac{1}{2}$ for $t>t_0$, 0 for $t=t_0$, and $-\frac{1}{2}$ for $t<t_0$. Let $K[\phi]$ be the linear functional

$$K[\phi]=\int_{-\infty}^{\infty}\phi(t)f(t)\,dt.$$

Show that $K'[\phi]\equiv K[-\phi']=\phi(t_0)$. Thus, $\delta(t-t_0)$ may be viewed as the derivative of $f(t)$.

2.13 The convolution integral

Consider the initial-value problem

$$a\frac{d^2y}{dt^2}+b\frac{dy}{dt}+cy=f(t);\qquad y(0)=y_0,\quad y'(0)=y_0'. \tag{1}$$

Let $Y(s)=\mathcal{L}\{y(t)\}$ and $F(s)=\mathcal{L}\{f(t)\}$. Taking Laplace transforms of both sides of the differential equation gives

$$a\left[s^2Y(s)-sy_0-y_0'\right]+b\left[sY(s)-y_0\right]+cY(s)=F(s)$$

and this implies that

$$Y(s)=\frac{as+b}{as^2+bs+c}y_0+\frac{a}{as^2+bs+c}y_0'+\frac{F(s)}{as^2+bs+c}.$$

Now, let

$$y_1(t)=\mathcal{L}^{-1}\left\{\frac{as+b}{as^2+bs+c}\right\}$$

and

$$y_2(t)=\mathcal{L}^{-1}\left\{\frac{a}{as^2+bs+c}\right\}.$$

Setting $f(t)=0$, $y_0=1$ and $y_0'=0$, we see that $y_1(t)$ is the solution of the homogeneous equation which satisfies the initial conditions $y_1(0)=1$, $y_1'(0)=0$. Similarly, by setting $f(t)=0$, $y_0=0$ and $y_0'=1$, we see that $y_2(t)$ is the solution of the homogeneous equation which satisfies the initial conditions $y_2(0)=0$, $y_2'(0)=1$. This implies that

$$\psi(t)=\mathcal{L}^{-1}\left\{\frac{F(s)}{as^2+bs+c}\right\}$$

is the particular solution of the nonhomogeneous equation which satisfies the initial conditions $\psi(0)=0$, $\psi'(0)=0$. Thus, the problem of finding a particular solution $\psi(t)$ of the nonhomogeneous equation is now reduced to the problem of finding the inverse Laplace transform of the function $F(s)/(as^2+bs+c)$. If we look carefully at this function, we see that it is the product of two Laplace transforms; that is

$$\frac{F(s)}{as^2+bs+c}=\mathcal{L}\{f(t)\}\times\mathcal{L}\left\{\frac{y_2(t)}{a}\right\}.$$

It is natural to ask whether there is any simple relationship between $\psi(t)$ and the functions $f(t)$ and $y_2(t)/a$. It would be nice, of course, if $\psi(t)$ were the product of $f(t)$ with $y_2(t)/a$, but this is obviously false. However, there is an extremely interesting way of combining two functions f and g together to form a new function $f*g$, which resembles multiplication, and for which

$$\mathcal{L}\{(f*g)(t)\}=\mathcal{L}\{f(t)\}\times\mathcal{L}\{g(t)\}.$$

This combination of f and g appears quite often in applications, and is known as the *convolution* of f with g.

Definition. The *convolution* $(f*g)(t)$ of f with g is defined by the equation

$$(f*g)(t)=\int_0^t f(t-u)\,g(u)\,du. \tag{2}$$

For example, if $f(t)=\sin 2t$ and $g(t)=e^{t^2}$, then

$$(f*g)(t)=\int_0^t \sin 2(t-u)e^{u^2}\,du.$$

The convolution operator $*$ clearly bears some resemblance to the multiplication operator since we multiply the value of f at the point $t-u$ by the value of g at the point u, and then integrate this product with respect to u. Therefore, it should not be too surprising to us that the convolution operator satisfies the following properties.

Property 1. The convolution operator obeys the commutative law of multiplication; that is, $(f*g)(t)=(g*f)(t)$.

PROOF. By definition,

$$(f*g)(t)=\int_0^t f(t-u)\,g(u)\,du.$$

Let us make the substitution $t-u=s$ in this integral. Then,

$$\begin{aligned}(f*g)(t)&=-\int_t^0 f(s)\,g(t-s)\,ds\\&=\int_0^t g(t-s)f(s)\,ds\equiv(g*f)(t).\end{aligned}$$

□

Property 2. The convolution operator satisfies the distributive law of multiplication; that is,

$$f*(g+h)=f*g+f*h.$$

PROOF. See Exercise 19. □

Property 3. The convolution operator satisfies the associative law of multiplication; that is, $(f*g)*h=f*(g*h)$.

PROOF. See Exercise 20. □

Property 4. The convolution of any function f with the zero function is zero.

PROOF. Obvious. □

On the other hand, the convolution operator differs from the multiplication operator in that $f*1\neq f$ and $f*f\neq f^2$. Indeed, the convolution of a function f with itself may even be negative.

Example 1. Compute the convolution of $f(t)=t^2$ with $g(t)=1$.
Solution. From Property 1,

$$(f*g)(t)=(g*f)(t)=\int_0^t 1\cdot u^2\,du=\frac{t^3}{3}.$$

Example 2. Compute the convolution of $f(t)=\cos t$ with itself, and show that it is not always positive.
Solution. By definition,

$$\begin{aligned}(f*f)(t)&=\int_0^t \cos(t-u)\cos u\,du\\&=\int_0^t(\cos t\cos^2u+\sin t\sin u\cos u)\,du\\&=\cos t\int_0^t\frac{1+\cos 2u}{2}\,du+\sin t\int_0^t\sin u\cos u\,du\\&=\cos t\left[\frac{t}{2}+\frac{\sin 2t}{4}\right]+\frac{\sin^3 t}{2}\\&=\frac{t\cos t+\sin t\cos^2 t+\sin^3 t}{2}\\&=\frac{t\cos t+\sin t(\cos^2 t+\sin^2 t)}{2}\\&=\frac{t\cos t+\sin t}{2}.\end{aligned}$$

This function, clearly, is negative for

$$(2n+1)\pi\leqslant t\leqslant(2n+1)\pi+\tfrac{1}{2}\pi,\qquad n=0,1,2,\ldots.$$

We now show that the Laplace transform of $f*g$ is the product of the Laplace transform of f with the Laplace transform of g.

Theorem 9. $\mathcal{L}\{(f*g)(t)\}=\mathcal{L}\{f(t)\}\times\mathcal{L}\{g(t)\}$.

PROOF. By definition,

$$\mathcal{L}\{(f*g)(t)\}=\int_0^\infty e^{-st}\left[\int_0^t f(t-u)\,g(u)\,du\right]dt.$$

This iterated integral equals the double integral

$$\iint_R e^{-st}f(t-u)\,g(u)\,du\,dt$$

where R is the triangular region described in Figure 1. Integrating first with respect to t, instead of u, gives

$$\mathcal{L}\{(f*g)(t)\}=\int_0^\infty g(u)\left[\int_u^\infty e^{-st}f(t-u)\,dt\right]du.$$

Setting $t-u=\xi$, we see that

$$\int_u^\infty e^{-st}f(t-u)\,dt=\int_0^\infty e^{-s(u+\xi)}f(\xi)\,d\xi.$$

Hence,

$$\begin{aligned}\mathcal{L}\{(f*g)(t)\}&=\int_0^\infty g(u)\left[\int_0^\infty e^{-su}e^{-s\xi}f(\xi)\,d\xi\right]du\\&=\left[\int_0^\infty g(u)e^{-su}\,du\right]\left[\int_0^\infty e^{-s\xi}f(\xi)\,d\xi\right]\\&\equiv\mathcal{L}\{f(t)\}\times\mathcal{L}\{g(t)\}.\end{aligned}$$

□

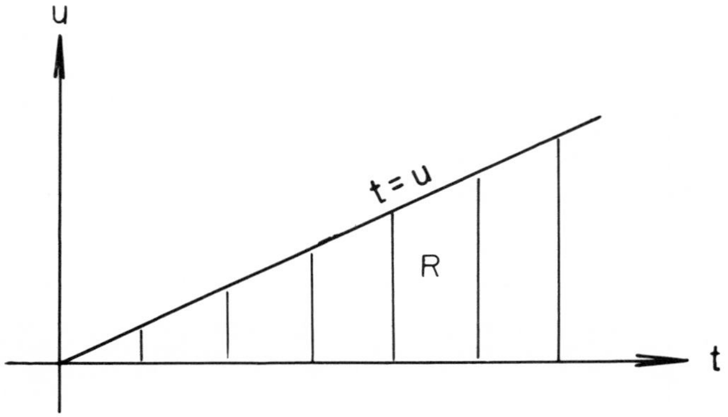

Figure 1

Example 3. Find the inverse Laplace transform of the function

$$\frac{a}{s^2(s^2+a^2)}.$$

Solution. Observe that

$$\frac{1}{s^2}=\mathcal{L}\{t\} \quad\text{and}\quad \frac{a}{s^2+a^2}=\mathcal{L}\{\sin at\}.$$

Hence, by Theorem 9

$$\mathcal{L}^{-1}\left\{\frac{a}{s^2(s^2+a^2)}\right\}=\int_0^t (t-u)\sin au\,du$$

$$=\frac{at-\sin at}{a^2}.$$

Example 4. Find the inverse Laplace transform of the function

$$\frac{1}{s(s^2+2s+2)}.$$

Solution. Observe that

$$\frac{1}{s}=\mathcal{L}\{1\} \quad\text{and}\quad \frac{1}{s^2+2s+2}=\frac{1}{(s+1)^2+1}=\mathcal{L}\{e^{-t}\sin t\}.$$

Hence, by Theorem 9,

$$\mathcal{L}^{-1}\left\{\frac{1}{s(s^2+2s+2)}\right\}=\int_0^t e^{-u}\sin u\,du$$

$$=\frac{1}{2}\left[1-e^{-t}(\cos t+\sin t)\right].$$

Remark. Let $y_2(t)$ be the solution of the homogeneous equation $ay''+by'+cy=0$ which satisfies the initial conditions $y_2(0)=0$, $y_2'(0)=1$. Then,

$$\psi(t)=f(t)*\frac{y_2(t)}{a} \tag{3}$$

is the particular solution of the nonhomogeneous equation $ay''+by'+cy=f(t)$ which satisfies the initial conditions $\psi(0)=\psi'(0)=0$. Equation (3) is often much simpler to use than the variation of parameters formula derived in Section 2.4.

Exercises

Compute the convolution of each of the following pairs of functions.

1. e^{at}, e^{bt}, $a\neq b$

2. e^{at}, e^{at}

3. $\cos at$, $\cos bt$

4. $\sin at$, $\sin bt$, $a\neq b$

5. $\sin at$, $\sin at$

6. t, $\sin t$

Use Theorem 9 to invert each of the following Laplace transforms.

7. $\dfrac{1}{s^2(s^2+1)}$ **8.** $\dfrac{s}{(s+1)(s^2+4)}$ **9.** $\dfrac{s}{(s^2+1)^2}$

10. $\dfrac{1}{s(s^2+1)}$ **11.** $\dfrac{1}{s^2(s+1)^2}$ **12.** $\dfrac{1}{(s^2+1)^2}$

Use Theorem 9 to find the solution $y(t)$ of each of the following integro-differential equations.

13. $y(t)=4t-3\int_0^t y(u)\sin(t-u)\,du$

14. $y(t)=4t-3\int_0^t y(t-u)\sin u\,du$

15. $y'(t)=\sin t+\int_0^t y(t-u)\cos u\,du,\ y(0)=0$

16. $y(t)=4t^2-\int_0^t y(u)e^{-(t-u)}\,du$

17. $y'(t)+2y+\int_0^t y(u)\,du=\sin t,\ y(0)=1$

18. $y(t)=t-e^t\int_0^t y(u)e^{-u}\,du$

19. Prove that $f*(g+h)=f*g+f*h$.

20. Prove that $(f*g)*h=f*(g*h)$.

2.14 The method of elimination for systems

The theory of second-order linear differential equations can also be used to find the solutions of two simultaneous first-order equations of the form

$$\begin{aligned} x' &= \frac{dx}{dt} = a(t)x+b(t)y+f(t) \\ y' &= \frac{dy}{dt} = c(t)x+d(t)y+g(t). \end{aligned} \tag{1}$$

The key idea is to eliminate one of the variables, say y, and then find x as the solution of a second-order linear differential equation. This technique is known as the *method of elimination*, and we illustrate it with the following two examples.

Example 1. Find all solutions of the simultaneous equations

$$\begin{aligned} x' &= 2x+y+t \\ y' &= x+3y+1. \end{aligned} \tag{2}$$

Solution. First, we solve for

$$y = x' - 2x - t \tag{3}$$

from the first equation of (2). Differentiating this equation gives

$$y' = x'' - 2x' - 1 = x + 3y + 1.$$

Then, substituting for y from (3) gives

$$x'' - 2x' - 1 = x + 3(x' - 2x - t) + 1$$

so that

$$x'' - 5x' + 5x = 2 - 3t. \tag{4}$$

Equation (4) is a second-order linear equation and its solution is

$$x(t) = e^{5t/2}\left[c_1 e^{\sqrt{5}\,t/2} + c_2 e^{-\sqrt{5}\,t/2}\right] - \frac{(1+3t)}{5}$$

for some constants c_1 and c_2. Finally, plugging this expression into (3) gives

$$y(t) = e^{5t/2}\left[\frac{1+\sqrt{5}}{2} c_1 e^{\sqrt{5}\,t/2} + \frac{1-\sqrt{5}}{2} c_2 e^{-\sqrt{5}\,t/2}\right] + \frac{t-1}{5}.$$

Example 2. Find the solution of the initial-value problem

$$\begin{aligned} x' &= 3x - y, \qquad & x(0) &= 3 \\ y' &= x + y, \qquad & y(0) &= 0. \end{aligned} \tag{5}$$

Solution. From the first equation of (5),

$$y = 3x - x'. \tag{6}$$

Differentiating this equation gives

$$y' = 3x' - x'' = x + y.$$

Then, substituting for y from (6) gives

$$3x' - x'' = x + 3x - x'$$

so that

$$x'' - 4x' + 4x = 0.$$

This implies that

$$x(t) = (c_1 + c_2 t)e^{2t}$$

for some constants c_1, c_2, and plugging this expression into (6) gives

$$y(t) = (c_1 - c_2 + c_2 t)e^{2t}.$$

The constants c_1 and c_2 are determined from the initial conditions

$$\begin{aligned} x(0) &= 3 = c_1 \\ y(0) &= 0 = c_1 - c_2. \end{aligned}$$

Hence $c_1 = 3$, $c_2 = 3$ and

$$x(t) = 3(1+t)e^{2t}, y(t) = 3te^{2t}$$

is the solution of (5).

Remark. The simultaneous equations (1) are usually referred to as a first-order *system* of equations. Systems of equations are treated fully in Chapters 3 and 4.

EXERCISES

Find all solutions of each of the following systems of equations.

1. $x' = 6x - 3y$
$y' = 2x + y$

2. $x' = -2x + y + t$
$y' = -4x + 3y - 1$

3. $x' = -3x + 2y$
$y' = -x - y$

4. $x' = x + y + e^t$
$y' = x - y - e^t$

Find the solution of each of the following initial-value problems.

5. $x' = x + y, \quad x(0) = 2$
$y' = 4x + y, \quad y(0) = 3$

6. $x' = x - 3y, \quad x(0) = 0$
$y' = -2x + 2y, \quad y(0) = 5$

7. $x' = x - y, \quad x(0) = 1$
$y' = 5x - 3y, \quad y(0) = 2$

8. $x' = 3x - 2y, \quad x(0) = 1$
$y' = 4x - y, \quad y(0) = 5$

9. $x' = 4x + 5y + 4e^t\cos t, \quad x(0) = 0$
$y' = -2x - 2y, \quad y(0) = 0$

10. $x' = 3x - 4y + e^t, \quad x(0) = 1$
$y' = x - y + e^t, \quad y(0) = 1$

11. $x' = 2x - 5y + \sin t, \quad x(0) = 0$
$y' = x - 2y + \tan t, \quad y(0) = 0$

12. $x' = y + f_1(t), \quad x(0) = 0$
$y' = -x + f_2(t), \quad y(0) = 0$

2.15 A few words about higher-order equations

In this section we briefly discuss higher-order linear differential equations.

Definition. The equation

$$L[y] = a_n(t)\frac{d^ny}{dt^n} + a_{n-1}(t)\frac{d^{n-1}y}{dt^{n-1}} + \dots + a_0(t)y = 0, \quad a_n(t) \neq 0 \qquad (1)$$

is called the general nth order homogeneous linear equation. The differential equation (1) together with the initial conditions

$$y(t_0) = y_0, \quad y'(t_0) = y_0', \dots, y^{(n-1)}(t_0) = y_0^{(n-1)} \qquad (1')$$

is called an initial-value problem. The theory for Equation (1) is completely analogous to the theory for the second-order linear homogeneous equation which we studied in Sections 2.1 and 2.2. Therefore, we will state the relevant theorems without proof. Complete proofs can be obtained by generalizing the methods used in Sections 2.1 and 2.2, or by using the methods to be developed in Chapter 3.

Theorem 10. *Let $y_1(t),\dots,y_n(t)$ be n independent solutions of* (1); *that is, no solution $y_j(t)$ is a linear combination of the other solutions. Then, every solution $y(t)$ of* (1) *is of the form*

$$y(t)=c_1y_1(t)+\dots+c_ny_n(t) \tag{2}$$

for some choice of constants $c_1,\dots,c_n$. For this reason, we say that (2) *is the general solution of* (1).

To find n independent solutions of (1) when the coefficients $a_0, a_1,\dots,a_n$ do not depend on t, we compute

$$L[e^{rt}]=(a_nr^n+a_{n-1}r^{n-1}+\dots+a_0)e^{rt}. \tag{3}$$

This implies that e^{rt} is a solution of (1) if, and only if, r is a root of the characteristic equation

$$a_nr^n+a_{n-1}r^{n-1}+\dots+a_0=0. \tag{4}$$

Thus, if Equation (4) has n distinct roots $r_1,\dots,r_n$, then the general solution of (1) is $y(t)=c_1e^{r_1t}+\dots+c_ne^{r_nt}$. If $r_j=\alpha_j+i\beta_j$ is a complex root of (4), then

$$u(t)=\text{Re}\{e^{r_jt}\}=e^{\alpha_jt}\cos\beta_jt$$

and

$$v(t)=\text{Im}\{e^{r_jt}\}=e^{\alpha_jt}\sin\beta_jt$$

are two real-valued solutions of (1). Finally, if r_1 is a root of multiplicity k; that is, if

$$a_nr^n+\dots+a_0=(r-r_1)^kq(r)$$

where $q(r_1)\neq 0$, then e^{r_1t}, $te^{r_1t},\dots,t^{k-1}e^{r_1t}$ are k independent solutions of (1). We prove this last assertion in the following manner. Observe from (3) that

$$L[e^{rt}]=(r-r_1)^kq(r)e^{rt}$$

if r_1 is a root of multiplicity k. Therefore,

$$\begin{aligned}L[t^je^{r_1t}]&=L\left[\frac{\partial^j}{\partial r^j}e^{rt}\right]\Bigg|_{r=r_1}\\&=\frac{\partial^j}{\partial r^j}L[e^{rt}]\Bigg|_{r=r_1}\\&=\frac{\partial^j}{\partial r^j}(r-r_1)^kq(r)e^{rt}\Bigg|_{r=r_1}\\&=0,\ \text{for } 1\leqslant j<k.\end{aligned}$$

Example 1. Find the general solution of the equation

$$\frac{d^4y}{dt^4}+y=0. \tag{5}$$

Solution. The characteristic equation of (5) is $r^4+1=0$. We find the roots of this equation by noting that

$$-1=e^{i\pi}=e^{3\pi i}=e^{5\pi i}=e^{7\pi i}.$$

Hence,

$$r_1=e^{i\pi/4}=\cos\frac{\pi}{4}+i\sin\frac{\pi}{4}=\frac{1}{\sqrt{2}}(1+i),$$

$$r_2=e^{3\pi i/4}=\cos\frac{3\pi}{4}+i\sin\frac{3\pi}{4}=-\frac{1}{\sqrt{2}}(1-i),$$

$$r_3=e^{5\pi i/4}=\cos\frac{5\pi}{4}+i\sin\frac{5\pi}{4}=-\frac{1}{\sqrt{2}}(1+i),$$

and

$$r_4=e^{7\pi i/4}=\cos\frac{7\pi}{4}+i\sin\frac{7\pi}{4}=\frac{1}{\sqrt{2}}(1-i)$$

are 4 roots of the equation $r^4+1=0$. The roots r_3 and r_4 are the complex conjugates of r_2 and r_1, respectively. Thus,

$$e^{r_1t}=e^{t/\sqrt{2}}\left[\cos\frac{t}{\sqrt{2}}+i\sin\frac{t}{\sqrt{2}}\right]$$

and

$$e^{r_2t}=e^{-t/\sqrt{2}}\left[\cos\frac{t}{\sqrt{2}}+i\sin\frac{t}{\sqrt{2}}\right]$$

are 2 complex-valued solutions of (5), and this implies that

$$y_1(t)=e^{t/\sqrt{2}}\cos\frac{t}{\sqrt{2}},\qquad y_2(t)=e^{t/\sqrt{2}}\sin\frac{t}{\sqrt{2}},$$

$$y_3(t)=e^{-t/\sqrt{2}}\cos\frac{t}{\sqrt{2}},\quad\text{and}\quad y_4(t)=e^{-t/\sqrt{2}}\sin\frac{t}{\sqrt{2}}$$

are 4 real-valued solutions of (5). These solutions are clearly independent. Hence, the general solution of (5) is

$$y(t)=e^{t/\sqrt{2}}\left[a_1\cos\frac{t}{\sqrt{2}}+b_1\sin\frac{t}{\sqrt{2}}\right]$$
$$+e^{-t/\sqrt{2}}\left[a_2\cos\frac{t}{\sqrt{2}}+b_2\sin\frac{t}{\sqrt{2}}\right].$$

Example 2. Find the general solution of the equation

$$\frac{d^4y}{dt^4}-3\frac{d^3y}{dt^3}+3\frac{d^2y}{dt^2}-\frac{dy}{dt}=0. \tag{6}$$

Solution. The characteristic equation of (6) is

$$\begin{aligned}0&=r^4-3r^3+3r^2-r=r(r^3-3r^2+3r-1)\\&=r(r-1)^3.\end{aligned}$$

Its roots are $r_1=0$ and $r_2=1$, with $r_2=1$ a root of multiplicity three. Hence, the general solution of (6) is

$$y(t)=c_1+(c_2+c_3t+c_4t^2)e^t.$$

The theory for the nonhomogeneous equation

$$L[y]=a_n(t)\frac{d^ny}{dt^n}+\ldots+a_0(t)y=f(t),\quad a_n(t)\neq 0 \tag{7}$$

is also completely analogous to the theory for the second-order nonhomogeneous equation. The following results are the analogs of Lemma 1 and Theorem 5 of Section 2.3.

Lemma 1. *The difference of any two solutions of the nonhomogeneous equation* (7) *is a solution of the homogeneous equation* (1).

Theorem 11. *Let $\psi(t)$ be a particular solution of the nonhomogeneous equation* (7), *and let $y_1(t),\ldots,y_n(t)$ be n independent solutions of the homogeneous equation* (1). *Then, every solution $y(t)$ of* (7) *is of the form*

$$y(t)=\psi(t)+c_1y_1(t)+\ldots+c_ny_n(t)$$

for some choice of constants $c_1, c_2,\ldots,c_n$.

The method of judicious guessing also applies to the nth-order equation

$$a_n\frac{d^ny}{dt^n}+\ldots+a_0y=\left[b_0+b_1t+\ldots+b_kt^k\right]e^{\alpha t}. \tag{8}$$

It is easily verified that Equation (8) has a particular solution $\psi(t)$ of the form

$$\psi(t)=\left[c_0+c_1t+\ldots+c_kt^k\right]e^{\alpha t}$$

if $e^{\alpha t}$ is not a solution of the homogeneous equation, and

$$\psi(t)=t^j\left[c_0+c_1t+\ldots+c_kt^k\right]e^{\alpha t}$$

if $t^{j-1}e^{\alpha t}$ is a solution of the homogeneous equation, but $t^je^{\alpha t}$ is not.

Example 3. Find a particular solution $\psi(t)$ of the equation

$$L[y]=\frac{d^3y}{dt^3}+3\frac{d^2y}{dt^2}+3\frac{dy}{dt}+y=e^t. \tag{9}$$

Solution. The characteristic equation

$$r^3+3r^2+3r+1=(r+1)^3$$

has $r=-1$ as a triple root. Hence, e^t is not a solution of the homogeneous equation, and Equation (9) has a particular solution $\psi(t)$ of the form

$$\psi(t)=Ae^t.$$

Computing $L[\psi](t)=8Ae^t$, we see that $A=\frac{1}{8}$. Consequently, $\psi(t)=\frac{1}{8}e^t$ is a particular solution of (9).

There is also a variation of parameters formula for the nonhomogeneous equation (7). Let $v(t)$ be the solution of the homogeneous equation (1) which satisfies the initial conditions $v(t_0)=0$, $v'(t_0)=0,\ldots,v^{(n-2)}(t_0)=0$, $v^{(n-1)}(t_0)=1$. Then,

$$\psi(t)=\int_{t_0}^{t}\frac{v(t-s)}{a_n(s)}f(s)\,ds$$

is a particular solution of the nonhomogeneous equation (7). We will prove this assertion in Section 3.10. (This can also be proven using the method of Laplace transforms; see Section 2.13.)

Exercises

Find the general solution of each of the following equations.

1. $y'''-2y''-y'+2y=0$

2. $y'''-6y''+5y'+12y=0$

3. $y^{(iv)}-5y'''+6y''+4y'-8y=0$

4. $y'''-y''+y'-y=0$

Solve each of the following initial-value problems.

5. $y^{(iv)}+4y'''+14y''-20y'+25y=0;\quad y(0)=y'(0)=y''(0)=0,\ y'''(0)=0$

6. $y^{(iv)}-y=0;\quad y(0)=1,\ y'(0)=y''(0)=0,\ y'''(0)=-1$

7. $y^{(v)}-2y^{(iv)}+y'''=0;\quad y(0)=y'(0)=y''(0)=y'''(0)=0,\ y^{(iv)}(0)=-1$

8. Given that $y_1(t)=e^t\cos t$ is a solution of

$$y^{(iv)}-2y'''+y''+2y'-2y=0, \tag{*}$$

find the general solution of (*). *Hint*: Use this information to find the roots of the characteristic equation of (*).

Find a particular solution of each of the following equations.

9. $y'''+y'=\tan t$

10. $y^{(iv)}-y=g(t)$

11. $y^{(iv)}+y=g(t)$

12. $y'''+y'=2t^2+4\sin t$

13. $y''' - 4y' = t + \cos t + 2e^{-2t}$

14. $y^{(iv)} - y = t + \sin t$

15. $y^{(iv)} + 2y'' + y = t^2 \sin t$

16. $y^{(vi)} + y'' = t^2$

17. $y''' + y'' + y' + y = t + e^{-t}$

18. $y^{(iv)} + 4y''' + 6y'' + 4y' + y = t^3 e^{-t}$

Hint for (18): Make the substitution $y = e^{-t}v$ and solve for v. Otherwise, it will take an awfully long time to do this problem.

3 Systems of differential equations

3.1 Algebraic properties of solutions of linear systems

In this chapter we will consider simultaneous first-order differential equations in several variables, that is, equations of the form

$$\begin{aligned}\frac{dx_1}{dt} &= f_1(t, x_1, \ldots, x_n),\\ \frac{dx_2}{dt} &= f_2(t, x_1, \ldots, x_n),\\ &\vdots\\ \frac{dx_n}{dt} &= f_n(t, x_1, \ldots, x_n).\end{aligned} \tag{1}$$

A solution of (1) is n functions $x_1(t), \ldots, x_n(t)$ such that $dx_j(t)/dt = f_j(t, x_1(t), \ldots, x_n(t))$, $j = 1, 2, \ldots, n$. For example, $x_1(t) = t$ and $x_2(t) = t^2$ is a solution of the simultaneous first-order differential equations

$$\frac{dx_1}{dt} = 1 \quad \text{and} \quad \frac{dx_2}{dt} = 2x_1$$

since $dx_1(t)/dt = 1$ and $dx_2(t)/dt = 2t = 2x_1(t)$.

In addition to Equation (1), we will often impose initial conditions on the functions $x_1(t), \ldots, x_n(t)$. These will be of the form

$$x_1(t_0) = x_1^0, \quad x_2(t_0) = x_2^0, \ldots, x_n(t_0) = x_n^0. \tag{1'}$$

Equation (1), together with the initial conditions $(1)'$, is referred to as an initial-value problem. A solution of this initial-value problem is n functions $x_1(t), \ldots, x_n(t)$ which satisfy (1) and the initial conditions

$$x_1(t_0) = x_1^0, \ldots, x_n(t_0) = x_n^0.$$

For example, $x_1(t) = e^t$ and $x_2(t) = 1 + e^{2t}/2$ is a solution of the initial-value problem

$$\frac{dx_1}{dt} = x_1, \qquad x_1(0) = 1,$$
$$\frac{dx_2}{dt} = x_1^2, \qquad x_2(0) = \frac{3}{2},$$

since $dx_1(t)/dt = e^t = x_1(t)$, $dx_2(t)/dt = e^{2t} = x_1^2(t)$, $x_1(0) = 1$ and $x_2(0) = \frac{3}{2}$.

Equation (1) is usually referred to as a system of n first-order differential equations. Equations of this type arise quite often in biological and physical applications and frequently describe very complicated systems since the rate of change of the variable x_j depends not only on t and x_j, but on the value of all the other variables as well. One particular example is the blood glucose model we studied in Section 2.7. In this model, the rates of change of g and h (respectively, the deviations of the blood glucose and net hormonal concentrations from their optimal values) are given by the equations

$$\frac{dg}{dt} = -m_1 g - m_2 h + J(t), \qquad \frac{dh}{dt} = -m_3 h + m_4 g.$$

This is a system of two first-order equations for the functions $g(t)$ and $h(t)$.

First-order systems of differential equations also arise from higher-order equations for a single variable $y(t)$. Every nth-order differential equation for the single variable y can be converted into a system of n first-order equations for the variables

$$x_1(t) = y, \quad x_2(t) = \frac{dy}{dt}, \ldots, x_n(t) = \frac{d^{n-1}y}{dt^{n-1}}.$$

Examples 1 and 2 illustrate how this works.

Example 1. Convert the differential equation

$$a_n(t)\frac{d^n y}{dt^n} + a_{n-1}(t)\frac{d^{n-1}y}{dt^{n-1}} + \ldots + a_0 y = 0$$

into a system of n first-order equations.

Solution. Let $x_1(t) = y$, $x_2(t) = dy/dt, \ldots,$ and $x_n(t) = d^{n-1}y/dt^{n-1}$. Then,

$$\frac{dx_1}{dt} = x_2, \quad \frac{dx_2}{dt} = x_3, \ldots, \frac{dx_{n-1}}{dt} = x_n,$$

and

$$\frac{dx_n}{dt} = -\frac{a_{n-1}(t)x_n + a_{n-2}(t)x_{n-1} + \ldots + a_0 x_1}{a_n(t)}.$$

Example 2. Convert the initial-value problem

$$\frac{d^3y}{dt^3} + \left(\frac{dy}{dt}\right)^2 + 3y = e^t; \qquad y(0) = 1, \quad y'(0) = 0, \quad y''(0) = 0$$

into an initial-value problem for the variables y, dy/dt, and d^2y/dt^2.

Solution. Set $x_1(t)=y$, $x_2(t)=dy/dt$, and $x_3(t)=d^2y/dt^2$. Then,

$$\frac{dx_1}{dt}=x_2, \qquad \frac{dx_2}{dt}=x_3, \qquad \frac{dx_3}{dt}=e^t-x_2^2-3x_1.$$

Moreover, the functions x_1, x_2, and x_3 satisfy the initial conditions $x_1(0)=1$, $x_2(0)=0$, and $x_3(0)=0$.

If each of the functions $f_1,\ldots,f_n$ in (1) is a linear function of the dependent variables $x_1,\ldots,x_n$, then the system of equations is said to be linear. The most general system of n first-order linear equations has the form

$$\begin{aligned}\frac{dx_1}{dt}&=a_{11}(t)x_1+\ldots+a_{1n}(t)x_n+g_1(t)\\ &\vdots\\ \frac{dx_n}{dt}&=a_{n1}(t)x_1+\ldots+a_{nn}(t)x_n+g_n(t).\end{aligned} \tag{2}$$

If each of the functions $g_1,\ldots,g_n$ is identically zero, then the system (2) is said to be homogeneous; otherwise it is nonhomogeneous. In this chapter, we only consider the case where the coefficients a_{ij} do not depend on t.

Now, even the homogeneous linear system with constant coefficients

$$\begin{aligned}\frac{dx_1}{dt}&=a_{11}x_1+\ldots+a_{1n}x_n\\ &\vdots\\ \frac{dx_n}{dt}&=a_{n1}x_1+\ldots+a_{nn}x_n\end{aligned} \tag{3}$$

is quite cumbersome to handle. This is especially true if n is large. Therefore, we seek to write these equations in as concise a manner as possible. To this end we introduce the concepts of *vectors* and *matrices*.

Definition. A *vector*

$$\mathbf{x}=\begin{bmatrix}x_1\\ x_2\\ \vdots\\ x_n\end{bmatrix}$$

is a shorthand notation for the sequence of numbers $x_1,\ldots,x_n$. The numbers $x_1,\ldots,x_n$ are called the *components* of $\mathbf{x}$. If $x_1=x_1(t),\ldots,$ and $x_n=x_n(t)$, then

$$\mathbf{x}(t)=\begin{bmatrix}x_1(t)\\ \vdots\\ x_n(t)\end{bmatrix}$$

is called a vector-valued function. Its derivative $d\mathbf{x}(t)/dt$ is the vector-valued function

$$\begin{bmatrix} \dfrac{dx_1(t)}{dt} \\ \vdots \\ \dfrac{dx_n(t)}{dt} \end{bmatrix}.$$

Definition. A *matrix*

$$\mathbf{A}=\begin{bmatrix} a_{11} & a_{12} & \dots & a_{1n} \\ a_{21} & a_{22} & \dots & a_{2n} \\ \vdots & \vdots & & \vdots \\ a_{m1} & a_{m2} & \dots & a_{mn} \end{bmatrix}$$

is a shorthand notation for the array of numbers a_{ij} arranged in m rows and n columns. The element lying in the ith row and jth column is denoted by a_{ij}, the first subscript identifying its row and the second subscript identifying its column. $\mathbf{A}$ is said to be a square matrix if $m=n$.

Next, we define the product of a matrix $\mathbf{A}$ with a vector $\mathbf{x}$.

Definition. Let $\mathbf{A}$ be an $n\times n$ matrix with elements a_{ij} and let $\mathbf{x}$ be a vector with components $x_1,\dots,x_n$. We define the product of $\mathbf{A}$ with $\mathbf{x}$, denoted by $\mathbf{Ax}$, as the vector whose ith component is

$$a_{i1}x_1+a_{i2}x_2+\dots+a_{in}x_n, \qquad i=1,2,\dots,n.$$

In other words, the ith component of $\mathbf{Ax}$ is the sum of the product of corresponding terms of the ith row of $\mathbf{A}$ with the vector $\mathbf{x}$. Thus,

$$\mathbf{Ax}=\begin{bmatrix} a_{11} & a_{12} & \dots & a_{1n} \\ a_{21} & a_{22} & \dots & a_{2n} \\ \vdots & \vdots & & \vdots \\ a_{n1} & a_{n2} & \dots & a_{nn} \end{bmatrix}\begin{bmatrix} x_1 \\ x_2 \\ \vdots \\ x_n \end{bmatrix}$$

$$=\begin{bmatrix} a_{11}x_1+a_{12}x_2+\dots+a_{1n}x_n \\ a_{21}x_1+a_{22}x_2+\dots+a_{2n}x_n \\ \vdots \\ a_{n1}x_1+a_{n2}x_2+\dots+a_{nn}x_n \end{bmatrix}.$$

For example,

$$\begin{bmatrix} 1 & 2 & 4 \\ -1 & 0 & 6 \\ 1 & 1 & 1 \end{bmatrix}\begin{bmatrix} 3 \\ 2 \\ 1 \end{bmatrix}=\begin{bmatrix} 3+4+4 \\ -3+0+6 \\ 3+2+1 \end{bmatrix}=\begin{bmatrix} 11 \\ 3 \\ 6 \end{bmatrix}.$$

Finally, we observe that the left-hand sides of (3) are the components of the vector $d\mathbf{x}/dt$, while the right-hand sides of (3) are the components of the vector $\mathbf{Ax}$. Hence, we can write (3) in the concise form

$$\dot{\mathbf{x}}=\frac{d\mathbf{x}}{dt}=\mathbf{Ax},\quad \text{where}\quad \mathbf{x}=\begin{pmatrix} x_1\\ \vdots\\ x_n\end{pmatrix}\quad\text{and}\quad \mathbf{A}=\begin{pmatrix} a_{11} & a_{12} & \dots & a_{1n}\\ a_{21} & a_{22} & \dots & a_{2n}\\ \vdots & \vdots & & \vdots\\ a_{n1} & a_{n2} & \dots & a_{nn}\end{pmatrix}. \tag{4}$$

Moreover, if $x_1(t),\dots,x_n(t)$ satisfy the initial conditions

$$x_1(t_0)=x_1^0,\dots,x_n(t_0)=x_n^0,$$

then $\mathbf{x}(t)$ satisfies the initial-value problem

$$\dot{\mathbf{x}}=\mathbf{Ax},\qquad \mathbf{x}(t_0)=\mathbf{x}^0,\quad\text{where}\quad \mathbf{x}^0=\begin{pmatrix} x_1^0\\ \vdots\\ x_n^0\end{pmatrix}. \tag{5}$$

For example, the system of equations

$$\frac{dx_1}{dt}=3x_1-7x_2+9x_3$$
$$\frac{dx_2}{dt}=15x_1+x_2-x_3$$
$$\frac{dx_3}{dt}=7x_1+6x_3$$

can be written in the concise form

$$\dot{\mathbf{x}}=\begin{pmatrix} 3 & -7 & 9\\ 15 & 1 & -1\\ 7 & 0 & 6\end{pmatrix}\mathbf{x},\qquad \mathbf{x}=\begin{pmatrix} x_1\\ x_2\\ x_3\end{pmatrix},$$

and the initial-value problem

$$\frac{dx_1}{dt}=x_1-x_2+x_3,\qquad x_1(0)=1$$
$$\frac{dx_2}{dt}=3x_2-x_3,\qquad x_2(0)=0$$
$$\frac{dx_3}{dt}=x_1+7x_3,\qquad x_3(0)=-1$$

can be written in the concise form

$$\dot{\mathbf{x}}=\begin{pmatrix} 1 & -1 & 1\\ 0 & 3 & -1\\ 1 & 0 & 7\end{pmatrix}\mathbf{x},\qquad \mathbf{x}(0)=\begin{pmatrix} 1\\ 0\\ -1\end{pmatrix}.$$

Now that we have succeeded in writing (3) in the more manageable form (4), we can tackle the problem of finding all of its solutions. Since these equations are linear, we will try and play the same game that we played, with so much success, with the second-order linear homogeneous equation. To wit, we will show that a constant times a solution and the sum of two solutions are again solutions of (4). Then, we will try and show that we can find every solution of (4) by taking all linear combinations of a finite number of solutions. Of course, we must first define what we mean by a constant times **x** and the sum of **x** and **y** if **x** and **y** are vectors with n components.

Definition. Let c be a number and **x** a vector with n components $x_1,\dots,x_n$. We define $c\mathbf{x}$ to be the vector whose components are $cx_1,\dots,cx_n$, that is

$$c\mathbf{x}=c\begin{pmatrix}x_1\\x_2\\\vdots\\x_n\end{pmatrix}=\begin{pmatrix}cx_1\\cx_2\\\vdots\\cx_n\end{pmatrix}.$$

For example, if

$$c=2\quad\text{and}\quad\mathbf{x}=\begin{pmatrix}3\\1\\7\end{pmatrix},\quad\text{then}\quad 2\mathbf{x}=2\begin{pmatrix}3\\1\\7\end{pmatrix}=\begin{pmatrix}6\\2\\14\end{pmatrix}.$$

This process of multiplying a vector **x** by a number c is called scalar multiplication.

Definition. Let **x** and **y** be vectors with components $x_1,\dots,x_n$ and $y_1,\dots,y_n$ respectively. We define $\mathbf{x}+\mathbf{y}$ to be the vector whose components are $x_1+y_1,\dots,x_n+y_n$, that is

$$\mathbf{x}+\mathbf{y}=\begin{pmatrix}x_1\\x_2\\\vdots\\x_n\end{pmatrix}+\begin{pmatrix}y_1\\y_2\\\vdots\\y_n\end{pmatrix}=\begin{pmatrix}x_1+y_1\\x_2+y_2\\\vdots\\x_n+y_n\end{pmatrix}.$$

For example, if

$$\mathbf{x}=\begin{pmatrix}1\\6\\3\\2\end{pmatrix}\quad\text{and}\quad\mathbf{y}=\begin{pmatrix}-1\\-6\\7\\9\end{pmatrix},$$

then

$$\mathbf{x}+\mathbf{y}=\begin{pmatrix}1\\6\\3\\2\end{pmatrix}+\begin{pmatrix}-1\\-6\\7\\9\end{pmatrix}=\begin{pmatrix}0\\0\\10\\11\end{pmatrix}.$$

This process of adding two vectors together is called vector addition.

Having defined the processes of scalar multiplication and vector addition, we can now state the following theorem.

Theorem 1. *Let* $\mathbf{x}(t)$ *and* $\mathbf{y}(t)$ *be two solutions of* (4). *Then* (*a*) $c\mathbf{x}(t)$ *is a solution, for any constant* c, *and* (*b*) $\mathbf{x}(t)+\mathbf{y}(t)$ *is again a solution.*

Theorem 1 can be proven quite easily with the aid of the following lemma.

Lemma. *Let* $\mathbf{A}$ *be an* $n\times n$ *matrix. For any vectors* $\mathbf{x}$ *and* $\mathbf{y}$ *and constant* c, (*a*) $\mathbf{A}(c\mathbf{x})=c\mathbf{A}\mathbf{x}$ *and* (*b*) $\mathbf{A}(\mathbf{x}+\mathbf{y})=\mathbf{A}\mathbf{x}+\mathbf{A}\mathbf{y}$.

PROOF OF LEMMA.
(a) We prove that two vectors are equal by showing that they have the same components. To this end, observe that the ith component of the vector $c\mathbf{A}\mathbf{x}$ is

$$ca_{i1}x_1+ca_{i2}x_2+\ldots+ca_{in}x_n=c(a_{i1}x_1+\ldots+a_{in}x_n),$$

and the ith component of the vector $\mathbf{A}(c\mathbf{x})$ is

$$a_{i1}(cx_1)+a_{i2}(cx_2)+\ldots+a_{in}(cx_n)=c(a_{i1}x_1+\ldots+a_{in}x_n).$$

Hence $\mathbf{A}(c\mathbf{x})=c\mathbf{A}\mathbf{x}$.
(b) The ith component of the vector $\mathbf{A}(\mathbf{x}+\mathbf{y})$ is

$$a_{i1}(x_1+y_1)+\ldots+a_{in}(x_n+y_n)=(a_{i1}x_1+\ldots+a_{in}x_n)+(a_{i1}y_1+\ldots+a_{in}y_n).$$

But this is also the ith component of the vector $\mathbf{A}\mathbf{x}+\mathbf{A}\mathbf{y}$ since the ith component of $\mathbf{A}\mathbf{x}$ is $a_{i1}x_1+\ldots+a_{in}x_n$ and the ith component of $\mathbf{A}\mathbf{y}$ is $a_{i1}y_1+\ldots+a_{in}y_n$. Hence $\mathbf{A}(\mathbf{x}+\mathbf{y})=\mathbf{A}\mathbf{x}+\mathbf{A}\mathbf{y}$. □

PROOF OF THEOREM 1.
(a). If $\mathbf{x}(t)$ is a solution of (4), then

$$\frac{d}{dt}c\mathbf{x}(t)=c\frac{d\mathbf{x}(t)}{dt}=c\mathbf{A}\mathbf{x}(t)=\mathbf{A}(c\mathbf{x}(t)).$$

Hence, $c\mathbf{x}(t)$ is also a solution of (4).
(b). If $\mathbf{x}(t)$ and $\mathbf{y}(t)$ are solutions of (4) then

$$\frac{d}{dt}(\mathbf{x}(t)+\mathbf{y}(t))=\frac{d\mathbf{x}(t)}{dt}+\frac{d\mathbf{y}(t)}{dt}=\mathbf{A}\mathbf{x}(t)+\mathbf{A}\mathbf{y}(t)=\mathbf{A}(\mathbf{x}(t)+\mathbf{y}(t)).$$

Hence, $\mathbf{x}(t)+\mathbf{y}(t)$ is also a solution of (4). □

An immediate corollary of Theorem 1 is that any linear combination of solutions of (4) is again a solution of (4). That is to say, if $\mathbf{x}^1(t),\ldots,\mathbf{x}^j(t)$ are j solutions of (4), then $c_1\mathbf{x}^1(t)+\ldots+c_j\mathbf{x}^j(t)$ is again a solution for any choice of constants $c_1, c_2, \ldots, c_j$. For example, consider the system of equations

$$\frac{dx_1}{dt} = x_2, \quad \frac{dx_2}{dt} = -4x_1, \quad \text{or} \quad \frac{d\mathbf{x}}{dt} = \begin{pmatrix} 0 & 1 \\ -4 & 0 \end{pmatrix}\mathbf{x}, \quad \mathbf{x} = \begin{pmatrix} x_1 \\ x_2 \end{pmatrix}. \tag{6}$$

This system of equations was derived from the second-order scalar equation $(d^2y/dt^2) + 4y = 0$ by setting $x_1 = y$ and $x_2 = dy/dt$. Since $y_1(t) = \cos 2t$ and $y_2(t) = \sin 2t$ are two solutions of the scalar equation, we know that

$$\begin{aligned} \mathbf{x}(t) = \begin{pmatrix} x_1(t) \\ x_2(t) \end{pmatrix} &= c_1 \begin{pmatrix} \cos 2t \\ -2\sin 2t \end{pmatrix} + c_2 \begin{pmatrix} \sin 2t \\ 2\cos 2t \end{pmatrix} \\ &= \begin{pmatrix} c_1 \cos 2t + c_2 \sin 2t \\ -2c_1 \sin 2t + 2c_2 \cos 2t \end{pmatrix} \end{aligned}$$

is a solution of (6) for any choice of constants c_1 and c_2.

The next step in our game plan is to show that every solution of (4) can be expressed as a linear combination of finitely many solutions. Equivalently, we seek to determine how many solutions we must find before we can generate all the solutions of (4). There is a branch of mathematics known as linear algebra, which addresses itself to exactly this question, and it is to this area that we now turn our attention.

Exercises

In each of Exercises 1–3 convert the given differential equation for the single variable y into a system of first-order equations.

1. $\dfrac{d^3y}{dt^3} + \left(\dfrac{dy}{dt}\right)^2 = 0$ **2.** $\dfrac{d^3y}{dt^3} + \cos y = e^t$ **3.** $\dfrac{d^4y}{dt^4} + \dfrac{d^2y}{dt^2} = 1$

4. Convert the pair of second-order equations

$$\frac{d^2y}{dt^2} + 3\frac{dz}{dt} + 2y = 0, \qquad \frac{d^2z}{dt^2} + 3\frac{dy}{dt} + 2z = 0$$

into a system of 4 first-order equations for the variables

$$x_1 = y, \qquad x_2 = y', \qquad x_3 = z, \quad \text{and} \quad x_4 = z'.$$

5. (a) Let $y(t)$ be a solution of the equation $y'' + y' + y = 0$. Show that

$$\mathbf{x}(t) = \begin{pmatrix} y(t) \\ y'(t) \end{pmatrix}$$

is a solution of the system of equations

$$\dot{\mathbf{x}} = \begin{pmatrix} 0 & 1 \\ -1 & -1 \end{pmatrix}\mathbf{x}.$$

(b) Let

$$\mathbf{x}(t)=\begin{pmatrix} x_1(t) \\ x_2(t) \end{pmatrix}$$

be a solution of the system of equations

$$\dot{\mathbf{x}}=\begin{pmatrix} 0 & 1 \\ -1 & -1 \end{pmatrix}\mathbf{x}.$$

Show that $y=x_1(t)$ is a solution of the equation $y''+y'+y=0$.

In each of Exercises 6–9, write the given system of differential equations and initial values in the form $\dot{\mathbf{x}}=\mathbf{A}\mathbf{x}, \mathbf{x}(t_0)=\mathbf{x}^0$.

6. $\dot{x}_1=3x_1-7x_2, \quad x_1(0)=1$
$\dot{x}_2=4x_1, \quad x_2(0)=1$

7. $\dot{x}_1=5x_1+5x_2, \quad x_1(3)=0$
$\dot{x}_2=-x_1+7x_2, \quad x_2(3)=6$

8. $\dot{x}_1=x_1+x_2-x_3, \quad x_1(0)=0$
$\dot{x}_2=3x_1-x_2+4x_3, \quad x_2(0)=1$
$\dot{x}_3=-x_1-x_2, \quad x_3(0)=-1$

9. $\dot{x}_1=-x_3, \quad x_1(-1)=2$
$\dot{x}_2=x_1, \quad x_2(-1)=3$
$\dot{x}_3=-x_2, \quad x_3(-1)=4$

10. Let

$$\mathbf{x}=\begin{pmatrix} 1 \\ 3 \\ 2 \end{pmatrix} \quad \text{and} \quad \mathbf{y}=\begin{pmatrix} -1 \\ 0 \\ 4 \end{pmatrix}.$$

Compute $\mathbf{x}+\mathbf{y}$ and $3\mathbf{x}-2\mathbf{y}$.

11. Let

$$\mathbf{A}=\begin{pmatrix} 1 & 2 & -1 \\ 3 & 0 & 4 \\ -1 & -1 & 2 \end{pmatrix}.$$

Compute $\mathbf{A}\mathbf{x}$ if

$$\text{(a) } \mathbf{x}=\begin{pmatrix} 1 \\ 0 \\ 0 \end{pmatrix}, \quad \text{(b) } \mathbf{x}=\begin{pmatrix} 0 \\ 1 \\ 0 \end{pmatrix}, \quad \text{(c) } \mathbf{x}=\begin{pmatrix} 0 \\ 0 \\ 1 \end{pmatrix}.$$

12. Let $\mathbf{A}$ be any $n\times n$ matrix and let $\mathbf{e}^j$ be the vector whose jth component is 1 and whose remaining components are zero. Verify that the vector $\mathbf{A}\mathbf{e}^j$ is the jth column of $\mathbf{A}$.

13. Let

$$\mathbf{A}=\begin{pmatrix} -1 & 6 & 0 \\ 2 & 1 & 3 \\ -1 & 0 & 2 \end{pmatrix}.$$

Compute $\mathbf{A}\mathbf{x}$ if

$$\text{(a) } \mathbf{x}=\begin{pmatrix} 1 \\ 2 \\ 4 \end{pmatrix}, \quad \text{(b) } \mathbf{x}=\begin{pmatrix} 1 \\ -1 \\ -1 \end{pmatrix}, \quad \text{(c) } \mathbf{x}=\begin{pmatrix} 1 \\ 1 \\ 1 \end{pmatrix}, \quad \text{(d) } \mathbf{x}=\begin{pmatrix} 1 \\ 0 \\ 1 \end{pmatrix}.$$

14. Let $\mathbf{A}$ be a 3×3 matrix with the property that

$$\mathbf{A}\begin{pmatrix} 1 \\ 1 \\ 1 \end{pmatrix}=\begin{pmatrix} 3 \\ 2 \\ 6 \end{pmatrix} \quad \text{and} \quad \mathbf{A}\begin{pmatrix} 1 \\ -1 \\ -1 \end{pmatrix}=\begin{pmatrix} 1 \\ 2 \\ 3 \end{pmatrix}.$$

Compute

$$\mathbf{A}\begin{pmatrix} 3 \\ -1 \\ -1 \end{pmatrix}.$$

Hint: Write

$$\begin{pmatrix} 3 \\ -1 \\ -1 \end{pmatrix}$$

as a linear combination of

$$\begin{pmatrix} 1 \\ 1 \\ 1 \end{pmatrix} \quad \text{and} \quad \begin{pmatrix} 1 \\ -1 \\ -1 \end{pmatrix}.$$

15. Let **A** be a 2×2 matrix with the property that

$$\mathbf{A}\begin{pmatrix} 1 \\ 1 \end{pmatrix}=\begin{pmatrix} 4 \\ 2 \end{pmatrix} \quad \text{and} \quad \mathbf{A}\begin{pmatrix} 1 \\ -1 \end{pmatrix}=\begin{pmatrix} -3 \\ -6 \end{pmatrix}.$$

Find **A**. *Hint*: The easy way is to use Exercise 12.

3.2 Vector spaces

In the previous section we defined, in a natural manner, a process of adding two vectors $\mathbf{x}$ and $\mathbf{y}$ together to form a new vector $\mathbf{z}=\mathbf{x}+\mathbf{y}$, and a process of multiplying a vector $\mathbf{x}$ by a scalar c to form a new vector $\mathbf{u}=c\mathbf{x}$. The former process was called vector addition and the latter process was called scalar multiplication. Our study of linear algebra begins with the more general premise that we have a set $\mathbf{V}$ of elements $\mathbf{x},\mathbf{y},\mathbf{z},\ldots$ and that we have one process that combines two elements $\mathbf{x}$ and $\mathbf{y}$ of $\mathbf{V}$ to form a third element $\mathbf{z}$ in $\mathbf{V}$ and a second process that combines a number c and an element $\mathbf{x}$ in $\mathbf{V}$ to form a new element $\mathbf{u}$ in $\mathbf{V}$. We will denote the first process by addition; that is, we will write $\mathbf{z}=\mathbf{x}+\mathbf{y}$, and the second process by scalar multiplication; that is, we will write $\mathbf{u}=c\mathbf{x}$, if they satisfy the usual axioms of addition and multiplication. These axioms are:

(i) $\mathbf{x}+\mathbf{y}=\mathbf{y}+\mathbf{x}$ (commutative law)
(ii) $\mathbf{x}+(\mathbf{y}+\mathbf{z})=(\mathbf{x}+\mathbf{y})+\mathbf{z}$ (associative law)
(iii) There is a unique element in $\mathbf{V}$, called the zero element, and denoted by $\mathbf{0}$, having the property that $\mathbf{x}+\mathbf{0}=\mathbf{x}$ for all $\mathbf{x}$ in $\mathbf{V}$.
(iv) For each element $\mathbf{x}$ in $\mathbf{V}$ there is a unique element, denoted by $-\mathbf{x}$ and called minus $\mathbf{x}$, such that $\mathbf{x}+(-\mathbf{x})=\mathbf{0}$.
(v) $1\cdot\mathbf{x}=\mathbf{x}$ for all $\mathbf{x}$ in $\mathbf{V}$.
(vi) $(ab)\mathbf{x}=a(b\mathbf{x})$ for any numbers a,b and any element $\mathbf{x}$ in $\mathbf{V}$.
(vii) $a(\mathbf{x}+\mathbf{y})=a\mathbf{x}+a\mathbf{y}$
(viii) $(a+b)\mathbf{x}=a\mathbf{x}+b\mathbf{x}$.

A set $\mathbf{V}$, together with processes addition and multiplication satisfying (i)–(viii) is said to be a *vector space* and its elements are called *vectors*. The numbers a,b will usually be real numbers, except in certain special cases where they will be complex numbers.

Remark 1. Implicit in axioms (i)–(viii) is the fact that if $\mathbf{x}$ and $\mathbf{y}$ are in $\mathbf{V}$, then the linear combination $a\mathbf{x}+b\mathbf{y}$ is again in $\mathbf{V}$ for any choice of constants a and b.

Remark 2. In the previous section we defined a vector $\mathbf{x}$ as a sequence of n numbers. In the more general context of this section, a quantity $\mathbf{x}$ is a vector by dint of its being in a vector space. That is to say, a quantity $\mathbf{x}$ is a vector if it belongs to a set of elements $\mathbf{V}$ which is equipped with two processes (addition and scalar multiplication) which satisfy (i)–(viii). As we shall see in Example 3 below, the set of all sequences

$$\mathbf{x}=\begin{bmatrix} x_1 \\ x_2 \\ \vdots \\ x_n \end{bmatrix}$$

of n real numbers is a vector space (with the usual operations of vector addition and scalar multiplication defined in Section 3.1). Thus, our two definitions are consistent.

Example 1. Let $\mathbf{V}$ be the set of all functions $x(t)$ which satisfy the differential equation

$$\frac{d^2x}{dt^2}-x=0 \tag{1}$$

with the sum of two functions and the product of a function by a number being defined in the usual manner. That is to say,

$$(f_1+f_2)(t)=f_1(t)+f_2(t)$$

and

$$(cf)(t)=cf(t).$$

It is trivial to verify that $\mathbf{V}$ is a vector space. Observe first that if x^1 and x^2 are in $\mathbf{V}$, then every linear combination $a_1x^1+a_2x^2$ is in $\mathbf{V}$, since the differential equation (1) is linear. Moreover, axioms (i), (ii), and (v)–(viii) are automatically satisfied since all we are doing at any time t in function addition and multiplication of a function by a number is adding or multiplying two numbers together. The zero vector in $\mathbf{V}$ is the function whose value at any time t is zero; this function is in $\mathbf{V}$ since $x(t)\equiv 0$ is a solution of (1). Finally, the negative of any function in $\mathbf{V}$ is again in $\mathbf{V}$, since the negative of any solution of (1) is again a solution of (1).

Example 2. Let **V** be the set of all solutions $x(t)$ of the differential equation $(d^2x/dt^2)-6x^2=0$, with the sum of two functions and the product of a function by a number being defined in the usual manner. **V** is not a vector space since the sum of any two solutions, while being defined, is not necessarily in **V**. Similarly, the product of a solution by a constant is not necessarily in **V**. For example, the function $x(t)=1/t^2$ is in **V** since it satisfies the differential equation, but the function $2x(t)=2/t^2$ is not in **V** since it does not satisfy the differential equation.

Example 3. Let **V** be the set of all sequences

$$\mathbf{x}=\begin{bmatrix} x_1 \\ \vdots \\ x_n \end{bmatrix}$$

of n real numbers. Define $\mathbf{x}+\mathbf{y}$ and $c\mathbf{x}$ as the vector addition and scalar multiplication defined in Section 3.1. It is trivial to verify that **V** is a vector space under these operations. The zero vector is the sequence

$$\begin{bmatrix} 0 \\ 0 \\ \vdots \\ 0 \end{bmatrix}$$

and the vector $-\mathbf{x}$ is the vector

$$\begin{bmatrix} -x_1 \\ \vdots \\ -x_n \end{bmatrix}.$$

This space is usually called n dimensional Euclidean space and is denoted by R^n.

Example 4. Let **V** be the set of all sequences

$$\mathbf{x}=\begin{bmatrix} x_1 \\ \vdots \\ x_n \end{bmatrix}$$

of n complex numbers $x_1,\ldots,x_n$. Define $\mathbf{x}+\mathbf{y}$ and $c\mathbf{x}$, for any complex number c, as the vector addition and scalar multiplication defined in Section 3.1. Again, it is trivial to verify that **V** is a vector space under these operations. This space is usually called complex n dimensional space and is denoted by C^n.

Example 5. Let **V** be the set of all $n \times n$ matrices **A**. Define the sum of two matrices **A** and **B** to be the matrix obtained by adding together corresponding elements of **A** and **B**, and define the matrix $c\mathbf{A}$ to be the matrix obtained by multiplying every element of **A** by c. In other words,

$$\begin{bmatrix} a_{11} & a_{12} & \dots & a_{1n} \\ a_{21} & a_{22} & \dots & a_{2n} \\ \vdots & \vdots & & \vdots \\ a_{n1} & a_{n2} & \dots & a_{nn} \end{bmatrix} + \begin{bmatrix} b_{11} & b_{12} & \dots & b_{1n} \\ b_{21} & b_{22} & \dots & b_{2n} \\ \vdots & \vdots & & \vdots \\ b_{n1} & b_{n2} & \dots & b_{nn} \end{bmatrix} =$$

$$\begin{bmatrix} a_{11}+b_{11} & a_{12}+b_{12} & \dots & a_{1n}+b_{1n} \\ a_{21}+b_{21} & a_{22}+b_{22} & \dots & a_{2n}+b_{2n} \\ \vdots & \vdots & & \vdots \\ a_{n1}+b_{n1} & a_{n2}+b_{n2} & \dots & a_{nn}+b_{nn} \end{bmatrix}$$

and

$$c\begin{bmatrix} a_{11} & a_{12} & \dots & a_{1n} \\ a_{21} & a_{22} & \dots & a_{2n} \\ \vdots & \vdots & & \vdots \\ a_{n1} & a_{n2} & \dots & a_{nn} \end{bmatrix} = \begin{bmatrix} ca_{11} & ca_{12} & \dots & ca_{1n} \\ ca_{21} & ca_{22} & \dots & ca_{2n} \\ \vdots & \vdots & & \vdots \\ ca_{n1} & ca_{n2} & \dots & ca_{nn} \end{bmatrix}.$$

Axioms (i), (ii), and (v)–(viii) are automatically satisfied since all we are doing in adding two matrices together or multiplying a matrix by a number is adding or multiplying two numbers together. The zero vector, or the matrix **0**, is the matrix whose every element is the number zero, and the negative of any matrix **A** is the matrix

$$\begin{bmatrix} -a_{11} & \dots & -a_{1n} \\ \vdots & & \vdots \\ -a_{n1} & \dots & -a_{nn} \end{bmatrix}.$$

Hence **V** is a vector space under these operations of matrix addition and scalar multiplication.

Example 6. We now present an example of a set of elements which comes close to being a vector space, but which doesn't quite make it. The purpose of this example is to show that the elements of **V** can be just about anything, and the operation of addition can be a rather strange process. Let **V** be the set consisting of three animals, a cat, a dog, and a mouse. Whenever any two of these animals meet, one eats up the other and changes into a different animal. The rules of eating are as follows.

(1) If a dog meets a cat, then the dog eats up the cat and changes into a mouse.
(2) If a dog meets another dog, then one dog eats up the other and changes into a cat.
(3) If a dog meets a mouse, then the dog eats up the mouse and remains unchanged.
(4) If a cat meets another cat, then one cat eats up the other and changes into a dog.
(5) If a cat meets a mouse, then the cat eats up the mouse and remains unchanged.
(6) If a mouse meets another mouse, then one mouse eats up the other and remains unchanged.

Clearly, "eating" is a process which combines two elements of **V** to form a third element in **V**. If we call this eating process addition, and denote it by $+$, then rules 1–6 can be written concisely in the form

1. $D+C=M$	2. $D+D=C$	3. $D+M=D$
4. $C+C=D$	5. $C+M=C$	6. $M+M=M$.

This operation of eating satisfies all the axioms of addition. To see this, note that axiom (i) is satisfied since the eating formulae do not depend on the order of the two animals involved. This is to say, $D+C=C+D$, etc. Moreover, the result of any addition is again an animal in **V**. This would not be the case, for example, if a dog ate up a cat and changed into a hippopotamus. The associative law (ii) is also satisfied, but it has to be verified explicitly. For example, suppose that we have an encounter between two cats and a dog. It is not obvious, a priori, that it does not make a difference whether the two cats meet first and their resultant meets the dog or whether one cat meets the dog and their resultant meets the other cat. To check that this is so we compute

$$(C+C)+D=D+D=C$$

and

$$(C+D)+C=M+C=C.$$

In a similar manner we can show that the result of any encounter between three animals is independent of the order in which they meet. Next, observe that the zero element in **V** is the mouse, since every animal is unchanged after eating a mouse. Finally, "minus a dog" is a cat (since $D+C=M$), "minus a cat" is a dog and "minus a mouse" is a mouse. However, **V** is *not* a vector space since there is no operation of scalar multiplication defined. Moreover, it is clearly impossible to define the quantities aC and aD, for all real numbers a, so as to satisfy axioms (v)–(viii).

Example 7. Let **V** be the set of all vector-valued solutions

$$\mathbf{x}(t)=\begin{bmatrix} x_1(t) \\ \vdots \\ x_n(t) \end{bmatrix}$$

of the vector differential equation

$$\dot{\mathbf{x}}=\mathbf{A}\mathbf{x}, \qquad \mathbf{A}=\begin{bmatrix} a_{11} & \dots & a_{1n} \\ \vdots & & \vdots \\ a_{n1} & \dots & a_{nn} \end{bmatrix}. \tag{2}$$

V is a vector space under the usual operations of vector addition and scalar multiplication. To wit, observe that axioms (i), (ii), and (v)–(viii) are automatically satisfied. Hence, we need only verify that

(a) The sum of any two solutions of (2) is again a solution.
(b) A constant times a solution of (2) is again a solution.
(c) The vector-valued function

$$\mathbf{x}(t)=\begin{bmatrix} x_1(t) \\ \vdots \\ x_n(t) \end{bmatrix}=\begin{bmatrix} 0 \\ \vdots \\ 0 \end{bmatrix}$$

is a solution of (2) (axiom (iii)).
(d) The negative of any solution of (2) is again a solution (axiom (iv)).

Now (a) and (b) are exactly Theorem 1 of the previous section, while (d) is a special case of (b). To verify (c) we observe that

$$\frac{d}{dt}\begin{bmatrix} 0 \\ \vdots \\ 0 \end{bmatrix}=\begin{bmatrix} 0 \\ \vdots \\ 0 \end{bmatrix} \quad \text{and} \quad \mathbf{A}\begin{bmatrix} 0 \\ \vdots \\ 0 \end{bmatrix}=\begin{bmatrix} 0 \\ \vdots \\ 0 \end{bmatrix}.$$

Hence the vector-valued function $\mathbf{x}(t)\equiv\mathbf{0}$ is always a solution of the differential equation (2).

Exercises

In each of Problems 1–6, determine whether the given set of elements

$$\mathbf{x}=\begin{bmatrix} x_1 \\ x_2 \\ x_3 \end{bmatrix}$$

form a vector space under the properties of vector addition and scalar multiplication defined in Section 3.1.

1. The set of all elements $\mathbf{x}=\begin{pmatrix} x_1\\ x_2\\ x_3 \end{pmatrix}$ where $3x_1-2x_2=0$

2. The set of all elements $\mathbf{x}=\begin{pmatrix} x_1\\ x_2\\ x_3 \end{pmatrix}$ where $x_1+x_2+x_3=0$

3. The set of all elements $\mathbf{x}=\begin{pmatrix} x_1\\ x_2\\ x_3 \end{pmatrix}$ where $x_1^2+x_2^2+x_3^2=1$

4. The set of all elements $\mathbf{x}=\begin{pmatrix} x_1\\ x_2\\ x_3 \end{pmatrix}$ where $x_1+x_2+x_3=1$

5. The set of elements $\mathbf{x}=\begin{pmatrix} 1\\ a\\ b \end{pmatrix}$ for all real numbers a and b

6. The set of all elements $\mathbf{x}=\begin{pmatrix} x_1\\ x_2\\ x_3 \end{pmatrix}$ where

$$x_1+x_2+x_3=0,\quad x_1-x_2+2x_3=0,\quad 3x_1-x_2+5x_3=0$$

In each of Problems 7–11 determine whether the given set of functions form a vector space under the usual operations of function addition and multiplication of a function by a constant.

7. The set of all polynomials of degree $\leqslant 4$

8. The set of all differentiable functions

9. The set of all differentiable functions whose derivative at $t=1$ is three

10. The set of all solutions of the differential equation $y''+y=\cos t$

11. The set of all functions $y(t)$ which have period 2π, that is $y(t+2\pi)=y(t)$

12. Show that the set of all vector-valued solutions

$$\mathbf{x}(t)=\begin{pmatrix} x_1(t)\\ x_2(t) \end{pmatrix}$$

of the system of differential equations

$$\frac{dx_1}{dt}=x_2+1,\qquad \frac{dx_2}{dt}=x_1+1$$

is not a vector space.

3.3 Dimension of a vector space

Let $\mathbf{V}$ be the set of all solutions $y(t)$ of the second-order linear homogeneous equation $(d^2y/dt^2)+p(t)(dy/dt)+q(t)y=0$. Recall that every solution $y(t)$ can be expressed as a linear combination of any two linearly independent solutions. Thus, if we knew two "independent" functions $y^1(t)$ and

$y^2(t)$ in **V**, then we could find every function in **V** by taking all linear combinations $c_1 y^1(t) + c_2 y^2(t)$ of y^1 and y^2. We would like to derive a similar property for solutions of the equation $\dot{\mathbf{x}} = \mathbf{A}\mathbf{x}$. To this end, we define the notion of a finite set of vectors generating the whole space, and the notion of independence of vectors in an arbitrary vector space **V**.

Definition. A set of vectors $\mathbf{x}^1, \mathbf{x}^2, \ldots, \mathbf{x}^n$ is said to *span* **V** if the set of all linear combinations $c_1\mathbf{x}^1 + c_2\mathbf{x}^2 + \ldots + c_n\mathbf{x}^n$ exhausts **V**. That is to say, the vectors $\mathbf{x}^1, \mathbf{x}^2, \ldots, \mathbf{x}^n$ span **V** if every element of **V** can be expressed as a linear combination of $\mathbf{x}^1, \mathbf{x}^2, \ldots, \mathbf{x}^n$.

Example 1. Let **V** be the set of all solutions of the differential equation $(d^2x/dt^2) - x = 0$. Let x^1 be the function whose value at any time t is e^t and let x^2 be the function whose value at any time t is e^{-t}. The functions x^1 and x^2 are in **V** since they satisfy the differential equation. Moreover, these functions also span **V** since every solution $x(t)$ of the differential equation can be written in the form

$$x(t) = c_1 e^t + c_2 e^{-t}$$

so that

$$x = c_1 x^1 + c_2 x^2.$$

Example 2. Let $\mathbf{V} = \mathsf{R}^n$ and let $\mathbf{e}^j$ denote the vector with a 1 in the jth place and zeros everywhere else, that is,

$$\mathbf{e}^1 = \begin{bmatrix} 1 \\ 0 \\ 0 \\ \vdots \\ 0 \end{bmatrix}, \quad \mathbf{e}^2 = \begin{bmatrix} 0 \\ 1 \\ 0 \\ \vdots \\ 0 \end{bmatrix}, \ldots, \mathbf{e}^n = \begin{bmatrix} 0 \\ 0 \\ \vdots \\ 0 \\ 1 \end{bmatrix}.$$

The set of vectors $\mathbf{e}^1, \mathbf{e}^2, \ldots, \mathbf{e}^n$ span R^n since any vector

$$\mathbf{x} = \begin{bmatrix} x_1 \\ x_2 \\ \vdots \\ x_n \end{bmatrix}$$

can be written in the form

$$\mathbf{x} = \begin{bmatrix} x_1 \\ 0 \\ \vdots \\ 0 \end{bmatrix} + \begin{bmatrix} 0 \\ x_2 \\ \vdots \\ 0 \end{bmatrix} + \ldots + \begin{bmatrix} 0 \\ 0 \\ \vdots \\ x_n \end{bmatrix} = x_1\mathbf{e}^1 + x_2\mathbf{e}^2 + \ldots + x_n\mathbf{e}^n.$$

Definition. A set of vectors $\mathbf{x}^1,\mathbf{x}^2,\ldots,\mathbf{x}^n$ in $\mathbf{V}$ is said to be *linearly dependent* if one of these vectors is a linear combination of the others. A very precise mathematical way of saying this is as follows. A set of vectors $\mathbf{x}^1,\mathbf{x}^2,\ldots,\mathbf{x}^n$ is said to be linearly dependent if there exist constants $c_1,c_2,\ldots,c_n$, *not all zero* such that

$$c_1\mathbf{x}^1+c_2\mathbf{x}^2+\ldots+c_n\mathbf{x}^n=\mathbf{0}.$$

These two definitions are equivalent, for if $\mathbf{x}^j$ is a linear combination of $\mathbf{x}^1,\ldots,\mathbf{x}^{j-1},\mathbf{x}^{j+1},\ldots,\mathbf{x}^n$, that is

$$\mathbf{x}^j=c_1\mathbf{x}^1+\ldots+c_{j-1}\mathbf{x}^{j-1}+c_{j+1}\mathbf{x}^{j+1}+\ldots+c_n\mathbf{x}^n,$$

then the linear combination

$$c_1\mathbf{x}^1+\ldots+c_{j-1}\mathbf{x}^{j-1}-\mathbf{x}^j+c_{j+1}\mathbf{x}^{j+1}+\ldots+c_n\mathbf{x}^n$$

equals zero and not all the constants are zero. Conversely, if $c_1\mathbf{x}^1+c_2\mathbf{x}^2+\ldots+c_n\mathbf{x}^n=\mathbf{0}$ and $c_j\neq 0$ for some j, then we can divide by c_j and solve for $\mathbf{x}^j$ as a linear combination of $\mathbf{x}^1,\ldots,\mathbf{x}^{j-1},\mathbf{x}^{j+1},\ldots,\mathbf{x}^n$. For example, if $c_1\neq 0$ then we can divide by c_1 to obtain that

$$\mathbf{x}^1=-\frac{c_2}{c_1}\mathbf{x}^2-\frac{c_3}{c_1}\mathbf{x}^3-\ldots-\frac{c_n}{c_1}\mathbf{x}^n.$$

Definition. If the vectors $\mathbf{x}^1,\mathbf{x}^2,\ldots,\mathbf{x}^n$ are not linearly dependent, that is, none of these vectors can be expressed as a linear combination of the others, then they are said to be *linearly independent*. The precise mathematical way of saying this is that the vectors $\mathbf{x}^1,\mathbf{x}^2,\ldots,\mathbf{x}^n$ are linearly independent if the equation

$$c_1\mathbf{x}^1+c_2\mathbf{x}^2+\ldots+c_n\mathbf{x}^n=\mathbf{0}$$

implies, of necessity, that all the constants $c_1,c_2,\ldots,c_n$ are zero.

In order to determine whether a set of vectors $\mathbf{x}^1,\mathbf{x}^2,\ldots,\mathbf{x}^n$ is linearly dependent or linearly independent, we write down the equation $c_1\mathbf{x}^1+c_2\mathbf{x}^2+\ldots+c_n\mathbf{x}^n=\mathbf{0}$ and see what this implies about the constants $c_1,c_2,\ldots,c_n$. If all these constants must be zero, then $\mathbf{x}^1,\mathbf{x}^2,\ldots,\mathbf{x}^n$ are linearly independent. On the other hand, if not all the constants $c_1,c_2,\ldots,c_n$ must be zero, then $\mathbf{x}^1,\mathbf{x}^2,\ldots,\mathbf{x}^n$ are linearly dependent.

Example 3. Let $\mathbf{V}=\mathsf{R}^3$ and let $\mathbf{x}^1$, $\mathbf{x}^2$, and $\mathbf{x}^3$ be the vectors

$$\mathbf{x}^1=\begin{pmatrix}1\\-1\\1\end{pmatrix},\quad \mathbf{x}^2=\begin{pmatrix}1\\2\\3\end{pmatrix}\quad\text{and}\quad \mathbf{x}^3=\begin{pmatrix}3\\0\\5\end{pmatrix}.$$

To determine whether these vectors are linearly dependent or linearly independent, we write down the equation $c_1\mathbf{x}^1 + c_2\mathbf{x}^2 + c_3\mathbf{x}^3 = \mathbf{0}$, that is

$$c_1\begin{bmatrix}1\\-1\\1\end{bmatrix} + c_2\begin{bmatrix}1\\2\\3\end{bmatrix} + c_3\begin{bmatrix}3\\0\\5\end{bmatrix} = \begin{bmatrix}0\\0\\0\end{bmatrix}.$$

The left-hand side of this equation is the vector

$$\begin{bmatrix}c_1+c_2+3c_3\\-c_1+2c_2\\c_1+3c_2+5c_3\end{bmatrix}.$$

Hence the constants c_1, c_2, and c_3 must satisfy the equations

$$c_1+c_2+3c_3=0, \tag{i}$$

$$-c_1+2c_2=0, \tag{ii}$$

$$c_1+3c_2+5c_3=0. \tag{iii}$$

Equation (ii) says that $c_1=2c_2$. Substituting this into Equations (i) and (iii) gives

$$3c_2+3c_3=0 \quad \text{and} \quad 5c_2+5c_3=0.$$

These equations have infinitely many solutions c_2, c_3 since they both reduce to the single equation $c_2+c_3=0$. One solution, in particular, is $c_2=-1$, $c_3=1$. Then, from Equation (ii), $c_1=-2$. Hence,

$$-2\begin{bmatrix}1\\-1\\1\end{bmatrix} - \begin{bmatrix}1\\2\\3\end{bmatrix} + \begin{bmatrix}3\\0\\5\end{bmatrix} = \begin{bmatrix}0\\0\\0\end{bmatrix}$$

and $\mathbf{x}^1$, $\mathbf{x}^2$ and $\mathbf{x}^3$ are linearly dependent vectors in R^3.

Example 4. Let $\mathbf{V}=\mathsf{R}^n$ and let $\mathbf{e}^1, \mathbf{e}^2, \ldots, \mathbf{e}^n$ be the vectors

$$\mathbf{e}^1 = \begin{bmatrix}1\\0\\0\\\vdots\\0\end{bmatrix}, \quad \mathbf{e}^2 = \begin{bmatrix}0\\1\\0\\\vdots\\0\end{bmatrix}, \ldots, \mathbf{e}^n = \begin{bmatrix}0\\0\\\vdots\\0\\1\end{bmatrix}.$$

To determine whether $\mathbf{e}^1, \mathbf{e}^2, \ldots, \mathbf{e}^n$ are linearly dependent or linearly independent, we write down the equation $c_1\mathbf{e}^1 + \ldots + c_n\mathbf{e}^n = \mathbf{0}$, that is

$$c_1\begin{bmatrix}1\\0\\0\\\vdots\\0\end{bmatrix} + c_2\begin{bmatrix}0\\1\\0\\\vdots\\0\end{bmatrix} + \ldots + c_n\begin{bmatrix}0\\0\\\vdots\\0\\1\end{bmatrix} = \begin{bmatrix}0\\0\\0\\\vdots\\0\end{bmatrix}.$$

The left-hand side of this equation is the vector

$$\begin{bmatrix} c_1 \\ c_2 \\ \vdots \\ c_n \end{bmatrix}.$$

Hence $c_1=0, c_2=0,\ldots,$ and $c_n=0$. Consequently, $\mathbf{e}^1,\mathbf{e}^2,\ldots,\mathbf{e}^n$ are linearly independent vectors in R^n.

Definition. The *dimension* of a vector space $\mathbf{V}$, denoted by $\dim\mathbf{V}$, is the fewest number of linearly independent vectors which span $\mathbf{V}$. $\mathbf{V}$ is said to be a finite dimensional space if its dimension is finite. On the other hand, $\mathbf{V}$ is said to be an infinite dimensional space if no set of finitely many elements span $\mathbf{V}$.

The dimension of a space $\mathbf{V}$ can be characterized as the fewest number of elements that we have to find in order to know all the elements of $\mathbf{V}$. In this sense, the definition of dimension captures our intuitive feeling. However, it is extremely difficult to compute the dimension of a space $\mathbf{V}$ from this definition alone. For example, let $\mathbf{V}=\mathsf{R}^n$. We have shown in Examples 2 and 4 that the vectors $\mathbf{e}^1,\mathbf{e}^2,\ldots,\mathbf{e}^n$ are linearly independent and span $\mathbf{V}$. Moreover, it seems intuitively obvious to us that we cannot generate R^n from fewer than n vectors. Thus, the dimension of R^n should be n. But how can we prove this rigorously? To wit, how can we prove that it is impossible to find a set of $(n-1)$ linearly independent vectors that span R^n? Thus, our definition of dimension is not, as yet, a very useful one. However, it will become extremely useful after we prove the following theorem.

Theorem 2. *If n linearly independent vectors span* $\mathbf{V}$, *then* $\dim\mathbf{V}=n$.

We will need two lemmas to prove Theorem 2. The first lemma concerns itself with the solutions of simultaneous linear equations and can be motivated as follows. Suppose that we are interested in determining n unknown numbers $x_1,x_2,\ldots,x_n$ uniquely. It seems pretty reasonable that we should be given n equations satisfied by these unknowns. If we are given too few equations then there may be many different solutions, that is, many different sets of values for $x_1,x_2,\ldots,x_n$ which satisfy the given equations. Lemma 1 proves this in the special case that we have m homogeneous linear equations for $n>m$ unknowns.

Lemma 1. *A set of m homogeneous linear equations for n unknowns $x_1,x_2,\ldots,x_n$ always admits a nontrivial solution if $m<n$. That is to say,*

the set of m equations in n unknowns

$$
\begin{aligned}
a_{11}x_1 + a_{12}x_2 + \dots + a_{1n}x_n &= 0\\
a_{21}x_1 + a_{22}x_2 + \dots + a_{2n}x_n &= 0\\
&\vdots\\
a_{m1}x_1 + a_{m2}x_2 + \dots + a_{mn}x_n &= 0
\end{aligned}
\tag{1}
$$

always has a solution $x_1, x_2, \dots, x_n$, *other than* $x_1 = \dots = x_n = 0$, *if* $m < n$.

Remark. Notice that $x_1 = 0, x_2 = 0, \dots, x_n = 0$ is certainly one solution of the system of equations (1). Thus, Lemma 1 is telling us that these equations have more than one solution.

PROOF OF LEMMA 1. We will prove Lemma 1 by induction on m. To this end, observe that the lemma is certainly true if $m = 1$, for in this case we have a single equation of the form $a_{11}x_1 + a_{12}x_2 + \dots + a_{1n}x_n = 0$, with $n \geqslant 2$. We can find a nontrivial solution of this equation, if $a_{11} = 0$, by taking $x_1 = 1, x_2 = 0, \dots, x_n = 0$. We can find a nontrivial solution of this equation, if $a_{11} \neq 0$, by taking $x_2 = 1, \dots, x_n = 1$ and $x_1 = -(a_{12} + \dots + a_{1n})/a_{11}$.

For the next step in our induction proof, we assume that Lemma 1 is true for some integer $m = k$ and show that this implies that Lemma 1 is true for $m = k+1$, and $k+1 < n$. To this end, consider the $k+1$ equations for the n unknowns $x_1, x_2, \dots, x_n$

$$
\begin{aligned}
a_{11}x_1 + a_{12}x_2 + \dots + a_{1n}x_n &= 0\\
a_{21}x_1 + a_{22}x_2 + \dots + a_{2n}x_n &= 0\\
&\vdots\\
a_{k+1,1}x_1 + a_{k+1,2}x_2 + \dots + a_{k+1,n}x_n &= 0
\end{aligned}
\tag{2}
$$

with $k+1 < n$. If $a_{11}, a_{21}, \dots, a_{k+1,1}$ are all zero, then $x_1 = 1, x_2 = 0, \dots, x_n = 0$ is clearly a non-trivial solution. Hence, we may assume that at least one of these coefficients is not zero. Without any loss of generality, we may assume that $a_{11} \neq 0$, for otherwise we can take the equation with the nonzero coefficient of x_1 and relabel it as the first equation. Then

$$x_1 = -\frac{a_{12}}{a_{11}}x_2 - \frac{a_{13}}{a_{11}}x_3 - \dots - \frac{a_{1n}}{a_{11}}x_n.$$

Substituting this value of x_1 in the second through the $(k+1)$st equations, we obtain the equivalent equations

$$
\begin{aligned}
a_{11}x_1 + a_{12}x_2 + a_{13}x_3 + \dots + a_{1n}x_n &= 0\\
b_{22}x_2 + b_{23}x_3 + \dots + b_{2n}x_n &= 0\\
&\vdots\\
b_{k2}x_2 + b_{k3}x_3 + \dots + b_{kn}x_n &= 0\\
b_{k+1,2}x_2 + b_{k+1,3}x_3 + \dots + b_{k+1,n}x_n &= 0
\end{aligned}
\tag{3}
$$

where $b_{ij} = a_{ij} - a_{i1}a_{1j}/a_{11}$. Now, the last k equations of (3) are k homogeneous linear equations for the $(n-1)$ unknowns $x_2,\dots,x_n$. Moreover, k is less than $n-1$ since $k+1$ is less than n. Hence, by the induction hypothesis, these equations have a nontrivial solution $x_2,\dots,x_n$. Once $x_2,\dots,x_n$ are known, we have as before $x_1 = -(a_{12}x_2 + \dots + a_{1n}x_n)/a_{11}$ from the first equation of (3). This establishes Lemma 1 for $m = k+1$, and therefore for all m, by induction. □

If a vector space $\mathbf{V}$ has dimension m, then it has m linearly independent vectors $\mathbf{x}^1,\dots,\mathbf{x}^m$ and every vector in the space can be written as a linear combination of the m vectors $\mathbf{x}^1,\mathbf{x}^2,\dots,\mathbf{x}^m$. It seems intuitively obvious to us in this case that there cannot be more than m linear independent vectors in $\mathbf{V}$. This is the content of Lemma 2.

Lemma 2. *In an m dimensional space, any set of $n > m$ vectors must be linearly dependent. In other words, the maximum number of linearly independent vectors in a finite dimensional space is the dimension of the space.*

Proof. Since $\mathbf{V}$ has dimension m, there exist m linearly independent vectors $\mathbf{x}^1,\mathbf{x}^2,\dots,\mathbf{x}^m$ which span $\mathbf{V}$. Let $\mathbf{y}^1,\mathbf{y}^2,\dots,\mathbf{y}^n$ be a set of n vectors in $\mathbf{V}$, with $n > m$. Since $\mathbf{x}^1,\mathbf{x}^2,\dots,\mathbf{x}^m$ span $\mathbf{V}$, all the $\mathbf{y}^j$ can be written as linear combinations of these vectors. That is to say, there exist constants a_{ij}, $1 \leqslant i \leqslant n$, $1 \leqslant j \leqslant m$ such that

$$
\begin{aligned}
\mathbf{y}^1 &= a_{11}\mathbf{x}^1 + a_{12}\mathbf{x}^2 + \dots + a_{1m}\mathbf{x}^m\\
\mathbf{y}^2 &= a_{21}\mathbf{x}^1 + a_{22}\mathbf{x}^2 + \dots + a_{2m}\mathbf{x}^m\\
&\vdots\\
\mathbf{y}^n &= a_{n1}\mathbf{x}^1 + a_{n2}\mathbf{x}^2 + \dots + a_{nm}\mathbf{x}^m.
\end{aligned}
\tag{4}
$$

To determine whether $\mathbf{y}^1, \mathbf{y}^2, \ldots, \mathbf{y}^n$ are linearly dependent or linearly independent, we consider the equation

$$c_1\mathbf{y}^1 + c_2\mathbf{y}^2 + \ldots + c_n\mathbf{y}^n = \mathbf{0}. \tag{5}$$

Using (4) we can rewrite (5) in the form

$$\begin{aligned}\mathbf{0} &= c_1\mathbf{y}^1 + c_2\mathbf{y}^2 + \ldots + c_n\mathbf{y}^n \\ &= (c_1a_{11} + \ldots + c_na_{n1})\mathbf{x}^1 + (c_1a_{12} + \ldots + c_na_{n2})\mathbf{x}^2 \\ &\quad + \ldots + (c_1a_{1m} + \ldots + c_na_{nm})\mathbf{x}^m.\end{aligned}$$

This equation states that a linear combination of $\mathbf{x}^1, \mathbf{x}^2, \ldots, \mathbf{x}^m$ is zero. Since $\mathbf{x}^1, \mathbf{x}^2, \ldots, \mathbf{x}^m$ are linearly independent, all these coefficients must be zero. Hence,

$$\begin{aligned} c_1a_{11} + c_2a_{21} + \ldots + c_na_{n1} &= 0 \\ c_1a_{12} + c_2a_{22} + \ldots + c_na_{n2} &= 0 \\ &\vdots \\ c_1a_{1m} + c_2a_{2m} + \ldots + c_na_{nm} &= 0. \end{aligned} \tag{6}$$

Now, observe that the system of Equations (6) is a set of m homogeneous linear equations for n unknowns $c_1, c_2, \ldots, c_n$, with $n > m$. By Lemma 1, these equations have a nontrivial solution. Thus, there exist constants $c_1, c_2, \ldots, c_n$, not all zero, such that $c_1\mathbf{y}^1 + c_2\mathbf{y}^2 + \ldots + c_n\mathbf{y}^n = \mathbf{0}$. Consequently, $\mathbf{y}^1, \mathbf{y}^2, \ldots, \mathbf{y}^n$ are linearly dependent. □

We are now in a position to prove Theorem 2.

Proof of Theorem 2. If n linearly independent vectors span $\mathbf{V}$, then, by the definition of dimension, $\dim \mathbf{V} \leqslant n$. By Lemma 2, $n \leqslant \dim \mathbf{V}$. Hence, $\dim \mathbf{V} = n$. □

Example 5. The dimension of R^n is n since $\mathbf{e}^1, \mathbf{e}^2, \ldots, \mathbf{e}^n$ are n linearly independent vectors which span R^n.

Example 6. Let $\mathbf{V}$ be the set of all 3×3 matrices

$$\mathbf{A} = \begin{bmatrix} a_{11} & a_{12} & a_{13} \\ a_{21} & a_{22} & a_{23} \\ a_{31} & a_{32} & a_{33} \end{bmatrix},$$

and let $\mathbf{E}_{ij}$ denote the matrix with a one in the ith row, jth column and zeros everywhere else. For example,

$$\mathbf{E}_{23} = \begin{bmatrix} 0 & 0 & 0 \\ 0 & 0 & 1 \\ 0 & 0 & 0 \end{bmatrix}.$$

To determine whether these matrices are linearly dependent or linearly independent, we consider the equation

$$\sum_{i,j=1}^{3} c_{ij}\mathbf{E}_{ij} = \mathbf{0} = \begin{bmatrix} 0 & 0 & 0 \\ 0 & 0 & 0 \\ 0 & 0 & 0 \end{bmatrix}. \tag{7}$$

Now, observe that the left-hand side of (7) is the matrix

$$c_{11}\begin{bmatrix} 1 & 0 & 0 \\ 0 & 0 & 0 \\ 0 & 0 & 0 \end{bmatrix} + c_{12}\begin{bmatrix} 0 & 1 & 0 \\ 0 & 0 & 0 \\ 0 & 0 & 0 \end{bmatrix} + \dots + c_{33}\begin{bmatrix} 0 & 0 & 0 \\ 0 & 0 & 0 \\ 0 & 0 & 1 \end{bmatrix} = \begin{bmatrix} c_{11} & c_{12} & c_{13} \\ c_{21} & c_{22} & c_{23} \\ c_{31} & c_{32} & c_{33} \end{bmatrix}.$$

Equating this matrix to the zero matrix gives $c_{11}=0, c_{12}=0,\dots,c_{33}=0$. Hence the 9 matrices $\mathbf{E}_{ij}$ are linearly independent. Moreover, these 9 matrices also span $\mathbf{V}$ since any matrix

$$\mathbf{A} = \begin{bmatrix} a_{11} & a_{12} & a_{13} \\ a_{21} & a_{22} & a_{23} \\ a_{31} & a_{32} & a_{33} \end{bmatrix}$$

can obviously be written in the form $\mathbf{A} = \sum_{i,j=1}^{3} a_{ij}\mathbf{E}_{ij}$. Hence $\dim \mathbf{V} = 9$.

Definition. If a set of linearly independent vectors span a vector space $\mathbf{V}$, then this set of vectors is said to be a *basis* for $\mathbf{V}$. A basis may also be called a *coordinate system*. For example, the vectors

$$\mathbf{e}^1 = \begin{bmatrix} 1 \\ 0 \\ 0 \\ 0 \end{bmatrix}, \quad \mathbf{e}^2 = \begin{bmatrix} 0 \\ 1 \\ 0 \\ 0 \end{bmatrix}, \quad \mathbf{e}^3 = \begin{bmatrix} 0 \\ 0 \\ 1 \\ 0 \end{bmatrix} \quad \text{and} \quad \mathbf{e}^4 = \begin{bmatrix} 0 \\ 0 \\ 0 \\ 1 \end{bmatrix}$$

are a basis for $\mathbf{R}^4$. If

$$\mathbf{x} = \begin{bmatrix} x_1 \\ x_2 \\ x_3 \\ x_4 \end{bmatrix},$$

then $\mathbf{x} = x_1\mathbf{e}^1 + x_2\mathbf{e}^2 + x_3\mathbf{e}^3 + x_4\mathbf{e}^4$, and relative to this basis the x_i are called "components" or "coordinates."

Corollary. *In a finite dimensional vector space, each basis has the same number of vectors, and this number is the dimension of the space.*

The following theorem is extremely useful in determining whether a set of vectors is a basis for $\mathbf{V}$.

Theorem 3. *Any n linearly independent vectors in an n dimensional space* $\mathbf{V}$ *must also span* $\mathbf{V}$. *That is to say, any n linearly independent vectors in an n dimensional space* $\mathbf{V}$ *are a basis for* $\mathbf{V}$.

PROOF. Let $\mathbf{x}^1,\mathbf{x}^2,\dots,\mathbf{x}^n$ be n linearly independent vectors in an n dimensional space $\mathbf{V}$. To show that they span $\mathbf{V}$, we must show that every $\mathbf{x}$ in $\mathbf{V}$ can be written as a linear combination of $\mathbf{x}^1,\mathbf{x}^2,\dots,\mathbf{x}^n$. To this end, pick any $\mathbf{x}$ in $\mathbf{V}$ and consider the set of vectors $\mathbf{x},\mathbf{x}^1,\mathbf{x}^2,\dots,\mathbf{x}^n$. This is a set of $(n+1)$ vectors in the n dimensional space $\mathbf{V}$; by Lemma 2, they must be linearly dependent. Consequently, there exist constants $c,c_1,c_2,\dots,c_n$, not all zero, such that

$$c\mathbf{x}+c_1\mathbf{x}^1+c_2\mathbf{x}^2+\dots+c_n\mathbf{x}^n=\mathbf{0}. \tag{8}$$

Now $c\neq 0$, for otherwise the set of vectors $\mathbf{x}^1,\mathbf{x}^2,\dots,\mathbf{x}^n$ would be linearly dependent. Therefore, we can divide both sides of (8) by c to obtain that

$$\mathbf{x}=-\frac{c_1}{c}\mathbf{x}^1-\frac{c_2}{c}\mathbf{x}^2-\dots-\frac{c_n}{c}\mathbf{x}^n.$$

Hence, any n linearly independent vectors in an n dimensional space $\mathbf{V}$ must also span $\mathbf{V}$. □

Example 7. Prove that the vectors

$$\mathbf{x}^1=\begin{pmatrix}1\\1\end{pmatrix}\quad\text{and}\quad\mathbf{x}^2=\begin{pmatrix}1\\-1\end{pmatrix}$$

form a basis for R^2.

Solution. To determine whether $\mathbf{x}^1$ and $\mathbf{x}^2$ are linearly dependent or linearly independent, we consider the equation

$$c_1\mathbf{x}^1+c_2\mathbf{x}^2=c_1\begin{pmatrix}1\\1\end{pmatrix}+c_2\begin{pmatrix}1\\-1\end{pmatrix}=\begin{pmatrix}0\\0\end{pmatrix}. \tag{9}$$

Equation (9) implies that $c_1+c_2=0$ and $c_1-c_2=0$. Adding these two equations gives $c_1=0$ while subtracting these two equations gives $c_2=0$. Consequently, $\mathbf{x}^1$ and $\mathbf{x}^2$ are two linearly independent vectors in the two dimensional space R^2. Hence, by Theorem 3, they must also span $\mathbf{V}$.

EXERCISES

In each of Exercises 1–4, determine whether the given set of vectors is linearly dependent or linearly independent.

1. $\begin{pmatrix}1\\1\\1\end{pmatrix},\ \begin{pmatrix}1\\-1\\1\end{pmatrix}$ and $\begin{pmatrix}-4\\0\\-4\end{pmatrix}$ **2.** $\begin{pmatrix}1\\1\\0\end{pmatrix},\ \begin{pmatrix}0\\1\\1\end{pmatrix}$ and $\begin{pmatrix}1\\0\\1\end{pmatrix}$

3. $\begin{pmatrix}-1\\1\\1\end{pmatrix}$, $\begin{pmatrix}1\\-1\\1\end{pmatrix}$ and $\begin{pmatrix}1\\1\\-1\end{pmatrix}$ **4.** $\begin{pmatrix}1\\2\\6\end{pmatrix}$, $\begin{pmatrix}3\\-13\\7\end{pmatrix}$, $\begin{pmatrix}-1\\0\\1\end{pmatrix}$ and $\begin{pmatrix}-1\\-1\\-1\end{pmatrix}$

5. Let **V** be the set of all 2×2 matrices. Determine whether the following sets of matrices are linearly dependent or linearly independent in **V**.

(a) $\begin{pmatrix}1&0\\1&0\end{pmatrix}$, $\begin{pmatrix}1&0\\0&1\end{pmatrix}$, $\begin{pmatrix}0&1\\1&0\end{pmatrix}$ and $\begin{pmatrix}0&1\\0&1\end{pmatrix}$

(b) $\begin{pmatrix}1&0\\1&0\end{pmatrix}$, $\begin{pmatrix}1&0\\0&1\end{pmatrix}$, $\begin{pmatrix}0&1\\1&0\end{pmatrix}$ and $\begin{pmatrix}2&-2\\-1&1\end{pmatrix}$.

6. Let **V** be the space of all polynomials in t of degree $\leqslant 2$.
(a) Show that $\dim \mathbf{V}=3$.
(b) Let p_1, p_2 and p_3 be the three polynomials whose values at any time t are $(t-1)^2$, $(t-2)^2$, and $(t-1)(t-2)$ respectively. Show that p_1, p_2, and p_3 are linearly independent. Hence, conclude from Theorem 3 that p_1, p_2, and p_3 form a basis for **V**.

7. Let **V** be the set of all solutions of the differential equation $d^2y/dt^2-y=0$.
(a) Show that **V** is a vector space.
(b) Find a basis for **V**.

8. Let **V** be the set of all solutions of the differential equation $(d^3y/dt^3)+y=0$ which satisfy $y(0)=0$. Show that **V** is a vector space and find a basis for it.

9. Let **V** be the set of all polynomials $p(t)=a_0+a_1t+a_2t^2$ which satisfy

$$p(0)+2p'(0)+3p''(0)=0.$$

Show that **V** is a vector space and find a basis for it.

10. Let **V** be the set of all solutions

$$\mathbf{x}=\begin{bmatrix}x_1(t)\\x_2(t)\\x_3(t)\end{bmatrix}$$

of the differential equation

$$\dot{\mathbf{x}}=\begin{pmatrix}0&1&0\\0&0&1\\6&-11&6\end{pmatrix}\mathbf{x}.$$

Show that

$$\mathbf{x}^1(t)=\begin{bmatrix}e^t\\e^t\\e^t\end{bmatrix},\qquad \mathbf{x}^2(t)=\begin{bmatrix}e^{2t}\\2e^{2t}\\4e^{2t}\end{bmatrix},\quad\text{and}\quad \mathbf{x}^3(t)=\begin{bmatrix}e^{3t}\\3e^{3t}\\9e^{3t}\end{bmatrix}$$

form a basis for **V**.

11. Let $\mathbf{V}$ be a vector space. We say that $\mathbf{W}$ is a subspace of $\mathbf{V}$ if $\mathbf{W}$ is a subset of $\mathbf{V}$ which is itself a vector space. Let $\mathbf{W}$ be the subset of R^3 which consists of all vectors

$$\mathbf{x}=\begin{pmatrix} x_1\\ x_2\\ x_3\end{pmatrix}$$

which satisfy the equations

$$\begin{aligned} x_1+x_2+2x_3&=0\\ 2x_1-x_2+\ x_3&=0\\ 6x_1\qquad\ +6x_3&=0. \end{aligned}$$

Show that $\mathbf{W}$ is a subspace of R^3 and find a basis for it.

12. Prove that any n vectors which span an n dimensional vector space $\mathbf{V}$ must be linearly independent. *Hint*: Show that any set of linearly dependent vectors contains a linearly independent subset which also spans $\mathbf{V}$.

13. Let $\mathbf{v}^1,\mathbf{v}^2,\ldots,\mathbf{v}^n$ be n vectors in a vector space $\mathbf{V}$. Let $\mathbf{W}$ be the subset of $\mathbf{V}$ which consists of all linear combinations $c_1\mathbf{v}^1+c_2\mathbf{v}^2+\ldots+c_n\mathbf{v}^n$ of $\mathbf{v}^1,\mathbf{v}^2,\ldots,\mathbf{v}^n$. Show that $\mathbf{W}$ is a subspace of $\mathbf{V}$, and that $\dim\mathbf{W}\leqslant n$.

14. Let $\mathbf{V}$ be the set of all functions $f(t)$ which are analytic for $|t|<1$, that is, $f(t)$ has a power series expansion $f(t)=a_0+a_1t+a_2t^2+\ldots$ which converges for $|t|<1$. Show that $\mathbf{V}$ is a vector space, and that its dimension is infinite. *Hint*: $\mathbf{V}$ contains all polynomials.

15. Let $\mathbf{v}^1,\mathbf{v}^2,\ldots,\mathbf{v}^m$ be m linearly independent vectors in an n dimensional vector space $\mathbf{V}$, with $n>m$. Show that we can find vectors $\mathbf{v}^{m+1},\ldots,\mathbf{v}^n$ so that $\mathbf{v}^1,\mathbf{v}^2,\ldots,\mathbf{v}^m,\mathbf{v}^{m+1},\ldots,\mathbf{v}^n$ form a basis for $\mathbf{V}$. That is to say, any set of m linearly independent vectors in an $n>m$ dimensional space $\mathbf{V}$ can be completed to form a basis for $\mathbf{V}$.

16. Find a basis for R^3 which includes the vectors

$$\begin{pmatrix}1\\1\\0\end{pmatrix}\quad\text{and}\quad\begin{pmatrix}1\\3\\4\end{pmatrix}.$$

17. (a) Show that

$$\mathbf{v}^1=\begin{pmatrix}1\\0\\0\end{pmatrix},\qquad \mathbf{v}^2=\frac{1}{\sqrt{2}}\begin{pmatrix}0\\1\\1\end{pmatrix},\quad\text{and}\quad \mathbf{v}^3=\frac{1}{\sqrt{2}}\begin{pmatrix}0\\-1\\1\end{pmatrix}$$

are linearly independent in R^3.

(b) Let

$$\mathbf{x}=\begin{pmatrix}x_1\\x_2\\x_3\end{pmatrix}=x_1\mathbf{e}^1+x_2\mathbf{e}^2+x_3\mathbf{e}^3.$$

Since $\mathbf{v}^1, \mathbf{v}^2$, and $\mathbf{v}^3$ are linearly independent they are a basis and $\mathbf{x} = y_1\mathbf{v}^1 + y_2\mathbf{v}^2 + y_3\mathbf{v}^3$. What is the relationship between the original coordinates x_i and the new coordinates y_j?

(c) Express the relations between coordinates in the form $\mathbf{x} = \mathbf{B}\mathbf{y}$. Show that the columns of $\mathbf{B}$ are $\mathbf{v}^1$, $\mathbf{v}^2$, and $\mathbf{v}^3$.

3.4 Applications of linear algebra to differential equations

Recall that an important tool in solving the second-order linear homogeneous equation $(d^2y/dt^2) + p(t)(dy/dt) + q(t)y = 0$ was the existence–uniqueness theorem stated in Section 2.1. In a similar manner, we will make extensive use of Theorem 4 below in solving the homogeneous linear system of differential equations

$$\frac{d\mathbf{x}}{dt} = \mathbf{A}\mathbf{x}, \qquad \mathbf{x} = \begin{pmatrix} x_1 \\ \vdots \\ x_n \end{pmatrix}, \qquad \mathbf{A} = \begin{pmatrix} a_{11} & \dots & a_{1n} \\ \vdots & & \vdots \\ a_{n1} & \dots & a_{nn} \end{pmatrix}. \tag{1}$$

Theorem 4 (Existence–uniqueness theorem). *There exists one, and only one, solution of the initial-value problem*

$$\frac{d\mathbf{x}}{dt} = \mathbf{A}\mathbf{x}, \qquad \mathbf{x}(t_0) = \mathbf{x}^0 = \begin{pmatrix} x_1^0 \\ x_2^0 \\ \vdots \\ x_n^0 \end{pmatrix}. \tag{2}$$

Moreover, this solution exists for $-\infty < t < \infty$.

Theorem 4 is an extremely powerful theorem, and has many implications. In particular, if $\mathbf{x}(t)$ is a nontrivial solution, then $\mathbf{x}(t) \neq \mathbf{0}$ for any t. (If $\mathbf{x}(t^*) = \mathbf{0}$ for some t^*, then $\mathbf{x}(t)$ must be identically zero, since it, and the trivial solution, satisfy the same differential equation and have the same value at $t = t^*$.)

We have already shown (see Example 7, Section 3.2) that the space $\mathbf{V}$ of all solutions of (1) is a vector space. Our next step is to determine the dimension of $\mathbf{V}$.

Theorem 5. *The dimension of the space* $\mathbf{V}$ *of all solutions of the homogeneous linear system of differential equations* (1) *is* n.

PROOF. We will exhibit a basis for $\mathbf{V}$ which contains n elements. To this end, let $\boldsymbol{\phi}^j(t), j=1,\ldots,n$ be the solution of the initial-value problem

$$\frac{d\mathbf{x}}{dt}=\mathbf{A}\mathbf{x}, \qquad \mathbf{x}(0)=\mathbf{e}^j=\begin{bmatrix}0\\ \vdots\\ 0\\ 1\\ 0\\ \vdots\\ 0\end{bmatrix} -j\text{th row.} \tag{3}$$

For example, $\boldsymbol{\phi}^1(t)$ is the solution of the differential equation (1) which satisfies the initial condition

$$\boldsymbol{\phi}^1(0)=\mathbf{e}^1=\begin{bmatrix}1\\ 0\\ \vdots\\ 0\end{bmatrix}.$$

Note from Theorem 4 that $\boldsymbol{\phi}^j(t)$ exists for all t and is unique. To determine whether $\boldsymbol{\phi}^1,\boldsymbol{\phi}^2,\ldots,\boldsymbol{\phi}^n$ are linearly dependent or linearly independent vectors in $\mathbf{V}$, we consider the equation

$$c_1\boldsymbol{\phi}^1+c_2\boldsymbol{\phi}^2+\ldots+c_n\boldsymbol{\phi}^n=\mathbf{0} \tag{4}$$

where the zero on the right-hand side of (4) stands for the zero vector in $\mathbf{V}$ (that is, the vector whose every component is the zero function). We want to show that (4) implies $c_1=c_2=\ldots=c_n=0$. Evaluating both sides of (4) at $t=0$ gives

$$c_1\boldsymbol{\phi}^1(0)+c_2\boldsymbol{\phi}^2(0)+\ldots+c_n\boldsymbol{\phi}^n(0)=\begin{bmatrix}0\\ 0\\ \vdots\\ 0\end{bmatrix}=\mathbf{0}$$

or

$$c_1\mathbf{e}^1+c_2\mathbf{e}^2+\ldots+c_n\mathbf{e}^n=\mathbf{0}.$$

Since we know that $\mathbf{e}^1,\mathbf{e}^2,\ldots,\mathbf{e}^n$ are linearly independent in R^n, $c_1=c_2=\ldots=c_n=0$. We conclude, therefore, that $\boldsymbol{\phi}^1,\boldsymbol{\phi}^2,\ldots,\boldsymbol{\phi}^n$ are linearly independent vectors in $\mathbf{V}$.

Next, we claim that $\boldsymbol{\phi}^1,\boldsymbol{\phi}^2,\ldots,\boldsymbol{\phi}^n$ also span $\mathbf{V}$. To prove this, we must show that any vector $\mathbf{x}$ in $\mathbf{V}$ (that is, any solution $\mathbf{x}(t)$ of (1)) can be written as a linear combination of $\boldsymbol{\phi}^1,\boldsymbol{\phi}^2,\ldots,\boldsymbol{\phi}^n$. To this end, pick any $\mathbf{x}$ in $\mathbf{V}$, and let

$$\mathbf{c} = \begin{bmatrix} c_1 \\ c_2 \\ \vdots \\ c_n \end{bmatrix}$$

be the value of $\mathbf{x}$ at $t=0$ ($\mathbf{x}(0)=\mathbf{c}$). With these constants $c_1, c_2, \ldots, c_n$, construct the vector-valued function

$$\boldsymbol{\phi}(t) = c_1\boldsymbol{\phi}^1(t) + c_2\boldsymbol{\phi}^2(t) + \ldots + c_n\boldsymbol{\phi}^n(t).$$

We know that $\boldsymbol{\phi}(t)$ satisfies (1) since it is a linear combination of solutions. Moreover,

$$\boldsymbol{\phi}(0) = c_1\boldsymbol{\phi}^1(0) + c_2\boldsymbol{\phi}^2(0) + \ldots + c_n\boldsymbol{\phi}^n(0)$$

$$= c_1\begin{bmatrix} 1 \\ 0 \\ \vdots \\ 0 \end{bmatrix} + c_2\begin{bmatrix} 0 \\ 1 \\ \vdots \\ 0 \end{bmatrix} + \ldots + c_n\begin{bmatrix} 0 \\ 0 \\ \vdots \\ 1 \end{bmatrix} = \begin{bmatrix} c_1 \\ c_2 \\ \vdots \\ c_n \end{bmatrix} = \mathbf{x}(0).$$

Now, observe that $\mathbf{x}(t)$ and $\boldsymbol{\phi}(t)$ satisfy the same homogeneous linear system of differential equations, and that $\mathbf{x}(t)$ and $\boldsymbol{\phi}(t)$ have the same value at $t=0$. Consequently, by Theorem 4, $\mathbf{x}(t)$ and $\boldsymbol{\phi}(t)$ must be identical, that is

$$\mathbf{x}(t) \equiv \boldsymbol{\phi}(t) = c_1\boldsymbol{\phi}^1(t) + c_2\boldsymbol{\phi}^2(t) + \ldots + c_n\boldsymbol{\phi}^n(t).$$

Thus, $\boldsymbol{\phi}^1, \boldsymbol{\phi}^2, \ldots, \boldsymbol{\phi}^n$ also span $\mathbf{V}$. Therefore, by Theorem 2 of Section 3.3, $\dim \mathbf{V} = n$. □

Theorem 5 states that the space $\mathbf{V}$ of all solutions of (1) has dimension n. Hence, we need only guess, or by some means find, n linearly independent solutions of (1). Theorem 6 below establishes a test for linear independence of solutions. It reduces the problem of determining whether n solutions $\mathbf{x}^1, \mathbf{x}^2, \ldots, \mathbf{x}^n$ are linearly independent to the much simpler problem of determining whether their values $\mathbf{x}^1(t_0), \mathbf{x}^2(t_0), \ldots, \mathbf{x}^n(t_0)$ at an appropriate time t_0 are linearly independent vectors in R^n.

Theorem 6 (Test for linear independence). *Let* $\mathbf{x}^1, \mathbf{x}^2, \ldots, \mathbf{x}^k$ *be k solutions of* $\dot{\mathbf{x}} = \mathbf{A}\mathbf{x}$. *Select a convenient* t_0. *Then,* $\mathbf{x}^1, \ldots, \mathbf{x}^k$ *are linear independent solutions if, and only if,* $\mathbf{x}^1(t_0), \mathbf{x}^2(t_0), \ldots, \mathbf{x}^k(t_0)$ *are linearly independent vectors in* R^n.

Proof. Suppose that $\mathbf{x}^1, \mathbf{x}^2, \ldots, \mathbf{x}^k$ are linearly dependent solutions. Then, there exist constants $c_1, c_2, \ldots, c_k$, not all zero, such that

$$c_1\mathbf{x}^1 + c_2\mathbf{x}^2 + \ldots + c_k\mathbf{x}^k = \mathbf{0}.$$

Evaluating this equation at $t=t_0$ gives

$$c_1\mathbf{x}^1(t_0)+c_2\mathbf{x}^2(t_0)+\dots+c_k\mathbf{x}^k(t_0)=\begin{bmatrix}0\\0\\\vdots\\0\end{bmatrix}.$$

Hence $\mathbf{x}^1(t_0),\mathbf{x}^2(t_0),\dots,\mathbf{x}^k(t_0)$ are linearly dependent vectors in R^n.

Conversely, suppose that the values of $\mathbf{x}^1,\mathbf{x}^2,\dots,\mathbf{x}^k$ at some time t_0 are linearly dependent vectors in R^n. Then, there exist constants $c_1,c_2,\dots,c_k$, not all zero, such that

$$c_1\mathbf{x}^1(t_0)+c_2\mathbf{x}^2(t_0)+\dots+c_k\mathbf{x}^k(t_0)=\begin{bmatrix}0\\0\\\vdots\\0\end{bmatrix}=\mathbf{0}.$$

With this choice of constants $c_1,c_2,\dots,c_k$, construct the vector-valued function

$$\boldsymbol{\phi}(t)=c_1\mathbf{x}^1(t)+c_2\mathbf{x}^2(t)+\dots+c_k\mathbf{x}^k(t).$$

This function satisfies (1) since it is a linear combination of solutions. Moreover, $\boldsymbol{\phi}(t_0)=\mathbf{0}$. Hence, by Theorem 4, $\boldsymbol{\phi}(t)=\mathbf{0}$ for all t. This implies that $\mathbf{x}^1,\mathbf{x}^2,\dots,\mathbf{x}^k$ are linearly dependent solutions. $\square$

Example 1. Consider the system of differential equations

$$\begin{aligned}\frac{dx_1}{dt}&=x_2\\ \frac{dx_2}{dt}&=-x_1-2x_2\end{aligned}\qquad\text{or}\qquad \frac{d\mathbf{x}}{dt}=\begin{pmatrix}0&1\\-1&-2\end{pmatrix}\mathbf{x},\quad \mathbf{x}=\begin{pmatrix}x_1\\x_2\end{pmatrix}. \tag{5}$$

This system of equations arose from the single second-order equation

$$\frac{d^2y}{dt^2}+2\frac{dy}{dt}+y=0 \tag{6}$$

by setting $x_1=y$ and $x_2=dy/dt$. Since $y_1(t)=e^{-t}$ and $y_2(t)=te^{-t}$ are two solutions of (6), we see that

$$\mathbf{x}^1(t)=\begin{pmatrix}e^{-t}\\-e^{-t}\end{pmatrix}$$

and

$$\mathbf{x}^2(t)=\begin{pmatrix}te^{-t}\\(1-t)e^{-t}\end{pmatrix}$$

are two solutions of (5). To determine whether $\mathbf{x}^1$ and $\mathbf{x}^2$ are linearly dependent or linearly independent, we check whether their initial values

$$\mathbf{x}^1(0)=\begin{pmatrix}1\\-1\end{pmatrix}$$

and

$$\mathbf{x}^2(0)=\begin{pmatrix}0\\1\end{pmatrix}$$

are linearly dependent or linearly independent vectors in R^2. Thus, we consider the equation

$$c_1\mathbf{x}^1(0)+c_2\mathbf{x}^2(0)=\begin{pmatrix}c_1\\-c_1+c_2\end{pmatrix}=\begin{pmatrix}0\\0\end{pmatrix}.$$

This equation implies that both c_1 and c_2 are zero. Hence, $\mathbf{x}^1(0)$ and $\mathbf{x}^2(0)$ are linearly independent vectors in R^2. Consequently, by Theorem 6, $\mathbf{x}^1(t)$ and $\mathbf{x}^2(t)$ are linearly independent solutions of (5), and every solution $\mathbf{x}(t)$ of (5) can be written in the form

$$\mathbf{x}(t)=\begin{pmatrix}x_1(t)\\x_2(t)\end{pmatrix}=c_1\begin{pmatrix}e^{-t}\\-e^{-t}\end{pmatrix}+c_2\begin{pmatrix}te^{-t}\\(1-t)e^{-t}\end{pmatrix}$$

$$=\begin{pmatrix}(c_1+c_2t)e^{-t}\\(c_2-c_1-c_2t)e^{-t}\end{pmatrix}. \tag{7}$$

Example 2. Solve the initial-value problem

$$\frac{d\mathbf{x}}{dt}=\begin{pmatrix}0&1\\-1&-2\end{pmatrix}\mathbf{x},\qquad \mathbf{x}(0)=\begin{pmatrix}1\\1\end{pmatrix}.$$

Solution. From Example 1, every solution $\mathbf{x}(t)$ must be of the form (7). The constants c_1 and c_2 are determined from the initial conditions

$$\begin{pmatrix}1\\1\end{pmatrix}=\mathbf{x}(0)=\begin{pmatrix}c_1\\c_2-c_1\end{pmatrix}.$$

Therefore, $c_1=1$ and $c_2=1+c_1=2$. Hence

$$\mathbf{x}(t)=\begin{pmatrix}x_1(t)\\x_2(t)\end{pmatrix}=\begin{pmatrix}(1+2t)e^{-t}\\(1-2t)e^{-t}\end{pmatrix}.$$

Up to this point in studying (1) we have found the concepts of linear algebra such as vector space, dependence, dimension, basis, etc., and vector-matrix notation useful, but we might well ask is all this other than simply an appropriate and convenient language. If it were nothing else it would be worth introducing. Good notations are important in expressing mathematical ideas. However, it is more. It is a body of theory with many applications.

In Sections 3.6–3.8 we will reduce the problem of finding all solutions of (1) to the much simpler algebraic problem of solving simultaneous linear equations of the form

$$\begin{aligned} a_{11}x_1 + a_{12}x_2 + \dots + a_{1n}x_n &= b_1 \\ a_{21}x_1 + a_{22}x_2 + \dots + a_{2n}x_n &= b_2 \\ &\vdots \\ a_{n1}x_1 + a_{n2}x_2 + \dots + a_{nn}x_n &= b_n \end{aligned}.$$

Therefore, we will now digress to study the theory of simultaneous linear equations. Here too we will see the role played by linear algebra.

Exercises

In each of Exercises 1–4 find a basis for the set of solutions of the given differential equation.

1. $\dot{\mathbf{x}} = \begin{pmatrix} 0 & 1 \\ -1 & -1 \end{pmatrix}\mathbf{x}$ (*Hint*: Find a second-order differential equation satisfied by $x_1(t)$.)

2. $\dot{\mathbf{x}} = \begin{pmatrix} 0 & 1 & 0 \\ 0 & 0 & 1 \\ 2 & -1 & 2 \end{pmatrix}\mathbf{x}$ (*Hint*: Find a third-order differential equation satisfied by $x_1(t)$.)

3. $\dot{\mathbf{x}} = \begin{pmatrix} 1 & 0 \\ 2 & 1 \end{pmatrix}\mathbf{x}$

4. $\dot{\mathbf{x}} = \begin{pmatrix} 1 & 0 & 0 \\ 1 & 1 & 0 \\ 1 & 1 & 1 \end{pmatrix}\mathbf{x}$

For each of the differential equations 5–9 determine whether the given solutions are a basis for the set of all solutions.

5. $\dot{\mathbf{x}} = \begin{pmatrix} 0 & -1 \\ -1 & 0 \end{pmatrix}\mathbf{x}$; $\mathbf{x}^1(t) = \begin{pmatrix} e^t \\ -e^t \end{pmatrix}$, $\mathbf{x}^2(t) = \begin{pmatrix} e^{-t} \\ e^{-t} \end{pmatrix}$

6. $\dot{\mathbf{x}} = \begin{pmatrix} 4 & -2 & 2 \\ -1 & 3 & 1 \\ 1 & -1 & 5 \end{pmatrix}\mathbf{x}$; $\mathbf{x}^1(t) = \begin{bmatrix} e^{2t} \\ e^{2t} \\ 0 \end{bmatrix}$, $\mathbf{x}^2(t) = \begin{bmatrix} 0 \\ e^{4t} \\ e^{4t} \end{bmatrix}$, $\mathbf{x}^3(t) = \begin{bmatrix} e^{6t} \\ 0 \\ e^{6t} \end{bmatrix}$

7. $\dot{\mathbf{x}} = \begin{pmatrix} -3 & -2 & 3 \\ 1 & 0 & -3 \\ 1 & -2 & -1 \end{pmatrix}\mathbf{x}$; $\mathbf{x}^1(t) = \begin{bmatrix} e^{-2t} \\ e^{-2t} \\ e^{-2t} \end{bmatrix}$, $\mathbf{x}^2(t) = \begin{bmatrix} e^{2t} \\ -e^{2t} \\ e^{2t} \end{bmatrix}$, $\mathbf{x}^3(t) = \begin{bmatrix} -e^{-4t} \\ e^{-4t} \\ e^{-4t} \end{bmatrix}$

8. $\dot{\mathbf{x}} = \begin{pmatrix} -5 & 2 & -2 \\ 1 & -4 & -1 \\ -1 & 1 & -6 \end{pmatrix}\mathbf{x}$; $\mathbf{x}^1(t) = \begin{bmatrix} e^{-3t} \\ e^{-3t} \\ 0 \end{bmatrix}$, $\mathbf{x}^2(t) = \begin{bmatrix} 0 \\ e^{-5t} \\ e^{-5t} \end{bmatrix}$

9. $\dot{\mathbf{x}}=\begin{pmatrix} -5 & 2 & -2 \\ 1 & -4 & -1 \\ -1 & 1 & -6 \end{pmatrix}\mathbf{x};\quad \mathbf{x}^1(t)=\begin{bmatrix} e^{-3t} \\ e^{-3t} \\ 0 \end{bmatrix},$

$$\mathbf{x}^2(t)=\begin{bmatrix} 0 \\ e^{-5t} \\ e^{-5t} \end{bmatrix},\quad \mathbf{x}^3(t)=\begin{bmatrix} e^{-3t}+e^{-7t} \\ e^{-3t} \\ e^{-7t} \end{bmatrix},\quad \mathbf{x}^4(t)=\begin{bmatrix} 2e^{-7t} \\ e^{-5t} \\ e^{-5t}+2e^{-7t} \end{bmatrix}$$

10. Determine the solutions $\boldsymbol{\phi}^1,\boldsymbol{\phi}^2,\dots,\boldsymbol{\phi}^n$ (see proof of Theorem 5) for the system of differential equations in (a) Problem 5; (b) Problem 6; (c) Problem 7.

11. Let $\mathbf{V}$ be the vector space of all continuous functions on $(-\infty,\infty)$ to R^n (the values of $\mathbf{x}(t)$ lie in R^n). Let $\mathbf{x}^1,\mathbf{x}^2,\dots,\mathbf{x}^n$ be functions in $\mathbf{V}$.
   (a) Show that $\mathbf{x}^1(t_0),\dots,\mathbf{x}^n(t_0)$ linearly independent vectors in R^n for some t_0 implies $\mathbf{x}^1,\mathbf{x}^2,\dots,\mathbf{x}^n$ are linearly independent functions in $\mathbf{V}$.
   (b) Is it true that $\mathbf{x}^1(t_0),\dots,\mathbf{x}^n(t_0)$ linearly dependent in R^n for some t_0 implies $\mathbf{x}^1,\mathbf{x}^2,\dots,\mathbf{x}^n$ are linearly dependent functions in $\mathbf{V}$? Justify your answer.

12. Let $\mathbf{u}$ be a vector in $\mathsf{R}^n\,(\mathbf{u}\neq 0)$.
   (a) Is $\mathbf{x}(t)=t\mathbf{u}$ a solution of a linear homogeneous differential equation $\dot{\mathbf{x}}=\mathbf{A}\mathbf{x}$?
   (b) Is $\mathbf{x}(t)=e^{\lambda t}\mathbf{u}$?
   (c) Is $\mathbf{x}(t)=(e^t-e^{-t})\mathbf{u}$?
   (d) Is $\mathbf{x}(t)=(e^t+e^{-t})\mathbf{u}$?
   (e) Is $\mathbf{x}(t)=(e^{\lambda_1 t}+e^{\lambda_2 t})\mathbf{u}$?
   (f) For what functions $\phi(t)$ can $\mathbf{x}(t)=\phi(t)\mathbf{u}$ be a solution of some $\dot{\mathbf{x}}=\mathbf{A}\mathbf{x}$?

3.5 The theory of determinants

In this section we briefly review some of the salient features of *determinants*. Complete proofs of all our assertions can be found in Sections 3.5–3.7 in the unabridged version of this text.

There are three special classes of matrices whose determinants are trivial to compute.

1. *Diagonal matrices*: A matrix

$$\mathbf{A}=\begin{bmatrix} a_{11} & 0 & 0 & \dots & 0 \\ 0 & a_{22} & 0 & \dots & 0 \\ \vdots & \vdots & \vdots & & \vdots \\ 0 & 0 & 0 & \dots & a_{nn} \end{bmatrix},$$

whose nondiagonal elements are all zero, is called a diagonal matrix. Its determinant is the product of the diagonal elements $a_{11},a_{22},\dots,a_{nn}$.

2. *Lower diagonal matrices*: A matrix

$$\mathbf{A}=\begin{bmatrix} a_{11} & 0 & \dots & 0 \\ a_{21} & a_{22} & \dots & 0 \\ \vdots & \vdots & & \vdots \\ a_{n1} & a_{n2} & \dots & a_{nn} \end{bmatrix}$$

whose elements above the main diagonal are all zero, is called a lower diagonal matrix, and its determinant too is the product of the diagonal elements $a_{11},\ldots,a_{nn}$.

3. *Upper diagonal matrices*: A matrix

$$\mathbf{A}=\begin{bmatrix} a_{11} & a_{12} & \cdots & a_{1n} \\ 0 & a_{22} & \cdots & a_{2n} \\ \vdots & \vdots & \vdots & \vdots \\ 0 & 0 & 0 & a_{nn} \end{bmatrix}$$

whose elements below the main diagonal are all zero is called an upper diagonal matrix and its determinant too is the product of the diagonal elements $a_{11},\ldots,a_{nn}$.

Manipulations inside A with respect to effect on det(A)

Property 1. If we interchange any two rows of $\mathbf{A}$, then we change the sign of its determinant.

Property 2. If any two rows of $\mathbf{A}$ are equal, then $\det\mathbf{A}=0$.

Property 3. $\det c\mathbf{A}=c^n\det\mathbf{A}$.

Property 4. Let $\mathbf{B}$ be the matrix obtained from $\mathbf{A}$ by multiplying its ith row by a constant c. Then, $\det\mathbf{B}=c\det\mathbf{A}$.

Property 5. Let $\mathbf{A}^T$ be the matrix obtained from $\mathbf{A}$ by switching rows and columns. The matrix $\mathbf{A}^T$ is called the transpose of $\mathbf{A}$. For example, if

$$\mathbf{A}=\begin{bmatrix} 1 & 3 & 2 \\ 6 & 9 & 4 \\ -1 & 2 & 7 \end{bmatrix}, \quad \text{then} \quad \mathbf{A}^T=\begin{bmatrix} 1 & 6 & -1 \\ 3 & 9 & 2 \\ 2 & 4 & 7 \end{bmatrix}.$$

A concise way of saying this is $(\mathbf{A}^T)_{ij}=a_{ji}$. Then,

$$\det\mathbf{A}^T=\det\mathbf{A}.$$

Remark. It follows immediately from Properties 1, 2, and 5 that we change the sign of the determinant when we interchange two columns of $\mathbf{A}$, and that $\det\mathbf{A}=0$ if two columns of $\mathbf{A}$ are equal.

Property 6. If we add a multiple of one row of $\mathbf{A}$ to another row of $\mathbf{A}$, then we do not change the value of its determinant.

Remark. Everything we say about rows applies to columns, since $\det\mathbf{A}^T=\det\mathbf{A}$. Thus, we do not change the value of the determinant when we add a multiple of one column of $\mathbf{A}$ to another column of $\mathbf{A}$.

Property 6 is extremely important because it enables us to reduce the problem of computing any determinant to the much simpler problem of computing the determinant of an upper diagonal matrix. To wit, if

$$\mathbf{A}=\begin{bmatrix} a_{11} & a_{12} & \cdots & a_{1n} \\ a_{21} & a_{22} & \cdots & a_{2n} \\ \vdots & \vdots & & \vdots \\ a_{n1} & a_{n2} & \cdots & a_{nn} \end{bmatrix}$$

and $a_{11}\neq 0$, then we can add suitable multiples of the first row of **A** to the remaining rows of **A** so as to make the resulting values of $a_{21},\ldots,a_{n1}$ all zero. Similarly, we can add multiples of the resulting second row of **A** to the rows beneath it so as to make the resulting values of $a_{32},\ldots,a_{n2}$ all zero, and so on. We illustrate this method with the following example.

Example 1. Compute

$$\det\begin{bmatrix} 1 & -1 & 2 & 3 \\ 2 & 2 & 0 & 2 \\ 4 & 1 & -1 & -1 \\ 1 & 2 & 3 & 0 \end{bmatrix}.$$

Solution. Subtracting twice the first row from the second row; four times the first row from the third row; and the first row from the last row gives

$$\det\begin{bmatrix} 1 & -1 & 2 & 3 \\ 2 & 2 & 0 & 2 \\ 4 & 1 & -1 & -1 \\ 1 & 2 & 3 & 0 \end{bmatrix}=\det\begin{bmatrix} 1 & -1 & 2 & 3 \\ 0 & 4 & -4 & -4 \\ 0 & 5 & -9 & -13 \\ 0 & 3 & 1 & -3 \end{bmatrix}$$

$$=4\det\begin{bmatrix} 1 & -1 & 2 & 3 \\ 0 & 1 & -1 & -1 \\ 0 & 5 & -9 & -13 \\ 0 & 3 & 1 & -3 \end{bmatrix}.$$

Next, we subtract five times the second row of this latter matrix from the third row, and three times the second row from the fourth row. Then

$$\det\begin{bmatrix} 1 & -1 & 2 & 3 \\ 2 & 2 & 0 & 2 \\ 4 & 1 & -1 & -1 \\ 1 & 2 & 3 & 0 \end{bmatrix}=4\det\begin{bmatrix} 1 & -1 & 2 & 3 \\ 0 & 1 & -1 & -1 \\ 0 & 0 & -4 & -8 \\ 0 & 0 & 4 & 0 \end{bmatrix}.$$

Finally, adding the third row of this matrix to the fourth row gives

$$\det\begin{bmatrix} 1 & -1 & 2 & 3 \\ 2 & 2 & 0 & 2 \\ 4 & 1 & -1 & -1 \\ 1 & 2 & 3 & 0 \end{bmatrix}=4\det\begin{bmatrix} 1 & -1 & 2 & 3 \\ 0 & 1 & -1 & -1 \\ 0 & 0 & -4 & -8 \\ 0 & 0 & 0 & -8 \end{bmatrix}$$

$$=4(-4)(-8)=128.$$

(Alternately, we could have interchanged the third and fourth columns of the matrix

$$\begin{bmatrix} 1 & -1 & 2 & 3 \\ 0 & 1 & -1 & -1 \\ 0 & 0 & -4 & -8 \\ 0 & 0 & 4 & 0 \end{bmatrix}$$

to yield the same result.)

Remark. In exactly the same manner as we reduced the matrix **A** to an upper diagonal matrix, we can reduce the system of equations

$$\begin{aligned} a_{11}x_1 + a_{12}x_2 + \dots + a_{1n}x_n &= b_1 \\ a_{21}x_1 + a_{22}x_2 + \dots + a_{2n}x_n &= b_2 \\ &\vdots \\ a_{n1}x_1 + a_{n2}x_2 + \dots + a_{nn}x_n &= b_n \end{aligned}$$

to an equivalent system of the form

Gaussian elimination

$$\begin{aligned} c_{11}x_1 + c_{12}x_2 + \dots + c_{1n}x_n &= d_1 \\ c_{22}x_2 + \dots + c_{2n}x_n &= d_2 \\ &\vdots \\ c_{nn}x_n &= d_n. \end{aligned}$$

We can then solve (if $c_{nn} \neq 0$) for x_n from the last equation, for x_{n-1} from the $(n-1)$st equation, and so on.

Example 2. Find all solutions of the system of equations

$$\begin{aligned} x_1 + x_2 + x_3 &= 1 \\ -x_1 + x_2 + x_3 &= 2 \\ 2x_1 - x_2 + x_3 &= 3. \end{aligned}$$

Solution. Adding the first equation to the second equation and subtracting twice the first equation from the third equation gives

$$\begin{aligned} x_1 + x_2 + x_3 &= 1 \\ 2x_2 + 2x_3 &= 3 \\ -3x_2 - x_3 &= 1. \end{aligned}$$

Next, adding $\frac{3}{2}$ the second equation to the third equation gives

$$\begin{aligned} x_1 + x_2 + x_3 &= 1 \\ 2x_2 + 2x_3 &= 3 \\ 2x_3 &= \tfrac{11}{2}. \end{aligned}$$

Consequently, $x_3 = \frac{11}{4}$, $x_2 = (3 - \frac{11}{2})/2 = -\frac{5}{4}$, and $x_1 = 1 + \frac{5}{4} - \frac{11}{4} = -\frac{1}{2}$.

Definition. Let **A** and **B** be $n\times n$ matrices with elements a_{ij} and b_{ij} respectively. We define their product **AB** as the $n\times n$ matrix **C** whose ij element c_{ij} is the product of the ith row of **A** with jth column of **B**. That is to say

$$c_{ij}=\sum_{k=1}^{n} a_{ik}b_{kj}.$$

Alternately, if we write **B** in the form $\mathbf{B}=(\mathbf{b}^1,\mathbf{b}^2,\dots,\mathbf{b}^n)$,where $\mathbf{b}^j$ is the jth column of **B**, then we can express the product $\mathbf{C}=\mathbf{AB}$ in the form $\mathbf{C}=(\mathbf{Ab}^1,\mathbf{Ab}^2,\dots,\mathbf{Ab}^n)$, since the ith component of the vector $\mathbf{Ab}^j$ is $\sum_{k=1}^{n}a_{ik}b_{kj}$.

Example 3. Let

$$\mathbf{A}=\begin{pmatrix} 3 & 1 & -1\\ 0 & 2 & 1\\ 1 & 1 & 1\end{pmatrix}$$

and

$$\mathbf{B}=\begin{pmatrix} 1 & -1 & 0\\ 2 & -1 & 1\\ -1 & 0 & 0\end{pmatrix}.$$

Compute **AB**.

Solution.

$$\begin{pmatrix} 3 & 1 & -1\\ 0 & 2 & 1\\ 1 & 1 & 1\end{pmatrix}\begin{pmatrix} 1 & -1 & 0\\ 2 & -1 & 1\\ -1 & 0 & 0\end{pmatrix}=\begin{pmatrix} 3+2+1 & -3-1+0 & 0+1+0\\ 0+4-1 & 0-2+0 & 0+2+0\\ 1+2-1 & -1-1+0 & 0+1+0\end{pmatrix}$$

$$=\begin{pmatrix} 6 & -4 & 1\\ 3 & -2 & 2\\ 2 & -2 & 1\end{pmatrix}$$

Example 4. Let **A** and **B** be the matrices in Example 3. Compute **BA**.

Solution.

$$\begin{pmatrix} 1 & -1 & 0\\ 2 & -1 & 1\\ -1 & 0 & 0\end{pmatrix}\begin{pmatrix} 3 & 1 & -1\\ 0 & 2 & 1\\ 1 & 1 & 1\end{pmatrix}=\begin{pmatrix} 3+0+0 & 1-2+0 & -1-1+0\\ 6+0+1 & 2-2+1 & -2-1+1\\ -3+0+0 & -1+0+0 & 1+0+0\end{pmatrix}$$

$$=\begin{pmatrix} 3 & -1 & -2\\ 7 & 1 & -2\\ -3 & -1 & 1\end{pmatrix}$$

Remark 1. As Examples 3 and 4 indicate, it is generally not true that **AB**=

BA. It can be shown, though, that

$$\mathbf{A}(\mathbf{BC})=(\mathbf{AB})\mathbf{C}$$

for any three $n\times n$ matrices **A**, **B**, and **C**.

Remark 2. Let **I** denote the diagonal matrix

$$\begin{bmatrix} 1 & 0 & \dots & 0 \\ 0 & 1 & \dots & 0 \\ \vdots & \vdots & & \vdots \\ 0 & 0 & \dots & 1 \end{bmatrix}.$$

I is called the identity matrix since $\mathbf{IA}=\mathbf{AI}=\mathbf{A}$ (see Exercise 18) for any $n\times n$ matrix **A**.

The following two properties of determinants are extremely useful in many applications.

Property 7.

$$\det \mathbf{AB} = \det \mathbf{A}\times\det \mathbf{B}.$$

That is to say, the determinant of the product is the product of the determinants.

Property 8. Let $\mathbf{A}(i|j)$ denote the $(n-1)\times(n-1)$ matrix obtained from **A** by deleting the ith row and jth column of **A**. For example, if

$$\mathbf{A}=\begin{bmatrix} 1 & 2 & 6 \\ 0 & -1 & -3 \\ -4 & -5 & 0 \end{bmatrix}, \quad \text{then} \quad \mathbf{A}(2|3)=\begin{pmatrix} 1 & 2 \\ -4 & -5 \end{pmatrix}.$$

Let $c_{ij}=(-1)^{i+j}\det\mathbf{A}(i|j)$. Then

$$\det \mathbf{A} = \sum_{i=1}^{n} a_{ij}c_{ij}$$

for any choice of j between 1 and n. This process of computing determinants is known as "expansion by the elements of columns," and Property 8 states that it does not matter which column we choose to expand about. For example, let

$$\mathbf{A}=\begin{bmatrix} 1 & 3 & 2 & 6 \\ 9 & -1 & 0 & 7 \\ 2 & 1 & 3 & 4 \\ -1 & 6 & 3 & 5 \end{bmatrix}.$$

Expanding about the first, second, third, and fourth columns of **A**, respectively, gives

$$\begin{aligned}
\det \mathbf{A} &= \det\begin{bmatrix} -1 & 0 & 7 \\ 1 & 3 & 4 \\ 6 & 3 & 5 \end{bmatrix} - 9\det\begin{bmatrix} 3 & 2 & 6 \\ 1 & 3 & 4 \\ 6 & 3 & 5 \end{bmatrix} \\
&\quad + 2\det\begin{bmatrix} 3 & 2 & 6 \\ -1 & 0 & 7 \\ 6 & 3 & 5 \end{bmatrix} + \det\begin{bmatrix} 3 & 2 & 6 \\ -1 & 0 & 7 \\ 1 & 3 & 4 \end{bmatrix} \\
&= -3\det\begin{bmatrix} 9 & 0 & 7 \\ 2 & 3 & 4 \\ -1 & 3 & 5 \end{bmatrix} - \det\begin{bmatrix} 1 & 2 & 6 \\ 2 & 3 & 4 \\ -1 & 3 & 5 \end{bmatrix} \\
&\quad - \det\begin{bmatrix} 1 & 2 & 6 \\ 9 & 0 & 7 \\ -1 & 3 & 5 \end{bmatrix} + 6\det\begin{bmatrix} 1 & 2 & 6 \\ 9 & 0 & 7 \\ 2 & 3 & 4 \end{bmatrix} \\
&= 2\det\begin{bmatrix} 9 & -1 & 7 \\ 2 & 1 & 4 \\ -1 & 6 & 5 \end{bmatrix} + 3\det\begin{bmatrix} 1 & 3 & 6 \\ 9 & -1 & 7 \\ -1 & 6 & 5 \end{bmatrix} \\
&\quad - 3\det\begin{bmatrix} 1 & 3 & 6 \\ 9 & -1 & 7 \\ 2 & 1 & 4 \end{bmatrix} \\
&= -6\det\begin{bmatrix} 9 & -1 & 0 \\ 2 & 1 & 3 \\ -1 & 6 & 3 \end{bmatrix} + 7\det\begin{bmatrix} 1 & 3 & 2 \\ 2 & 1 & 3 \\ -1 & 6 & 3 \end{bmatrix} \\
&\quad - 4\det\begin{bmatrix} 1 & 3 & 2 \\ 9 & -1 & 0 \\ -1 & 6 & 3 \end{bmatrix} + 5\det\begin{bmatrix} 1 & 3 & 2 \\ 9 & -1 & 0 \\ 2 & 1 & 3 \end{bmatrix}.
\end{aligned}$$

Definition. Let **C** be the matrix whose ij element is

$$c_{ij} = (-1)^{i+j} \det \mathbf{A}(i|j)$$

The *classical adjoint* of **A**, denoted by adj **A**, is defined to be the transpose of **C**, that is,

$$\operatorname{adj} \mathbf{A} \equiv \mathbf{C}^T.$$

Property 9. For every $n \times n$ matrix **A**,

$$\operatorname{adj} \mathbf{A} \times \mathbf{A} = \mathbf{A} \times \operatorname{adj} \mathbf{A} = (\det \mathbf{A})\mathbf{I}. \tag{1}$$

Example 5. Let

$$\mathbf{A}=\begin{bmatrix}1 & 0 & -1\\ 1 & 1 & -1\\ 1 & 2 & 1\end{bmatrix}.$$

Compute adj $\mathbf{A}$ and verify directly the identity (1).
Solution.

$$\operatorname{adj}\mathbf{A}=\mathbf{C}^T=\begin{bmatrix}3 & -2 & 1\\ -2 & 2 & -2\\ 1 & 0 & 1\end{bmatrix}^T=\begin{bmatrix}3 & -2 & 1\\ -2 & 2 & 0\\ 1 & -2 & 1\end{bmatrix}$$

so that

$$\operatorname{adj}\mathbf{A}\times\mathbf{A}=\begin{bmatrix}3 & -2 & 1\\ -2 & 2 & 0\\ 1 & -2 & 1\end{bmatrix}\begin{bmatrix}1 & 0 & -1\\ 1 & 1 & -1\\ 1 & 2 & 1\end{bmatrix}=\begin{bmatrix}2 & 0 & 0\\ 0 & 2 & 0\\ 0 & 0 & 2\end{bmatrix}=2\mathbf{I}$$

and

$$\mathbf{A}\times\operatorname{adj}\mathbf{A}=\begin{bmatrix}1 & 0 & -1\\ 1 & 1 & -1\\ 1 & 2 & 1\end{bmatrix}\begin{bmatrix}3 & -2 & 1\\ -2 & 2 & 0\\ 1 & -2 & 1\end{bmatrix}=\begin{bmatrix}2 & 0 & 0\\ 0 & 2 & 0\\ 0 & 0 & 2\end{bmatrix}=2\mathbf{I}.$$

Now, $\det\mathbf{A}=1-2+1+2=2$. Hence,

$$\operatorname{adj}\mathbf{A}\times\mathbf{A}=\mathbf{A}\times\operatorname{adj}\mathbf{A}=(\det\mathbf{A})\mathbf{I}.$$

We consider now the system of equations

$$\mathbf{A}\mathbf{x}=\mathbf{b},\qquad \mathbf{A}=\begin{bmatrix}a_{11} & \dots & a_{1n}\\ \vdots & & \vdots\\ a_{n1} & \dots & a_{nn}\end{bmatrix},\quad \mathbf{x}=\begin{bmatrix}x_1\\ \vdots\\ x_n\end{bmatrix},\quad \mathbf{b}=\begin{bmatrix}b_1\\ \vdots\\ b_n\end{bmatrix}. \tag{2}$$

If $\mathbf{A}$ were a non-zero number instead of a matrix, we would divide both sides of (2) by $\mathbf{A}$ to obtain that $\mathbf{x}=\mathbf{b}/\mathbf{A}$. This expression, of course, does not make sense if $\mathbf{A}$ is a matrix. However, there is a way of deriving the solution $\mathbf{x}=\mathbf{b}/\mathbf{A}$ which *does* generalize to the case where $\mathbf{A}$ is an $n\times n$ matrix. To wit, if the number $\mathbf{A}$ were unequal to zero, then we can multiply both sides of (2) by the number $\mathbf{A}^{-1}$ to obtain that

$$\mathbf{A}^{-1}\mathbf{A}\mathbf{x}=\mathbf{x}=\mathbf{A}^{-1}\mathbf{b}.$$

Now, if we can define $\mathbf{A}^{-1}$ as an $n\times n$ matrix when $\mathbf{A}$ is an $n\times n$ matrix, then the expression $\mathbf{A}^{-1}\mathbf{b}$ would make perfectly good sense. This leads us to ask the following two questions.

Question 1: Given an $n\times n$ matrix $\mathbf{A}$, does there exist another $n\times n$ matrix, which we will call $\mathbf{A}^{-1}$, with the property that

$$\mathbf{A}^{-1}\mathbf{A}=\mathbf{A}\mathbf{A}^{-1}=\mathbf{I}?$$

Question 2: If $\mathbf{A}^{-1}$ exists, is it unique? That is to say, can there exist two *distinct* matrices $\mathbf{B}$ and $\mathbf{C}$ with the property that

$$\mathbf{BA}=\mathbf{AB}=\mathbf{I} \quad \text{and} \quad \mathbf{CA}=\mathbf{AC}=\mathbf{I}?$$

The answers to these two questions are supplied in Theorems 7 and 8.

Theorem 7. *An $n\times n$ matrix $\mathbf{A}$ has at most one inverse.*

Theorem 8. $\mathbf{A}^{-1}$ *exists if, and only if,* $\det\mathbf{A}\neq 0$, *and in this case*

$$\mathbf{A}^{-1}=\frac{1}{\det\mathbf{A}}\operatorname{adj}\mathbf{A}. \tag{3}$$

Next, suppose that $\det\mathbf{A}\neq 0$. Then, $\mathbf{A}^{-1}$ exists, and multiplying both sides of (2) by this matrix gives

$$\mathbf{A}^{-1}\mathbf{A}\mathbf{x}=\mathbf{I}\mathbf{x}=\mathbf{x}=\mathbf{A}^{-1}\mathbf{b}.$$

Hence, if a solution exists, it must be $\mathbf{A}^{-1}\mathbf{b}$. Moreover, this vector is a solution of (2) since

$$\mathbf{A}(\mathbf{A}^{-1}\mathbf{b})=\mathbf{A}\mathbf{A}^{-1}\mathbf{b}=\mathbf{I}\mathbf{b}=\mathbf{b}.$$

Thus, Equation (2) has a unique solution $\mathbf{x}=\mathbf{A}^{-1}\mathbf{b}$ if $\det\mathbf{A}\neq 0$.

Example 6. Find all solutions of the equation

$$\begin{bmatrix} 1 & 1 & 1\\ 2 & -2 & 2\\ 3 & 3 & -3\end{bmatrix}\mathbf{x}=\begin{bmatrix}0\\0\\0\end{bmatrix},\qquad \mathbf{x}=\begin{bmatrix}x_1\\x_2\\x_3\end{bmatrix}. \tag{4}$$

Solution.

$$\det\begin{bmatrix} 1 & 1 & 1\\ 2 & -2 & 2\\ 3 & 3 & -3\end{bmatrix}=\det\begin{bmatrix} 1 & 1 & 1\\ 0 & -4 & 0\\ 0 & 0 & -6\end{bmatrix}=24.$$

Hence, Equation (4) has a *unique* solution $\mathbf{x}$. But

$$\mathbf{x}=\begin{bmatrix}0\\0\\0\end{bmatrix}$$

is obviously one solution. Therefore,

$$\mathbf{x}=\begin{bmatrix}0\\0\\0\end{bmatrix}$$

is the unique solution of (4).

Remark. It is often quite cumbersome and time consuming to compute the inverse of an $n\times n$ matrix $\mathbf{A}$ from (3). This is especially true for $n\geqslant 4$. An alternate, and much more efficient way of computing $\mathbf{A}^{-1}$, is by means of "elementary row operations."

Definition. An elementary row operation on a matrix **A** is either
(i) an interchange of two rows,
(ii) the multiplication of one row by a non-zero number,
or
(iii) the addition of a multiple of one row to another row.

It can be shown that every matrix **A**, with $\det \mathbf{A} \neq 0$, can be transformed into the identity **I** by a systematic sequence of these operations. Moreover, if the same sequence of operations is then performed upon **I**, it is transformed into $\mathbf{A}^{-1}$. We illustrate this method with the following example.

Example 7. Find the inverse of the matrix

$$\mathbf{A}=\begin{pmatrix} 1 & 0 & -1 \\ 1 & 1 & -1 \\ 1 & 2 & 1 \end{pmatrix}.$$

Solution. The matrix **A** can be transformed into **I** by the following sequence of elementary row operations. The result of each step appears below the operation performed.

(a) We obtain zeros in the off-diagonal positions in the first column by subtracting the first row from both the second and third rows.

$$\begin{pmatrix} 1 & 0 & -1 \\ 0 & 1 & 0 \\ 0 & 2 & 2 \end{pmatrix}$$

(b) We obtain zeros in the off-diagonal positions in the second column by adding (-2) times the second row to the third row.

$$\begin{pmatrix} 1 & 0 & -1 \\ 0 & 1 & 0 \\ 0 & 0 & 2 \end{pmatrix}$$

(c) We obtain a one in the diagonal position in the third column by multiplying the third row by $\frac{1}{2}$.

$$\begin{pmatrix} 1 & 0 & -1 \\ 0 & 1 & 0 \\ 0 & 0 & 1 \end{pmatrix}$$

(d) Finally, we obtain zeros in the off-diagonal positions in the third column by adding the third row to the first row.

$$\begin{pmatrix} 1 & 0 & 0 \\ 0 & 1 & 0 \\ 0 & 0 & 1 \end{pmatrix}$$

If we perform the same sequence of elementary row operations upon **I**, we obtain the following sequence of matrices:

$$\begin{bmatrix} 1 & 0 & 0 \\ 0 & 1 & 0 \\ 0 & 0 & 1 \end{bmatrix}, \begin{bmatrix} 1 & 0 & 0 \\ -1 & 1 & 0 \\ -1 & 0 & 1 \end{bmatrix}, \begin{bmatrix} 1 & 0 & 0 \\ -1 & 1 & 0 \\ 1 & -2 & 1 \end{bmatrix},$$

$$\begin{bmatrix} 1 & 0 & 0 \\ -1 & 1 & 0 \\ \frac{1}{2} & -1 & \frac{1}{2} \end{bmatrix}, \begin{bmatrix} \frac{3}{2} & -1 & \frac{1}{2} \\ -1 & 1 & 0 \\ \frac{1}{2} & -1 & \frac{1}{2} \end{bmatrix}.$$

The last of these matrices is $\mathbf{A}^{-1}$.

The following two theorems, which are equivalent, summarize everything we need to know concerning the solutions of (2).

Theorem 9. (a) *The equation* $\mathbf{Ax}=\mathbf{b}$ *has a unique solution if the columns of* $\mathbf{A}$ *are linearly independent.*

(b) *The equation* $\mathbf{Ax}=\mathbf{b}$ *has either no solution, or infinitely many solutions, if the columns of* $\mathbf{A}$ *are linearly dependent.*

Theorem 10. *The equation* $\mathbf{Ax}=\mathbf{b}$ *has a unique solution* $\mathbf{x}=\mathbf{A}^{-1}\mathbf{b}$ *if* $\det\mathbf{A}\neq 0$. *The equation* $\mathbf{Ax}=\mathbf{b}$ *has either no solutions, or infinitely many solutions if* $\det\mathbf{A}=0$.

Corollary. *The equation* $\mathbf{Ax}=\mathbf{0}$ *has a nontrivial solution (that is, a solution*

$$\mathbf{x}=\begin{bmatrix} x_1 \\ \vdots \\ x_n \end{bmatrix}$$

with not all the x_i *equal to zero) if, and only if,* $\det\mathbf{A}=0$.

PROOF. Observe that

$$\mathbf{x}=\begin{bmatrix} 0 \\ \vdots \\ 0 \end{bmatrix}$$

is always one solution of the equation $\mathbf{Ax}=\mathbf{0}$. Hence, it is the only solution if $\det\mathbf{A}\neq 0$. On the other hand, there exist infinitely many solutions if $\det\mathbf{A}=0$, and all but one of these are nontrivial. □

Finally, we derive the following simple, but extremely important theorem concerning the vector $\mathbf{Ax}$.

Theorem 11. *Let* $\mathbf{A}$ *be an* $n\times n$ *matrix with elements* a_{ij}, *and let* $\mathbf{x}$ *be a vector with components* $x_1,x_2,\ldots,x_n$. *Let*

$$\mathbf{a}^j=\begin{bmatrix} a_{1j}\\ \vdots \\ a_{nj}\end{bmatrix}$$

denote the jth column of $\mathbf{A}$. *Then*

$$\mathbf{Ax}=x_1\mathbf{a}^1+x_2\mathbf{a}^2+\ldots+x_n\mathbf{a}^n.$$

PROOF. We will show that the vectors $\mathbf{Ax}$ and $x_1\mathbf{a}^1+\ldots+x_n\mathbf{a}^n$ have the same components. To this end, observe that $(\mathbf{Ax})_j$, the jth component of $\mathbf{Ax}$, is $a_{j1}x_1+\ldots+a_{jn}x_n$, while the jth component of the vector $x_1\mathbf{a}^1+\ldots+x_n\mathbf{a}^n$ is

$$x_1\mathbf{a}^1_j+\ldots+x_n\mathbf{a}^n_j=x_1a_{j1}+\ldots+x_na_{jn}=(\mathbf{Ax})_j.$$

Hence, $\mathbf{A}x=x_1\mathbf{a}^1+x_2\mathbf{a}^2+\ldots+x_n\mathbf{a}^n$. □

EXERCISES

In each of Problems 1–6 compute the determinant of the given matrix.

1. $\begin{pmatrix} 1 & 0 & 2\\ 1 & 2 & 5\\ 6 & 8 & 0\end{pmatrix}$

2. $\begin{bmatrix} 1 & t & t^2\\ t & t^2 & 1\\ t^2 & t & 1\end{bmatrix}$

3. $\begin{bmatrix} 0 & a & b & 0\\ -a & 0 & c & 0\\ -b & -c & 0 & 0\\ 0 & 0 & 0 & 1\end{bmatrix}$

4. $\begin{bmatrix} 2 & -1 & 6 & 3\\ 1 & 0 & 1 & -1\\ 1 & 3 & 0 & 2\\ 1 & -1 & 0 & 0\end{bmatrix}$

5. $\begin{bmatrix} 0 & 2 & 3 & -1\\ 1 & 8 & -1 & 1\\ 1 & -1 & 0 & -1\\ 0 & 2 & 6 & 1\end{bmatrix}$

6. $\begin{bmatrix} 1 & 1 & 1 & 1 & 1\\ 1 & 0 & 0 & 0 & 2\\ 0 & 1 & 0 & 0 & 3\\ 0 & 0 & 1 & 0 & 4\\ 0 & 0 & 0 & 1 & 5\end{bmatrix}$

7. Without doing any computations, show that

$$\det\begin{bmatrix} a & b & c\\ d & e & f\\ g & h & k\end{bmatrix}=\det\begin{bmatrix} e & b & h\\ d & a & g\\ f & c & k\end{bmatrix}.$$

In each of Problems 8–13, find all solutions of the given system of equations.

8. $$\begin{aligned} x_1+x_2-x_3&=0\\ 2x_1\quad\ +x_3&=14\\ x_2+x_3&=13 \end{aligned}$$

9. $$\begin{aligned} x_1+x_2+x_3&=6\\ x_1-x_2-x_3&=-4\\ x_2+x_3&=-1 \end{aligned}$$

10. $$\begin{aligned} x_1+x_2+x_3&=0\\ x_1-x_2-x_3&=0\\ x_2+x_3&=0 \end{aligned}$$

11. $$\begin{aligned} x_1+\ x_2+\ x_3-x_4&=1\\ x_1+2x_2-2x_3+x_4&=1\\ x_1+3x_2-3x_3-x_4&=1\\ x_1+4x_2-4x_3-x_4&=1 \end{aligned}$$

12. $$\begin{aligned} x_1+x_2+2x_3-x_4&=1\\ x_1-x_2+2x_3+x_4&=2\\ x_1+x_2+2x_3-x_4&=1\\ -x_1-x_2-2x_3+x_4&=0 \end{aligned}$$

13. $$\begin{aligned} x_1-\ x_2+x_3+\ x_4&=0\\ x_1+2x_2-x_3+3x_4&=0\\ 3x_1+3x_2-x_3+7x_4&=0\\ -x_1+2x_2+x_3-\ x_4&=0 \end{aligned}$$

In each of Problems 14–20, compute **AB** and **BA** for the given matrices **A** and **B**.

14. $\mathbf{A}=\begin{pmatrix}3&2&0\\1&1&1\\2&6&0\end{pmatrix},\quad \mathbf{B}=\begin{pmatrix}-1&0&9\\2&1&-1\\0&6&2\end{pmatrix}$

15. $\mathbf{A}=\begin{pmatrix}1&2\\3&4\end{pmatrix},\quad \mathbf{B}=\begin{pmatrix}4&3\\2&1\end{pmatrix}$

16. $\mathbf{A}=\begin{pmatrix}1&0&1\\1&1&0\\0&1&1\end{pmatrix},\quad \mathbf{B}=\begin{pmatrix}0&1&1\\1&1&0\\1&0&1\end{pmatrix}$

17. $\mathbf{A}=\begin{bmatrix}1&1&1&1\\1&1&1&1\\1&1&1&1\\1&1&1&1\end{bmatrix},\quad \mathbf{B}=\begin{bmatrix}1&1&1&1\\2&2&2&2\\3&3&3&3\\4&4&4&4\end{bmatrix}$

18. Show that **IA** = **AI** = **A** for all matrices **A**.

19. Show that any two diagonal matrices **A** and **B** *commute*, that is **AB** = **BA**, if **A** and **B** are diagonal matrices.

20. Suppose that **AD** = **DA** for all matrices **A**. Prove that **D** is a multiple of the identity matrix.

In each of Problems 21–26, find the inverse, if it exists, of the given matrix.

21. $\begin{pmatrix}-2&3&2\\6&0&3\\4&1&-1\end{pmatrix}$

22. $\begin{pmatrix}\cos\theta&0&-\sin\theta\\0&1&0\\\sin\theta&0&\cos\theta\end{pmatrix}$

23. $\begin{pmatrix}1&1&1\\-1&i&-i\\2&1&1\end{pmatrix}$

24. $\begin{pmatrix}1&2&-1\\3&1&2\\2&3&-1\end{pmatrix}$

25. $\begin{pmatrix}1&1+i&1-i\\1&0&0\\0&1&1\end{pmatrix}$

26. $\begin{pmatrix}0&1&1\\-1&0&1\\-1&-1&0\end{pmatrix}$

27. Let

$$\mathbf{A}=\begin{pmatrix} a_{11} & a_{12} \\ a_{21} & a_{22} \end{pmatrix}.$$

Show that

$$\mathbf{A}^{-1}=\frac{1}{\det \mathbf{A}}\begin{pmatrix} a_{22} & -a_{12} \\ -a_{21} & a_{11} \end{pmatrix}$$

if $\det \mathbf{A}\neq 0$.

28. Show that $(\mathbf{AB})^{-1}=\mathbf{B}^{-1}\mathbf{A}^{-1}$ if $\det \mathbf{A}\times\det \mathbf{B}\neq 0$.

In each of Problems 29–32 show that $\mathbf{x}=\mathbf{0}$ is the unique solution of the given system of equations.

29.
$$\begin{aligned} x_1 - x_2 - x_3 &= 0 \\ 3x_1 - x_2 + 2x_3 &= 0 \\ 2x_1 + 2x_2 + 3x_3 &= 0 \end{aligned}$$

30.
$$\begin{aligned} x_1 + 2x_2 + 4x_3 &= 0 \\ x_2 + x_3 &= 0 \\ x_1 + x_2 + x_3 &= 0 \end{aligned}$$

31.
$$\begin{aligned} x_1 + 2x_2 - x_3 \qquad &= 0 \\ 2x_1 + 3x_2 + x_3 - x_4 &= 0 \\ -x_1 \qquad + 2x_3 + 2x_4 &= 0 \\ 3x_1 - x_2 + x_3 + 3x_4 &= 0 \end{aligned}$$

32.
$$\begin{aligned} x_1 + 2x_2 - x_3 + 3x_4 &= 0 \\ 2x_1 + 3x_2 \qquad - x_4 &= 0 \\ -x_1 + x_2 + 2x_3 + x_4 &= 0 \\ -x_2 + 2x_3 + 3x_4 &= 0 \end{aligned}$$

In each of Problems 33–38, find all solutions of the given system of equations.

33. $\begin{pmatrix} 1 & 2 & -3 \\ 2 & 3 & -1 \\ 1 & 3 & 10 \end{pmatrix}\mathbf{x}=\mathbf{0}$

34. $\begin{pmatrix} 1 & 2 & 1 \\ -3 & -2 & 1 \\ 6 & 8 & 2 \end{pmatrix}\mathbf{x}=\begin{pmatrix} 3 \\ 5 \\ 4 \end{pmatrix}$

35. $\begin{bmatrix} 1 & 1 & 1 & 1 \\ 1 & -1 & 1 & 1 \\ 1 & 1 & -1 & 1 \\ 1 & 1 & 1 & -1 \end{bmatrix}\mathbf{x}=\mathbf{0}$

36. $\begin{pmatrix} 2 & -1 & -1 \\ -5 & 3 & 1 \\ -1 & 1 & -1 \end{pmatrix}\mathbf{x}=\begin{pmatrix} -1 \\ 2 \\ 3 \end{pmatrix}$

37. $\begin{bmatrix} 1 & 2 & 3 & 4 \\ 0 & 1 & -1 & 6 \\ 2 & 0 & 0 & 1 \\ 2 & 4 & 6 & 8 \end{bmatrix}\mathbf{x}=\mathbf{0}$

38. $\begin{pmatrix} 1 & 1 & 1 \\ 1 & -1 & 1 \\ 1 & 1 & -1 \end{pmatrix}\mathbf{x}=\begin{pmatrix} 1 \\ 1 \\ 1 \end{pmatrix}$

3.6 The eigenvalue–eigenvector method of finding solutions

We return now to the first-order linear homogeneous differential equation

$$\dot{\mathbf{x}}=\mathbf{A}\mathbf{x},\quad \mathbf{x}=\begin{bmatrix} x_1 \\ \vdots \\ x_n \end{bmatrix},\quad \mathbf{A}=\begin{bmatrix} a_{11} & \cdots & a_{1n} \\ \vdots & & \vdots \\ a_{n1} & \cdots & a_{nn} \end{bmatrix}. \tag{1}$$

Our goal is to find n linearly independent solutions $\mathbf{x}^1(t),\ldots,\mathbf{x}^n(t)$. Now, recall that both the first-order and second-order linear homogeneous scalar equations have exponential functions as solutions. This suggests that we try $\mathbf{x}(t)=e^{\lambda t}\mathbf{v}$, where $\mathbf{v}$ is a constant vector, as a solution of (1). To this end, observe that

$$\frac{d}{dt}e^{\lambda t}\mathbf{v}=\lambda e^{\lambda t}\mathbf{v}$$

and

$$\mathbf{A}(e^{\lambda t}\mathbf{v})=e^{\lambda t}\mathbf{A}\mathbf{v}.$$

Hence, $\mathbf{x}(t)=e^{\lambda t}\mathbf{v}$ is a solution of (1) if, and only if, $\lambda e^{\lambda t}\mathbf{v}=e^{\lambda t}\mathbf{A}\mathbf{v}$. Dividing both sides of this equation by $e^{\lambda t}$ gives

$$\mathbf{A}\mathbf{v}=\lambda\mathbf{v}. \tag{2}$$

Thus, $\mathbf{x}(t)=e^{\lambda t}\mathbf{v}$ is a solution of (1) if, and only if, λ and $\mathbf{v}$ satisfy (2).

Definition. A nonzero vector $\mathbf{v}$ satisfying (2) is called an *eigenvector* of $\mathbf{A}$ with *eigenvalue* λ.

Remark. The vector $\mathbf{v}=\mathbf{0}$ is excluded because it is uninteresting. Obviously, $\mathbf{A0}=\lambda\cdot\mathbf{0}$ for any number λ.

An eigenvector of a matrix $\mathbf{A}$ is a rather special vector: under the transformation $\mathbf{x}\rightarrow\mathbf{A}\mathbf{x}$, it goes into a multiple λ of itself. Vectors which are transformed into multiples of themselves play an important role in many applications. To find such vectors, we rewrite Equation (2) in the form

$$\mathbf{0}=\mathbf{A}\mathbf{v}-\lambda\mathbf{v}=(\mathbf{A}-\lambda\mathbf{I})\mathbf{v}. \tag{3}$$

But, Equation (3) has a nonzero solution $\mathbf{v}$ only if $\det(\mathbf{A}-\lambda\mathbf{I})=0$. Hence the eigenvalues λ of $\mathbf{A}$ are the roots of the equation

$$0=\det(\mathbf{A}-\lambda\mathbf{I})=\det\begin{bmatrix} a_{11}-\lambda & a_{12} & \dots & a_{1n} \\ a_{21} & a_{22}-\lambda & \dots & a_{2n} \\ \vdots & \vdots & & \vdots \\ a_{n1} & a_{n2} & \dots & a_{nn}-\lambda \end{bmatrix},$$

and the eigenvectors of $\mathbf{A}$ are then the nonzero solutions of the equations $(\mathbf{A}-\lambda\mathbf{I})\mathbf{v}=\mathbf{0}$, for these values of λ.

The determinant of the matrix $\mathbf{A}-\lambda\mathbf{I}$ is clearly a polynomial in λ of degree n, with leading term $(-1)^n\lambda^n$. It is customary to call this polynomial the characteristic polynomial of $\mathbf{A}$ and to denote it by $p(\lambda)$. For each root λ_j of $p(\lambda)$, that is, for each number λ_j such that $p(\lambda_j)=0$, there exists at least one nonzero vector $\mathbf{v}^j$ such that $\mathbf{A}\mathbf{v}^j=\lambda_j\mathbf{v}^j$. Now, every polynomial of degree $n\geqslant 1$ has at least one (possibly complex) root. Therefore, every matrix has at least one eigenvalue, and consequently, at least one eigenvector. On the other hand, $p(\lambda)$ has at most n distinct roots. Therefore, every $n\times n$

matrix has at most n eigenvalues. Finally, observe that every $n \times n$ matrix has at most n linearly independent eigenvectors, since the space of all vectors

$$\mathbf{v} = \begin{bmatrix} v_1 \\ \vdots \\ v_n \end{bmatrix}$$

has dimension n.

Remark. Let $\mathbf{v}$ be an eigenvector of $\mathbf{A}$ with eigenvalue λ. Observe that

$$\mathbf{A}(c\mathbf{v}) = c\mathbf{A}\mathbf{v} = c\lambda\mathbf{v} = \lambda(c\mathbf{v})$$

for any constant c. Hence, any constant multiple $(c \neq 0)$ of an eigenvector of $\mathbf{A}$ is again an eigenvector of $\mathbf{A}$, with the same eigenvalue.

For each eigenvector $\mathbf{v}^j$ of $\mathbf{A}$ with eigenvalue λ_j, we have a solution $\mathbf{x}^j(t) = e^{\lambda_j t}\mathbf{v}^j$ of (1). If $\mathbf{A}$ has n linearly independent eigenvectors $\mathbf{v}^1, \dots, \mathbf{v}^n$ with eigenvalues $\lambda_1, \dots, \lambda_n$ respectively ($\lambda_1, \dots, \lambda_n$ need not be distinct), then $\mathbf{x}^j(t) = e^{\lambda_j t}\mathbf{v}^j$, $j = 1, \dots, n$ are n linearly independent solutions of (1). This follows immediately from Theorem 6 of Section 3.4 and the fact that $\mathbf{x}^j(0) = \mathbf{v}^j$. In this case, then, every solution $\mathbf{x}(t)$ of (1) is of the form

$$\mathbf{x}(t) = c_1 e^{\lambda_1 t}\mathbf{v}^1 + c_2 e^{\lambda_2 t}\mathbf{v}^2 + \dots + c_n e^{\lambda_n t}\mathbf{v}^n. \tag{4}$$

This is sometimes called the "general solution" of (1).

The situation is simplest when $\mathbf{A}$ has n distinct real eigenvalues $\lambda_1, \lambda_2, \dots, \lambda_n$ with eigenvectors $\mathbf{v}^1, \mathbf{v}^2, \dots, \mathbf{v}^n$ respectively, for in this case we are guaranteed that $\mathbf{v}^1, \mathbf{v}^2, \dots, \mathbf{v}^n$ are linearly independent. This is the content of Theorem 12.

Theorem 12. *Any k eigenvectors $\mathbf{v}^1, \dots, \mathbf{v}^k$ of $\mathbf{A}$ with distinct eigenvalues $\lambda_1, \dots, \lambda_k$ respectively, are linearly independent.*

PROOF. We will prove Theorem 12 by induction on k, the number of eigenvectors. Observe that this theorem is certainly true for $k = 1$. Next, we assume that Theorem 12 is true for $k = j$. That is to say, we assume that any set of j eigenvectors of $\mathbf{A}$ with distinct eigenvalues is linearly independent. We must show that any set of $j+1$ eigenvectors of $\mathbf{A}$ with distinct eigenvalues is also linearly independent. To this end, let $\mathbf{v}^1, \dots, \mathbf{v}^{j+1}$ be $j+1$ eigenvectors of $\mathbf{A}$ with distinct eigenvalues $\lambda_1, \dots, \lambda_{j+1}$ respectively. To determine whether these vectors are linearly dependent or linearly independent, we consider the equation

$$c_1\mathbf{v}^1 + c_2\mathbf{v}^2 + \dots + c_{j+1}\mathbf{v}^{j+1} = \mathbf{0}. \tag{5}$$

Applying **A** to both sides of (5) gives

$$\lambda_1 c_1 \mathbf{v}^1 + \lambda_2 c_2 \mathbf{v}^2 + \ldots + \lambda_{j+1} c_{j+1} \mathbf{v}^{j+1} = \mathbf{0}. \tag{6}$$

Thus, if we multiply both sides of (5) by λ_1 and subtract the resulting equation from (6), we obtain that

$$(\lambda_2 - \lambda_1) c_2 \mathbf{v}^2 + \ldots + (\lambda_{j+1} - \lambda_1) c_{j+1} \mathbf{v}^{j+1} = \mathbf{0}. \tag{7}$$

But $\mathbf{v}^2, \ldots, \mathbf{v}^{j+1}$ are j eigenvectors of **A** with distinct eigenvalues $\lambda_2, \ldots, \lambda_{j+1}$ respectively. By the induction hypothesis, they are linearly independent. Consequently,

$$(\lambda_2 - \lambda_1) c_2 = 0, \qquad (\lambda_3 - \lambda_1) c_3 = 0, \ldots, \quad \text{and} \quad (\lambda_{j+1} - \lambda_1) c_{j+1} = 0.$$

Since $\lambda_1, \lambda_2, \ldots, \lambda_{j+1}$ are distinct, we conclude that $c_2, c_3, \ldots, c_{j+1}$ are all zero. Equation (5) now forces c_1 to be zero. Hence, $\mathbf{v}^1, \mathbf{v}^2, \ldots, \mathbf{v}^{j+1}$ are linearly independent. By induction, therefore, every set of k eigenvectors of **A** with distinct eigenvalues is linearly independent. □

Example 1. Find all solutions of the equation

$$\dot{\mathbf{x}} = \begin{pmatrix} 1 & -1 & 4 \\ 3 & 2 & -1 \\ 2 & 1 & -1 \end{pmatrix} \mathbf{x}.$$

Solution. The characteristic polynomial of the matrix

$$\mathbf{A} = \begin{pmatrix} 1 & -1 & 4 \\ 3 & 2 & -1 \\ 2 & 1 & -1 \end{pmatrix}$$

is

$$\begin{aligned} p(\lambda) = \det(\mathbf{A} - \lambda \mathbf{I}) &= \det \begin{pmatrix} 1-\lambda & -1 & 4 \\ 3 & 2-\lambda & -1 \\ 2 & 1 & -1-\lambda \end{pmatrix} \\ &= -(1+\lambda)(1-\lambda)(2-\lambda) + 2 + 12 - 8(2-\lambda) + (1-\lambda) - 3(1+\lambda) \\ &= (1-\lambda)(\lambda-3)(\lambda+2). \end{aligned}$$

Thus the eigenvalues of **A** are $\lambda_1 = 1$, $\lambda_2 = 3$, and $\lambda_3 = -2$.

(i) $\lambda_1 = 1$: We seek a nonzero vector **v** such that

$$(\mathbf{A} - \mathbf{I})\mathbf{v} = \begin{pmatrix} 0 & -1 & 4 \\ 3 & 1 & -1 \\ 2 & 1 & -2 \end{pmatrix} \begin{pmatrix} v_1 \\ v_2 \\ v_3 \end{pmatrix} = \begin{pmatrix} 0 \\ 0 \\ 0 \end{pmatrix}.$$

This implies that

$$-v_2 + 4v_3 = 0, \qquad 3v_1 + v_2 - v_3 = 0, \quad \text{and} \quad 2v_1 + v_2 - 2v_3 = 0.$$

Solving for v_1 and v_2 in terms of v_3 from the first two equations gives $v_1 = -v_3$ and $v_2 = 4v_3$. Hence, each vector

$$\mathbf{v} = c\begin{bmatrix} -1 \\ 4 \\ 1 \end{bmatrix}$$

is an eigenvector of $\mathbf{A}$ with eigenvalue one. Consequently,

$$ce^t\begin{bmatrix} -1 \\ 4 \\ 1 \end{bmatrix}$$

is a solution of the differential equation for any constant c. For simplicity, we take

$$\mathbf{x}^1(t) = e^t\begin{bmatrix} -1 \\ 4 \\ 1 \end{bmatrix}.$$

(ii) $\lambda_2 = 3$: We seek a nonzero vector $\mathbf{v}$ such that

$$(\mathbf{A} - 3\mathbf{I})\mathbf{v} = \begin{bmatrix} -2 & -1 & 4 \\ 3 & -1 & -1 \\ 2 & 1 & -4 \end{bmatrix}\begin{bmatrix} v_1 \\ v_2 \\ v_3 \end{bmatrix} = \begin{bmatrix} 0 \\ 0 \\ 0 \end{bmatrix}.$$

This implies that

$$-2v_1 - v_2 + 4v_3 = 0, \qquad 3v_1 - v_2 - v_3 = 0, \quad \text{and} \quad 2v_1 + v_2 - 4v_3 = 0.$$

Solving for v_1 and v_2 in terms of v_3 from the first two equations gives $v_1 = v_3$ and $v_2 = 2v_3$. Consequently, each vector

$$\mathbf{v} = c\begin{bmatrix} 1 \\ 2 \\ 1 \end{bmatrix}$$

is an eigenvector of $\mathbf{A}$ with eigenvalue 3. Therefore,

$$\mathbf{x}^2(t) = e^{3t}\begin{bmatrix} 1 \\ 2 \\ 1 \end{bmatrix}$$

is a second solution of the differential equation.

(iii) $\lambda_3 = -2$: We seek a nonzero vector $\mathbf{v}$ such that

$$(\mathbf{A} + 2\mathbf{I})\mathbf{v} = \begin{bmatrix} 3 & -1 & 4 \\ 3 & 4 & -1 \\ 2 & 1 & 1 \end{bmatrix}\begin{bmatrix} v_1 \\ v_2 \\ v_3 \end{bmatrix} = \begin{bmatrix} 0 \\ 0 \\ 0 \end{bmatrix}.$$

This implies that

$$3v_1 - v_2 + 4v_3 = 0, \qquad 3v_1 + 4v_2 - v_3 = 0 \quad \text{and} \quad 2v_1 + v_2 + v_3 = 0.$$

Solving for v_1 and v_2 in terms of v_3 gives $v_1 = -v_3$ and $v_2 = v_3$. Hence, each vector

$$\mathbf{v}=c\begin{bmatrix}-1\\1\\1\end{bmatrix}$$

is an eigenvector of **A** with eigenvalue -2. Consequently,

$$\mathbf{x}^3(t)=e^{-2t}\begin{bmatrix}-1\\1\\1\end{bmatrix}$$

is a third solution of the differential equation. These solutions must be linearly independent, since **A** has distinct eigenvalues. Therefore, every solution $\mathbf{x}(t)$ must be of the form

$$\mathbf{x}(t)=c_1e^{t}\begin{bmatrix}-1\\4\\1\end{bmatrix}+c_2e^{3t}\begin{bmatrix}1\\2\\1\end{bmatrix}+c_3e^{-2t}\begin{bmatrix}-1\\1\\1\end{bmatrix}$$

$$=\begin{bmatrix}-c_1e^{t}+\ c_2e^{3t}-c_3e^{-2t}\\4c_1e^{t}+2c_2e^{3t}+c_3e^{-2t}\\c_1e^{t}+\ c_2e^{3t}+c_3e^{-2t}\end{bmatrix}.$$

Remark. If λ is an eigenvalue of **A**, then the n equations

$$a_{j1}v_1+\ldots+(a_{jj}-\lambda)v_j+\ldots+a_{jn}v_n=0,\qquad j=1,\ldots,n$$

are not independent; at least one of them is a linear combination of the others. Consequently, we have at most $n-1$ independent equations for the n unknowns $v_1,\ldots,v_n$. This implies that at least one of the unknowns $v_1,\ldots,v_n$ can be chosen arbitrarily.

Example 2. Solve the initial-value problem

$$\dot{\mathbf{x}}=\begin{pmatrix}1&12\\3&1\end{pmatrix}\mathbf{x},\qquad \mathbf{x}(0)=\begin{pmatrix}0\\1\end{pmatrix}.$$

Solution. The characteristic polynomial of the matrix

$$\mathbf{A}=\begin{pmatrix}1&12\\3&1\end{pmatrix}$$

is

$$p(\lambda)=\det\begin{pmatrix}1-\lambda&12\\3&1-\lambda\end{pmatrix}=(1-\lambda)^2-36=(\lambda-7)(\lambda+5).$$

Thus, the eigenvalues of A are $\lambda_1=7$ and $\lambda_2=-5$.

(i) $\lambda_1=7$: We seek a nonzero vector $\mathbf{v}$ such that

$$(\mathbf{A}-7\mathbf{I})\mathbf{v}=\begin{pmatrix} -6 & 12\\ 3 & -6\end{pmatrix}\begin{pmatrix} v_1\\ v_2\end{pmatrix}=\begin{pmatrix} 0\\ 0\end{pmatrix}.$$

This implies that $v_1=2v_2$. Consequently, every vector

$$\mathbf{v}=c\begin{pmatrix} 2\\ 1\end{pmatrix}$$

is an eigenvector of $\mathbf{A}$ with eigenvalue 7. Therefore,

$$\mathbf{x}^1(t)=e^{7t}\begin{pmatrix} 2\\ 1\end{pmatrix}$$

is a solution of the differential equation.

(ii) $\lambda_2=-5$: We seek a nonzero vector $\mathbf{v}$ such that

$$(\mathbf{A}+5\mathbf{I})\mathbf{v}=\begin{pmatrix} 6 & 12\\ 3 & 6\end{pmatrix}\begin{pmatrix} v_1\\ v_2\end{pmatrix}=\begin{pmatrix} 0\\ 0\end{pmatrix}.$$

This implies that $v_1=-2v_2$. Consequently,

$$\begin{pmatrix} -2\\ 1\end{pmatrix}$$

is an eigenvector of $\mathbf{A}$ with eigenvalue -5, and

$$\mathbf{x}^2(t)=e^{-5t}\begin{pmatrix} -2\\ 1\end{pmatrix}$$

is a second solution of the differential equation. These solutions are linearly independent since $\mathbf{A}$ has distinct eigenvalues. Hence, $\mathbf{x}(t)=c_1\mathbf{x}^1(t)+c_2\mathbf{x}^2(t)$. The constants c_1 and c_2 are determined from the initial condition

$$\begin{pmatrix} 0\\ 1\end{pmatrix}=c_1\mathbf{x}^1(0)+c_2\mathbf{x}^2(0)=\begin{pmatrix} 2c_1\\ c_1\end{pmatrix}+\begin{pmatrix} -2c_2\\ c_2\end{pmatrix}.$$

Thus, $2c_1-2c_2=0$ and $c_1+c_2=1$. The solution of these two equations is $c_1=\frac{1}{2}$ and $c_2=\frac{1}{2}$. Consequently,

$$\mathbf{x}(t)=\frac{1}{2}e^{7t}\begin{pmatrix} 2\\ 1\end{pmatrix}+\frac{1}{2}e^{-5t}\begin{pmatrix} -2\\ 1\end{pmatrix}=\begin{bmatrix} e^{7t}-e^{-5t}\\ \frac{1}{2}e^{7t}+\frac{1}{2}e^{-5t}\end{bmatrix}.$$

Example 3. Find all solutions of the equation

$$\dot{\mathbf{x}} = \mathbf{A}\mathbf{x} = \begin{bmatrix} 1 & 1 & 1 & 1 & 1 \\ 2 & 2 & 2 & 2 & 2 \\ 3 & 3 & 3 & 3 & 3 \\ 4 & 4 & 4 & 4 & 4 \\ 5 & 5 & 5 & 5 & 5 \end{bmatrix} \mathbf{x}.$$

Solution. It is not necessary to compute the characteristic polynomial of **A** in order to find the eigenvalues and eigenvectors of **A**. To wit, observe that

$$\mathbf{A}\mathbf{x} = \mathbf{A}\begin{bmatrix} x_1 \\ x_2 \\ x_3 \\ x_4 \\ x_5 \end{bmatrix} = (x_1 + x_2 + x_3 + x_4 + x_5)\begin{bmatrix} 1 \\ 2 \\ 3 \\ 4 \\ 5 \end{bmatrix}.$$

Hence, any vector **x** whose components add up to zero is an eigenvector of **A** with eigenvalue 0. In particular

$$\mathbf{v}^1 = \begin{bmatrix} 1 \\ 0 \\ 0 \\ 0 \\ -1 \end{bmatrix}, \quad \mathbf{v}^2 = \begin{bmatrix} 0 \\ 1 \\ 0 \\ 0 \\ -1 \end{bmatrix}, \quad \mathbf{v}^3 = \begin{bmatrix} 0 \\ 0 \\ 1 \\ 0 \\ -1 \end{bmatrix}, \quad \text{and} \quad \mathbf{v}^4 = \begin{bmatrix} 0 \\ 0 \\ 0 \\ 1 \\ -1 \end{bmatrix}$$

are four independent eigenvectors of **A** with eigenvalue zero. Moreover, observe that

$$\mathbf{v}^5 = \begin{bmatrix} 1 \\ 2 \\ 3 \\ 4 \\ 5 \end{bmatrix}$$

is an eigenvector of **A** with eigenvalue 15 since

$$\mathbf{A}\begin{bmatrix} 1 \\ 2 \\ 3 \\ 4 \\ 5 \end{bmatrix} = (1+2+3+4+5)\begin{bmatrix} 1 \\ 2 \\ 3 \\ 4 \\ 5 \end{bmatrix} = 15\begin{bmatrix} 1 \\ 2 \\ 3 \\ 4 \\ 5 \end{bmatrix}.$$

The five vectors $\mathbf{v}^1$, $\mathbf{v}^2$, $\mathbf{v}^3$, $\mathbf{v}^4$, and $\mathbf{v}^5$ are easily seen to be linearly independent. Hence, every solution $\mathbf{x}(t)$ is of the form

$$\mathbf{x}(t)=c_1\begin{bmatrix}1\\0\\0\\0\\-1\end{bmatrix}+c_2\begin{bmatrix}0\\1\\0\\0\\-1\end{bmatrix}+c_3\begin{bmatrix}0\\0\\1\\0\\-1\end{bmatrix}+c_4\begin{bmatrix}0\\0\\0\\1\\-1\end{bmatrix}+c_5e^{15t}\begin{bmatrix}1\\2\\3\\4\\5\end{bmatrix}.$$

Exercises

In each of Problems 1–6 find all solutions of the given differential equation.

1. $\dot{\mathbf{x}}=\begin{pmatrix}6 & -3\\ 2 & 1\end{pmatrix}\mathbf{x}$

2. $\dot{\mathbf{x}}=\begin{pmatrix}-2 & 1\\ -4 & 3\end{pmatrix}\mathbf{x}$

3. $\dot{\mathbf{x}}=\begin{pmatrix}3 & 2 & 4\\ 2 & 0 & 2\\ 4 & 2 & 3\end{pmatrix}\mathbf{x}$

4. $\dot{\mathbf{x}}=\begin{pmatrix}7 & -1 & 6\\ -10 & 4 & -12\\ -2 & 1 & -1\end{pmatrix}\mathbf{x}$

5. $\dot{\mathbf{x}}=\begin{pmatrix}-7 & 0 & 6\\ 0 & 5 & 0\\ 6 & 0 & 2\end{pmatrix}\mathbf{x}$

6. $\dot{\mathbf{x}}=\begin{bmatrix}1 & 2 & 3 & 6\\ 3 & 6 & 9 & 18\\ 5 & 10 & 15 & 30\\ 7 & 14 & 21 & 42\end{bmatrix}\mathbf{x}$

In each of Problems 7–12, solve the given initial-value problem.

7. $\dot{\mathbf{x}}=\begin{pmatrix}1 & 1\\ 4 & 1\end{pmatrix}\mathbf{x},\quad \mathbf{x}(0)=\begin{pmatrix}2\\3\end{pmatrix}$

8. $\dot{\mathbf{x}}=\begin{pmatrix}1 & -3\\ -2 & 2\end{pmatrix}\mathbf{x},\quad \mathbf{x}(0)=\begin{pmatrix}0\\5\end{pmatrix}$

9. $\dot{\mathbf{x}}=\begin{pmatrix}3 & 1 & -1\\ 1 & 3 & -1\\ 3 & 3 & -1\end{pmatrix}\mathbf{x},\quad \mathbf{x}(0)=\begin{pmatrix}1\\-2\\-1\end{pmatrix}$

10. $\dot{\mathbf{x}}=\begin{pmatrix}1 & -1 & 0\\ 1 & 2 & 1\\ 1 & 10 & 2\end{pmatrix}\mathbf{x},\quad \mathbf{x}(0)=\begin{pmatrix}-1\\-4\\13\end{pmatrix}$

11. $\dot{\mathbf{x}}=\begin{pmatrix}1 & -3 & 2\\ 0 & -1 & 0\\ 0 & -1 & -2\end{pmatrix}\mathbf{x},\quad \mathbf{x}(0)=\begin{pmatrix}-2\\0\\3\end{pmatrix}$

12. $\dot{\mathbf{x}}=\begin{pmatrix}3 & 1 & -2\\ -1 & 2 & 1\\ 4 & 1 & -3\end{pmatrix}\mathbf{x},\quad \mathbf{x}(0)=\begin{pmatrix}1\\4\\-7\end{pmatrix}$

13. (a) Show that $e^{\lambda(t-t_0)}\mathbf{v}$, t_0 constant, is a solution of $\dot{\mathbf{x}}=\mathbf{A}\mathbf{x}$ if $\mathbf{A}\mathbf{v}=\lambda\mathbf{v}$.
(b) Solve the initial-value problem

$$\dot{\mathbf{x}}=\begin{pmatrix}3 & 1 & -2\\ -1 & 2 & 1\\ 4 & 1 & -3\end{pmatrix}\mathbf{x},\quad \mathbf{x}(1)=\begin{pmatrix}1\\4\\-7\end{pmatrix}$$

(see Exercise 12).

14. Three solutions of the equation $\dot{\mathbf{x}}=\mathbf{A}\mathbf{x}$ are

$$\begin{bmatrix} e^{t}+e^{2t} \\ e^{2t} \\ 0 \end{bmatrix}, \quad \begin{bmatrix} e^{t}+e^{3t} \\ e^{3t} \\ e^{3t} \end{bmatrix} \quad \text{and} \quad \begin{bmatrix} e^{t}-e^{3t} \\ -e^{3t} \\ -e^{3t} \end{bmatrix}.$$

Find the eigenvalues and eigenvectors of $\mathbf{A}$.

15. Show that the eigenvalues of $\mathbf{A}^{-1}$ are the reciprocals of the eigenvalues of $\mathbf{A}$.

16. Show that the eigenvalues of $\mathbf{A}^n$ are the nth power of the eigenvalues of $\mathbf{A}$.

17. Show that $\lambda=0$ is an eigenvalue of $\mathbf{A}$ if $\det\mathbf{A}=0$.

18. Show, by example, that the eigenvalues of $\mathbf{A}+\mathbf{B}$ are not necessarily the sum of an eigenvalue of $\mathbf{A}$ and an eigenvalue of $\mathbf{B}$.

19. Show, by example, that the eigenvalues of $\mathbf{AB}$ are not necessarily the product of an eigenvalue of $\mathbf{A}$ with an eigenvalue of $\mathbf{B}$.

20. Show that the matrices $\mathbf{A}$ and $\mathbf{T}^{-1}\mathbf{AT}$ have the same characteristic polynomial.

21. Suppose that either $\mathbf{B}^{-1}$ or $\mathbf{A}^{-1}$ exists. Prove that $\mathbf{AB}$ and $\mathbf{BA}$ have the same eigenvalues. *Hint*: Use Exercise 20. (This result is true even if neither $\mathbf{B}^{-1}$ or $\mathbf{A}^{-1}$ exist; however, it is more difficult to prove then.)

3.7 Complex roots

If $\lambda=\alpha+i\beta$ is a complex eigenvalue of $\mathbf{A}$ with eigenvector $\mathbf{v}=\mathbf{v}^1+i\mathbf{v}^2$, then $\mathbf{x}(t)=e^{\lambda t}\mathbf{v}$ is a complex-valued solution of the differential equation

$$\dot{\mathbf{x}}=\mathbf{A}\mathbf{x}. \tag{1}$$

This complex-valued solution gives rise to *two* real-valued solutions, as we now show.

Lemma 1. *Let* $\mathbf{x}(t)=\mathbf{y}(t)+i\mathbf{z}(t)$ *be a complex-valued solution of* (1). *Then, both* $\mathbf{y}(t)$ *and* $\mathbf{z}(t)$ *are real-valued solutions of* (1).

PROOF. If $\mathbf{x}(t)=\mathbf{y}(t)+i\mathbf{z}(t)$ is a complex-valued solution of (1), then

$$\dot{\mathbf{y}}(t)+i\dot{\mathbf{z}}(t)=\mathbf{A}(\mathbf{y}(t)+i\mathbf{z}(t))=\mathbf{A}\mathbf{y}(t)+i\mathbf{A}\mathbf{z}(t). \tag{2}$$

Equating real and imaginary parts of (2) gives $\dot{\mathbf{y}}(t)=\mathbf{A}\mathbf{y}(t)$ and $\dot{\mathbf{z}}(t)=\mathbf{A}\mathbf{z}(t)$. Consequently, both $\mathbf{y}(t)=\mathrm{Re}\{\mathbf{x}(t)\}$ and $\mathbf{z}(t)=\mathrm{Im}\{\mathbf{x}(t)\}$ are real-valued solutions of (1). □

The complex-valued function $\mathbf{x}(t)=e^{(\alpha+i\beta)t}(\mathbf{v}^1+i\mathbf{v}^2)$ can be written in the form

$$\begin{aligned}\mathbf{x}(t)&=e^{\alpha t}(\cos\beta t+i\sin\beta t)(\mathbf{v}^1+i\mathbf{v}^2)\\ &=e^{\alpha t}\left[(\mathbf{v}^1\cos\beta t-\mathbf{v}^2\sin\beta t)+i(\mathbf{v}^1\sin\beta t+\mathbf{v}^2\cos\beta t)\right].\end{aligned}$$

Hence, if $\lambda=\alpha+i\beta$ is an eigenvalue of $\mathbf{A}$ with eigenvector $\mathbf{v}=\mathbf{v}^1+i\mathbf{v}^2$, then

$$\mathbf{y}(t)=e^{\alpha t}(\mathbf{v}^1\cos\beta t-\mathbf{v}^2\sin\beta t)$$

and

$$\mathbf{z}(t)=e^{\alpha t}(\mathbf{v}^1\sin\beta t+\mathbf{v}^2\cos\beta t)$$

are two real-valued solutions of (1). Moreover, these two solutions must be linearly independent (see Exercise 10).

Example 1. Solve the initial-value problem

$$\dot{\mathbf{x}}=\begin{pmatrix}1&0&0\\0&1&-1\\0&1&1\end{pmatrix}\mathbf{x},\qquad \mathbf{x}(0)=\begin{pmatrix}1\\1\\1\end{pmatrix}.$$

Solution. The characteristic polynomial of the matrix

$$\mathbf{A}=\begin{pmatrix}1&0&0\\0&1&-1\\0&1&1\end{pmatrix}$$

is

$$p(\lambda)=\det(\mathbf{A}-\lambda\mathbf{I})=\det\begin{pmatrix}1-\lambda&0&0\\0&1-\lambda&-1\\0&1&1-\lambda\end{pmatrix}$$
$$=(1-\lambda)^3+(1-\lambda)=(1-\lambda)(\lambda^2-2\lambda+2).$$

Hence the eigenvalues of $\mathbf{A}$ are

$$\lambda=1\quad\text{and}\quad\lambda=\frac{2\pm\sqrt{4-8}}{2}=1\pm i.$$

(i) $\lambda=1$: Clearly,

$$\begin{pmatrix}1\\0\\0\end{pmatrix}$$

is an eigenvector of $\mathbf{A}$ with eigenvalue 1. Hence

$$\mathbf{x}^1(t)=e^t\begin{pmatrix}1\\0\\0\end{pmatrix}$$

is one solution of the differential equation $\dot{\mathbf{x}}=\mathbf{A}\mathbf{x}$.

(ii) $\lambda=1+i$: We seek a nonzero vector $\mathbf{v}$ such that

$$[\mathbf{A}-(1+i)\mathbf{I}]\mathbf{v}=\begin{pmatrix}-i&0&0\\0&-i&-1\\0&1&-i\end{pmatrix}\begin{pmatrix}v_1\\v_2\\v_3\end{pmatrix}=\begin{pmatrix}0\\0\\0\end{pmatrix}.$$

This implies that $-iv_1=0$, $-iv_2-v_3=0$, and $v_2-iv_3=0$. The first equation says that $v_1=0$ and the second and third equations both say that $v_2=iv_3$. Consequently, each vector

$$\mathbf{v}=c\begin{pmatrix}0\\ i\\ 1\end{pmatrix}$$

is an eigenvector of $\mathbf{A}$ with eigenvalue $1+i$. Thus,

$$\mathbf{x}(t)=e^{(1+i)t}\begin{pmatrix}0\\ i\\ 1\end{pmatrix}$$

is a complex-valued solution of the differential equation $\dot{\mathbf{x}}=\mathbf{A}\mathbf{x}$. Now,

$$e^{(1+i)t}\begin{pmatrix}0\\ i\\ 1\end{pmatrix}=e^{t}(\cos t+i\sin t)\left[\begin{pmatrix}0\\ 0\\ 1\end{pmatrix}+i\begin{pmatrix}0\\ 1\\ 0\end{pmatrix}\right]$$

$$=e^{t}\left[\cos t\begin{pmatrix}0\\ 0\\ 1\end{pmatrix}-\sin t\begin{pmatrix}0\\ 1\\ 0\end{pmatrix}+i\sin t\begin{pmatrix}0\\ 0\\ 1\end{pmatrix}+i\cos t\begin{pmatrix}0\\ 1\\ 0\end{pmatrix}\right]$$

$$=e^{t}\begin{pmatrix}0\\ -\sin t\\ \cos t\end{pmatrix}+ie^{t}\begin{pmatrix}0\\ \cos t\\ \sin t\end{pmatrix}.$$

Consequently, by Lemma 1,

$$\mathbf{x}^2(t)=e^{t}\begin{pmatrix}0\\ -\sin t\\ \cos t\end{pmatrix}\quad\text{and}\quad \mathbf{x}^3(t)=e^{t}\begin{pmatrix}0\\ \cos t\\ \sin t\end{pmatrix}$$

are real-valued solutions. The three solutions $\mathbf{x}^1(t)$, $\mathbf{x}^2(t)$, and $\mathbf{x}^3(t)$ are linearly independent since their initial values

$$\mathbf{x}^1(0)=\begin{pmatrix}1\\ 0\\ 0\end{pmatrix},\qquad \mathbf{x}^2(0)=\begin{pmatrix}0\\ 1\\ 0\end{pmatrix},\quad\text{and}\quad \mathbf{x}^3(0)=\begin{pmatrix}0\\ 0\\ 1\end{pmatrix}$$

are linearly independent vectors in R^3. Therefore, the solution $\mathbf{x}(t)$ of our initial-value problem must have the form

$$\mathbf{x}(t)=c_1e^{t}\begin{pmatrix}1\\ 0\\ 0\end{pmatrix}+c_2e^{t}\begin{pmatrix}0\\ -\sin t\\ \cos t\end{pmatrix}+c_3e^{t}\begin{pmatrix}0\\ \cos t\\ \sin t\end{pmatrix}.$$

Setting $t=0$, we see that

$$\begin{pmatrix}1\\ 1\\ 1\end{pmatrix}=c_1\begin{pmatrix}1\\ 0\\ 0\end{pmatrix}+c_2\begin{pmatrix}0\\ 0\\ 1\end{pmatrix}+c_3\begin{pmatrix}0\\ 1\\ 0\end{pmatrix}=\begin{pmatrix}c_1\\ c_3\\ c_2\end{pmatrix}.$$

Consequently $c_1 = c_2 = c_3 = 1$ and

$$\mathbf{x}(t) = e^t \begin{bmatrix} 1 \\ 0 \\ 0 \end{bmatrix} + e^t \begin{bmatrix} 0 \\ -\sin t \\ \cos t \end{bmatrix} + e^t \begin{bmatrix} 0 \\ \cos t \\ \sin t \end{bmatrix}$$

$$= e^t \begin{bmatrix} 1 \\ \cos t - \sin t \\ \cos t + \sin t \end{bmatrix}.$$

Remark. If $\mathbf{v}$ is an eigenvector of $\mathbf{A}$ with eigenvalue λ, then $\bar{\mathbf{v}}$, the complex conjugate of $\mathbf{v}$, is an eigenvector of $\mathbf{A}$ with eigenvalue $\bar{\lambda}$. (Each component of $\bar{\mathbf{v}}$ is the complex conjugate of the corresponding component of $\mathbf{v}$.) To prove this, we take complex conjugates of both sides of the equation $\mathbf{A}\mathbf{v} = \lambda\mathbf{v}$ and observe that the complex conjugate of the vector $\mathbf{A}\mathbf{v}$ is $\mathbf{A}\bar{\mathbf{v}}$ if $\mathbf{A}$ is real. Hence, $\mathbf{A}\bar{\mathbf{v}} = \bar{\lambda}\bar{\mathbf{v}}$, which shows that $\bar{\mathbf{v}}$ is an eigenvector of $\mathbf{A}$ with eigenvalue $\bar{\lambda}$.

Exercises

In each of Problems 1–4 find the general solution of the given system of differential equations.

1. $\dot{\mathbf{x}} = \begin{pmatrix} -3 & 2 \\ -1 & -1 \end{pmatrix} \mathbf{x}$

2. $\dot{\mathbf{x}} = \begin{pmatrix} 1 & -5 & 0 \\ 1 & -3 & 0 \\ 0 & 0 & 1 \end{pmatrix} \mathbf{x}$

3. $\dot{\mathbf{x}} = \begin{pmatrix} 1 & 0 & 0 \\ 3 & 1 & -2 \\ 2 & 2 & 1 \end{pmatrix} \mathbf{x}$

4. $\dot{\mathbf{x}} = \begin{pmatrix} 1 & 0 & 1 \\ 0 & 1 & -1 \\ -2 & 0 & -1 \end{pmatrix} \mathbf{x}$

In each of Problems 5–8, solve the given initial-value problem.

5. $\dot{\mathbf{x}} = \begin{pmatrix} 1 & -1 \\ 5 & -3 \end{pmatrix} \mathbf{x}, \quad \mathbf{x}(0) = \begin{pmatrix} 1 \\ 2 \end{pmatrix}$

6. $\dot{\mathbf{x}} = \begin{pmatrix} 3 & -2 \\ 4 & -1 \end{pmatrix} \mathbf{x}, \quad \mathbf{x}(0) = \begin{pmatrix} 1 \\ 5 \end{pmatrix}$

7. $\dot{\mathbf{x}} = \begin{pmatrix} -3 & 0 & 2 \\ 1 & -1 & 0 \\ -2 & -1 & 0 \end{pmatrix} \mathbf{x}, \quad \mathbf{x}(0) = \begin{pmatrix} 0 \\ -1 \\ -2 \end{pmatrix}$

8. $\dot{\mathbf{x}} = \begin{bmatrix} 0 & 2 & 0 & 0 \\ -2 & 0 & 0 & 0 \\ 0 & 0 & 0 & -3 \\ 0 & 0 & 3 & 0 \end{bmatrix} \mathbf{x}, \quad \mathbf{x}(0) = \begin{bmatrix} 1 \\ 1 \\ 1 \\ 0 \end{bmatrix}$

9. Determine all vectors $\mathbf{x}^0$ such that the solution of the initial-value problem

$$\dot{\mathbf{x}} = \begin{pmatrix} 1 & 0 & -2 \\ 0 & 1 & 0 \\ 1 & -1 & -1 \end{pmatrix} \mathbf{x}, \qquad \mathbf{x}(0) = \mathbf{x}^0$$

is a periodic function of time.

10. Let $\mathbf{x}(t)=e^{\lambda t}\mathbf{v}$ be a solution of $\dot{\mathbf{x}}=\mathbf{A}\mathbf{x}$. Prove that $\mathbf{y}(t)=\text{Re}\{\mathbf{x}(t)\}$ and $\mathbf{z}(t)=\text{Im}\{\mathbf{x}(t)\}$ are linearly independent. *Hint*: Observe that $\mathbf{v}$ and $\bar{\mathbf{v}}$ are linearly independent in $\mathbf{C}^n$ since they are eigenvectors of $\mathbf{A}$ with distinct eigenvalues.

3.8 Equal roots

If the characteristic polynomial of $\mathbf{A}$ does not have n distinct roots, then $\mathbf{A}$ may not have n linearly independent eigenvectors. For example, the matrix

$$\mathbf{A}=\begin{pmatrix}1&1&0\\0&1&0\\0&0&2\end{pmatrix}$$

has only two distinct eigenvalues $\lambda_1=1$ and $\lambda_2=2$ and two linearly independent eigenvectors, which we take to be

$$\begin{pmatrix}1\\0\\0\end{pmatrix}\quad\text{and}\quad\begin{pmatrix}0\\0\\1\end{pmatrix}.$$

Consequently, the differential equation $\dot{\mathbf{x}}=\mathbf{A}\mathbf{x}$ has only two linearly independent solutions

$$e^{t}\begin{pmatrix}1\\0\\0\end{pmatrix}\quad\text{and}\quad e^{2t}\begin{pmatrix}0\\0\\1\end{pmatrix}$$

of the form $e^{\lambda t}\mathbf{v}$. Our problem, in this case, is to find a third linearly independent solution. More generally, suppose that the $n\times n$ matrix $\mathbf{A}$ has only $k<n$ linearly independent eigenvectors. Then, the differential equation $\dot{\mathbf{x}}=\mathbf{A}\mathbf{x}$ has only k linearly independent solutions of the form $e^{\lambda t}\mathbf{v}$. Our problem is to find an additional $n-k$ linearly independent solutions.

We approach this problem in the following ingenious manner. Recall that $x(t)=e^{at}c$ is a solution of the scalar differential equation $\dot{x}=ax$, for every constant c. Analogously, we would like to say that $\mathbf{x}(t)=e^{\mathbf{A}t}\mathbf{v}$ is a solution of the vector differential equation

$$\dot{\mathbf{x}}=\mathbf{A}\mathbf{x} \tag{1}$$

for every constant vector $\mathbf{v}$. However, $e^{\mathbf{A}t}$ is not defined if $\mathbf{A}$ is an $n\times n$ matrix. This is not a serious difficulty, though. There is a very natural way of defining $e^{\mathbf{A}t}$ so that it resembles the scalar exponential e^{at}; simply set

$$e^{\mathbf{A}t}\equiv\mathbf{I}+\mathbf{A}t+\frac{\mathbf{A}^2t^2}{2!}+\dots+\frac{\mathbf{A}^nt^n}{n!}+\dots. \tag{2}$$

It can be shown that the infinite series (2) converges for all t, and can be differentiated term by term. In particular

$$\frac{d}{dt}e^{\mathbf{A}t} = \mathbf{A} + \mathbf{A}^2 t + \ldots + \frac{\mathbf{A}^{n+1}}{n!}t^n + \ldots$$

$$= \mathbf{A}\left[\mathbf{I} + \mathbf{A}t + \ldots + \frac{\mathbf{A}^n t^n}{n!} + \ldots\right] = \mathbf{A}e^{\mathbf{A}t}.$$

This implies that $e^{\mathbf{A}t}\mathbf{v}$ is a solution of (1) for every constant vector $\mathbf{v}$, since

$$\frac{d}{dt}e^{\mathbf{A}t}\mathbf{v} = \mathbf{A}e^{\mathbf{A}t}\mathbf{v} = \mathbf{A}(e^{\mathbf{A}t}\mathbf{v}).$$

Remark. The matrix exponential $e^{\mathbf{A}t}$ and the scalar exponential e^{at} satisfy many similar properties. For example,

$$\left(e^{\mathbf{A}t}\right)^{-1} = e^{-\mathbf{A}t} \quad \text{and} \quad e^{\mathbf{A}(t+s)} = e^{\mathbf{A}t}e^{\mathbf{A}s}. \tag{3}$$

Indeed, the same proofs which show that $(e^{at})^{-1} = e^{-at}$ and $e^{a(t+s)} = e^{at}e^{as}$ can be used to establish the identities (3): we need only replace every a by $\mathbf{A}$ and every 1 by $\mathbf{I}$. However, $e^{\mathbf{A}t+\mathbf{B}t}$ equals $e^{\mathbf{A}t}e^{\mathbf{B}t}$ only if $\mathbf{AB} = \mathbf{BA}$ (see Exercise 15, Section 3.9).

There are several classes of matrices $\mathbf{A}$ (see Problems 9–11) for which the infinite series (2) can be summed exactly. In general, though, it does not seem possible to express $e^{\mathbf{A}t}$ in closed form. Yet, the remarkable fact is that we can always find n linearly independent vectors $\mathbf{v}$ for which the infinite series $e^{\mathbf{A}t}\mathbf{v}$ can be summed exactly. Moreover, once we know n linearly independent solutions of (1), we can even compute $e^{\mathbf{A}t}$ exactly. (This latter property will be proven in the next section.)

We now show how to find n linearly independent vectors $\mathbf{v}$ for which the infinite series $e^{\mathbf{A}t}\mathbf{v}$ can be summed exactly. Observe that

$$e^{\mathbf{A}t}\mathbf{v} = e^{(\mathbf{A}-\lambda\mathbf{I})t}e^{\lambda\mathbf{I}t}\mathbf{v}$$

for any constant λ, since $(\mathbf{A}-\lambda\mathbf{I})(\lambda\mathbf{I}) = (\lambda\mathbf{I})(\mathbf{A}-\lambda\mathbf{I})$. Moreover,

$$e^{\lambda\mathbf{I}t}\mathbf{v} = \left[\mathbf{I} + \lambda\mathbf{I}t + \frac{\lambda^2\mathbf{I}^2t^2}{2!} + \ldots\right]\mathbf{v}$$

$$= \left[1 + \lambda t + \frac{\lambda^2 t^2}{2!} + \ldots\right]\mathbf{v} = e^{\lambda t}\mathbf{v}.$$

Hence, $e^{\mathbf{A}t}\mathbf{v} = e^{\lambda t}e^{(\mathbf{A}-\lambda\mathbf{I})t}\mathbf{v}$.

Next, we make the crucial observation that if $\mathbf{v}$ satisfies $(\mathbf{A}-\lambda\mathbf{I})^m\mathbf{v} = \mathbf{0}$ for some integer m, then the infinite series $e^{(\mathbf{A}-\lambda\mathbf{I})t}\mathbf{v}$ terminates after m terms. If $(\mathbf{A}-\lambda\mathbf{I})^m\mathbf{v} = \mathbf{0}$, then $(\mathbf{A}-\lambda\mathbf{I})^{m+l}\mathbf{v}$ is also zero, for every positive integer l, since

$$(\mathbf{A}-\lambda\mathbf{I})^{m+l}\mathbf{v} = (\mathbf{A}-\lambda\mathbf{I})^l\left[(\mathbf{A}-\lambda\mathbf{I})^m\mathbf{v}\right] = \mathbf{0}.$$

Consequently,

$$e^{(\mathbf{A}-\lambda\mathbf{I})t}\mathbf{v}=\mathbf{v}+t(\mathbf{A}-\lambda\mathbf{I})\mathbf{v}+\ldots+\frac{t^{m-1}}{(m-1)!}(\mathbf{A}-\lambda\mathbf{I})^{m-1}\mathbf{v}$$

and

$$\begin{aligned}e^{\mathbf{A}t}\mathbf{v}&=e^{\lambda t}e^{(\mathbf{A}-\lambda\mathbf{I})t}\mathbf{v}\\&=e^{\lambda t}\left[\mathbf{v}+t(\mathbf{A}-\lambda\mathbf{I})\mathbf{v}+\ldots+\frac{t^{m-1}}{(m-1)!}(\mathbf{A}-\lambda\mathbf{I})^{m-1}\mathbf{v}\right].\end{aligned}$$

This suggests the following algorithm for finding n linearly independent solutions of (1).

(1) Find all the eigenvalues and eigenvectors of $\mathbf{A}$. If $\mathbf{A}$ has n linearly independent eigenvectors, then the differential equation $\dot{\mathbf{x}}=\mathbf{A}\mathbf{x}$ has n linearly independent solutions of the form $e^{\lambda t}\mathbf{v}$. (Observe that the infinite series $e^{(\mathbf{A}-\lambda\mathbf{I})t}\mathbf{v}$ terminates after one term if $\mathbf{v}$ is an eigenvector of $\mathbf{A}$ with eigenvalue λ.)

(2) Suppose that $\mathbf{A}$ has only $k<n$ linearly independent eigenvectors. Then, we have only k linearly independent solutions of the form $e^{\lambda t}\mathbf{v}$. To find additional solutions we pick an eigenvalue λ of $\mathbf{A}$ and find all vectors $\mathbf{v}$ for which $(\mathbf{A}-\lambda\mathbf{I})^2\mathbf{v}=\mathbf{0}$, but $(\mathbf{A}-\lambda\mathbf{I})\mathbf{v}\neq\mathbf{0}$. For each such vector $\mathbf{v}$

$$e^{\mathbf{A}t}\mathbf{v}=e^{\lambda t}e^{(\mathbf{A}-\lambda\mathbf{I})t}\mathbf{v}=e^{\lambda t}\left[\mathbf{v}+t(\mathbf{A}-\lambda\mathbf{I})\mathbf{v}\right]$$

is an additional solution of $\dot{\mathbf{x}}=\mathbf{A}\mathbf{x}$. We do this for all the eigenvalues λ of $\mathbf{A}$.

(3) If we still do not have enough solutions, then we find all vectors $\mathbf{v}$ for which $(\mathbf{A}-\lambda\mathbf{I})^3\mathbf{v}=\mathbf{0}$, but $(\mathbf{A}-\lambda\mathbf{I})^2\mathbf{v}\neq\mathbf{0}$. For each such vector $\mathbf{v}$,

$$e^{\mathbf{A}t}\mathbf{v}=e^{\lambda t}\left[\mathbf{v}+t(\mathbf{A}-\lambda\mathbf{I})\mathbf{v}+\frac{t^2}{2!}(\mathbf{A}-\lambda\mathbf{I})^2\mathbf{v}\right]$$

is an additional solution of $\dot{\mathbf{x}}=\mathbf{A}\mathbf{x}$.

(4) We keep proceeding in this manner until, hopefully, we obtain n linearly independent solutions.

The following lemma from linear algebra, which we accept without proof, guarantees that this algorithm always works. Moreover, it puts an upper bound on the number of steps we have to perform in this algorithm.

Lemma 1. *Let the characteristic polynomial of* $\mathbf{A}$ *have* k *distinct roots* $\lambda_1,\ldots,\lambda_k$ *with multiplicities* $n_1,\ldots,n_k$ *respectively. (This means that* $p(\lambda)$ *can be factored into the form* $(\lambda_1-\lambda)^{n_1}\ldots(\lambda_k-\lambda)^{n_k}$.) *Suppose that* $\mathbf{A}$ *has only* $\nu_j<n_j$ *linearly independent eigenvectors with eigenvalue* λ_j. *Then the equation* $(\mathbf{A}-\lambda_j\mathbf{I})^2\mathbf{v}=\mathbf{0}$ *has at least* ν_j+1 *independent solutions. More generally, if the equation* $(\mathbf{A}-\lambda_j\mathbf{I})^m\mathbf{v}$ *has only* $m_j<n_j$ *independent solutions, then the equation* $(\mathbf{A}-\lambda_j\mathbf{I})^{m+1}\mathbf{v}=\mathbf{0}$ *has at least* m_j+1 *independent solutions.*

Lemma 1 clearly implies that there exists an integer d_j with $d_j \leq n_j$, such that the equation $(\mathbf{A}-\lambda_j\mathbf{I})^{d_j}\mathbf{v}=\mathbf{0}$ has at least n_j linearly independent solutions. Thus, for each eigenvalue λ_j of $\mathbf{A}$, we can compute n_j linearly independent solutions of $\dot{\mathbf{x}}=\mathbf{A}\mathbf{x}$. All these solutions have the form

$$\mathbf{x}(t)=e^{\lambda_j t}\left[\mathbf{v}+t(\mathbf{A}-\lambda_j\mathbf{I})\mathbf{v}+\dots+\frac{t^{d_j-1}}{(d_j-1)!}(\mathbf{A}-\lambda_j\mathbf{I})^{d_j-1}\mathbf{v}\right].$$

In addition, it can be shown that the set of $n_1+\dots+n_k=n$ solutions thus obtained must be linearly independent.

Example 1. Find three linearly independent solutions of the differential equation

$$\dot{\mathbf{x}}=\begin{pmatrix}1&1&0\\0&1&0\\0&0&2\end{pmatrix}\mathbf{x}.$$

Solution. The characteristic polynomial of the matrix

$$\mathbf{A}=\begin{pmatrix}1&1&0\\0&1&0\\0&0&2\end{pmatrix}$$

is $(1-\lambda)^2(2-\lambda)$. Hence $\lambda=1$ is an eigenvalue of $\mathbf{A}$ with multiplicity two, and $\lambda=2$ is an eigenvalue of $\mathbf{A}$ with multiplicity one.

(i) $\lambda=1$: We seek all nonzero vectors $\mathbf{v}$ such that

$$(\mathbf{A}-\mathbf{I})\mathbf{v}=\begin{pmatrix}0&1&0\\0&0&0\\0&0&1\end{pmatrix}\begin{pmatrix}v_1\\v_2\\v_3\end{pmatrix}=\begin{pmatrix}0\\0\\0\end{pmatrix}.$$

This implies that $v_2=v_3=0$, and v_1 is arbitrary. Consequently,

$$\mathbf{x}^1(t)=e^t\begin{pmatrix}1\\0\\0\end{pmatrix}$$

is one solution of $\dot{\mathbf{x}}=\mathbf{A}\mathbf{x}$. Since $\mathbf{A}$ has only one linearly independent eigenvector with eigenvalue 1, we look for all solutions of the equation

$$(\mathbf{A}-\mathbf{I})^2\mathbf{v}=\begin{pmatrix}0&1&0\\0&0&0\\0&0&1\end{pmatrix}\begin{pmatrix}0&1&0\\0&0&0\\0&0&1\end{pmatrix}\mathbf{v}=\begin{pmatrix}0&0&0\\0&0&0\\0&0&1\end{pmatrix}\begin{pmatrix}v_1\\v_2\\v_3\end{pmatrix}=\begin{pmatrix}0\\0\\0\end{pmatrix}.$$

This implies that $v_3=0$ and both v_1 and v_2 are arbitrary. Now, the vector

$$\mathbf{v}=\begin{pmatrix}0\\1\\0\end{pmatrix}$$

satisfies $(\mathbf{A}-\mathbf{I})^2\mathbf{v}=\mathbf{0}$, but $(\mathbf{A}-\mathbf{I})\mathbf{v}\neq\mathbf{0}$. (We could just as well choose any

$$\mathbf{v}=\begin{bmatrix} v_1\\ v_2\\ 0\end{bmatrix}$$

for which $v_2\neq 0$.) Hence,

$$\mathbf{x}^2(t)=e^{\mathbf{A}t}\begin{bmatrix}0\\1\\0\end{bmatrix}=e^te^{(\mathbf{A}-\mathbf{I})t}\begin{bmatrix}0\\1\\0\end{bmatrix}$$

$$=e^t\left[\mathbf{I}+t(\mathbf{A}-\mathbf{I})\right]\begin{bmatrix}0\\1\\0\end{bmatrix}=e^t\left[\begin{bmatrix}0\\1\\0\end{bmatrix}+t\begin{bmatrix}0&1&0\\0&0&0\\0&0&1\end{bmatrix}\begin{bmatrix}0\\1\\0\end{bmatrix}\right]$$

$$=e^t\begin{bmatrix}0\\1\\0\end{bmatrix}+te^t\begin{bmatrix}1\\0\\0\end{bmatrix}=e^t\begin{bmatrix}t\\1\\0\end{bmatrix}$$

is a second linearly independent solution.

(ii) $\lambda=2$: We seek a nonzero vector $\mathbf{v}$ such that

$$(\mathbf{A}-2\mathbf{I})\mathbf{v}=\begin{bmatrix}-1&1&0\\0&-1&0\\0&0&0\end{bmatrix}\begin{bmatrix}v_1\\v_2\\v_3\end{bmatrix}=\begin{bmatrix}0\\0\\0\end{bmatrix}.$$

This implies that $v_1=v_2=0$ and v_3 is arbitrary. Hence

$$\mathbf{x}^3(t)=e^{2t}\begin{bmatrix}0\\0\\1\end{bmatrix}$$

is a third linearly independent solution.

Example 2. Solve the initial-value problem

$$\dot{\mathbf{x}}=\begin{bmatrix}2&1&3\\0&2&-1\\0&0&2\end{bmatrix}\mathbf{x},\qquad \mathbf{x}(0)=\begin{bmatrix}1\\2\\1\end{bmatrix}.$$

Solution. The characteristic polynomial of the matrix

$$\mathbf{A}=\begin{bmatrix}2&1&3\\0&2&-1\\0&0&2\end{bmatrix}$$

is $(2-\lambda)^3$. Hence $\lambda=2$ is an eigenvalue of $\mathbf{A}$ with multiplicity three. The eigenvectors of $\mathbf{A}$ satisfy the equation

$$(\mathbf{A}-2\mathbf{I})\mathbf{v}=\begin{bmatrix}0&1&3\\0&0&-1\\0&0&0\end{bmatrix}\begin{bmatrix}v_1\\v_2\\v_3\end{bmatrix}=\begin{bmatrix}0\\0\\0\end{bmatrix}.$$

This implies that $v_2 = v_3 = 0$ and v_1 is arbitrary. Hence

$$\mathbf{x}^1(t) = e^{2t}\begin{bmatrix}1\\0\\0\end{bmatrix}$$

is one solution of $\dot{\mathbf{x}} = \mathbf{A}\mathbf{x}$.

Since $\mathbf{A}$ has only one linearly independent eigenvector we look for all solutions of the equation

$$(\mathbf{A}-2\mathbf{I})^2\mathbf{v} = \begin{bmatrix}0&1&3\\0&0&-1\\0&0&0\end{bmatrix}\begin{bmatrix}0&1&3\\0&0&-1\\0&0&0\end{bmatrix}\mathbf{v} = \begin{bmatrix}0&0&-1\\0&0&0\\0&0&0\end{bmatrix}\begin{bmatrix}v_1\\v_2\\v_3\end{bmatrix} = \begin{bmatrix}0\\0\\0\end{bmatrix}.$$

This implies that $v_3 = 0$ and both v_1 and v_2 are arbitrary. Now, the vector

$$\mathbf{v} = \begin{bmatrix}0\\1\\0\end{bmatrix}$$

satisfies $(\mathbf{A}-2\mathbf{I})^2\mathbf{v} = \mathbf{0}$, but $(\mathbf{A}-2\mathbf{I})\mathbf{v} \neq \mathbf{0}$. Hence

$$\mathbf{x}^2(t) = e^{\mathbf{A}t}\begin{bmatrix}0\\1\\0\end{bmatrix} = e^{2t}e^{(\mathbf{A}-2\mathbf{I})t}\begin{bmatrix}0\\1\\0\end{bmatrix}$$

$$= e^{2t}\big[\mathbf{I} + t(\mathbf{A}-2\mathbf{I})\big]\begin{bmatrix}0\\1\\0\end{bmatrix} = e^{2t}\left[\mathbf{I} + t\begin{bmatrix}0&1&3\\0&0&-1\\0&0&0\end{bmatrix}\right]\begin{bmatrix}0\\1\\0\end{bmatrix}$$

$$= e^{2t}\left[\begin{bmatrix}0\\1\\0\end{bmatrix} + t\begin{bmatrix}1\\0\\0\end{bmatrix}\right] = e^{2t}\begin{bmatrix}t\\1\\0\end{bmatrix}$$

is a second solution of $\dot{\mathbf{x}} = \mathbf{A}\mathbf{x}$.

Since the equation $(\mathbf{A}-2\mathbf{I})^2\mathbf{v} = \mathbf{0}$ has only two linearly independent solutions, we look for all solutions of the equation

$$(\mathbf{A}-2\mathbf{I})^3\mathbf{v} = \begin{bmatrix}0&1&3\\0&0&-1\\0&0&0\end{bmatrix}^3\mathbf{v} = \begin{bmatrix}0&0&0\\0&0&0\\0&0&0\end{bmatrix}\begin{bmatrix}v_1\\v_2\\v_3\end{bmatrix} = \begin{bmatrix}0\\0\\0\end{bmatrix}.$$

Obviously, every vector $\mathbf{v}$ is a solution of this equation. The vector

$$\mathbf{v} = \begin{bmatrix}0\\0\\1\end{bmatrix}$$

does not satisfy $(\mathbf{A}-2\mathbf{I})^2\mathbf{v}=\mathbf{0}$. Hence

$$\mathbf{x}^3(t)=e^{\mathbf{A}t}\begin{pmatrix}0\\0\\1\end{pmatrix}=e^{2t}e^{(\mathbf{A}-2\mathbf{I})t}\begin{pmatrix}0\\0\\1\end{pmatrix}$$

$$=e^{2t}\left[\mathbf{I}+t(\mathbf{A}-2\mathbf{I})+\frac{t^2}{2}(\mathbf{A}-2\mathbf{I})^2\right]\begin{pmatrix}0\\0\\1\end{pmatrix}$$

$$=e^{2t}\left[\begin{pmatrix}0\\0\\1\end{pmatrix}+t\begin{pmatrix}3\\-1\\0\end{pmatrix}+\frac{t^2}{2}\begin{pmatrix}-1\\0\\0\end{pmatrix}\right]=e^{2t}\begin{pmatrix}3t-\frac{1}{2}t^2\\-t\\1\end{pmatrix}$$

is a third linearly independent solution. Therefore,

$$\mathbf{x}(t)=e^{2t}\left[c_1\begin{pmatrix}1\\0\\0\end{pmatrix}+c_2\begin{pmatrix}t\\1\\0\end{pmatrix}+c_3\begin{pmatrix}3t-\frac{1}{2}t^2\\-t\\1\end{pmatrix}\right].$$

The constants c_1, c_2, and c_3 are determined from the initial conditions

$$\begin{pmatrix}1\\2\\1\end{pmatrix}=c_1\begin{pmatrix}1\\0\\0\end{pmatrix}+c_2\begin{pmatrix}0\\1\\0\end{pmatrix}+c_3\begin{pmatrix}0\\0\\1\end{pmatrix}.$$

This implies that $c_1=1$, $c_2=2$, and $c_3=1$. Hence

$$\mathbf{x}(t)=e^{2t}\begin{pmatrix}1+5t-\frac{1}{2}t^2\\2-t\\1\end{pmatrix}.$$

For the matrix $\mathbf{A}$ in Example 2, $p(\lambda)=(2-\lambda)^3$ and $(2\mathbf{I}-\mathbf{A})^3=\mathbf{0}$. This is not an accident. Every matrix $\mathbf{A}$ satisfies its own characteristic equation. This is the content of the following theorem.

Theorem 13 (Cayley–Hamilton). *Let $p(\lambda)=p_0+p_1\lambda+\ldots+(-1)^n\lambda^n$ be the characteristic polynomial of* $\mathbf{A}$. *Then,*

$$p(\mathbf{A})\equiv p_0\mathbf{I}+p_1\mathbf{A}+\ldots+(-1)^n\mathbf{A}^n=\mathbf{0}.$$

FAKE PROOF. Setting $\lambda=\mathbf{A}$ in the equation $p(\lambda)=\det(\mathbf{A}-\lambda\mathbf{I})$ gives $p(\mathbf{A})=\det(\mathbf{A}-\mathbf{A}\mathbf{I})=\det\mathbf{0}=0$. The fallacy in this proof is that we cannot set $\lambda=\mathbf{A}$ in the expression $\det(\mathbf{A}-\lambda\mathbf{I})$ since we cannot subtract a matrix from the diagonal elements of $\mathbf{A}$. However, there is a very clever way to make this proof kosher. Let $\mathbf{C}(\lambda)$ be the classical adjoint (see Section 3.5) of the matrix $(\mathbf{A}-\lambda\mathbf{I})$. Then,

$$(\mathbf{A}-\lambda\mathbf{I})\mathbf{C}(\lambda)=p(\lambda)\mathbf{I}. \tag{4}$$

Each element of the matrix $\mathbf{C}(\lambda)$ is a polynomial in λ of degree at most $(n-1)$. Therefore, we can write $\mathbf{C}(\lambda)$ in the form

$$\mathbf{C}(\lambda)=\mathbf{C}_0+\mathbf{C}_1\lambda+\dots+\mathbf{C}_{n-1}\lambda^{n-1}$$

where $\mathbf{C}_0,\dots,\mathbf{C}_{n-1}$ are $n\times n$ matrices. For example,

$$\begin{aligned}\begin{pmatrix}\lambda+\lambda^2 & 2\lambda\\ \lambda^2 & 1-\lambda\end{pmatrix}&=\begin{pmatrix}0&0\\0&1\end{pmatrix}+\begin{pmatrix}\lambda & 2\lambda\\ 0 & -\lambda\end{pmatrix}+\begin{pmatrix}\lambda^2 & 0\\ \lambda^2 & 0\end{pmatrix}\\ &=\begin{pmatrix}0&0\\0&1\end{pmatrix}+\lambda\begin{pmatrix}1&2\\0&-1\end{pmatrix}+\lambda^2\begin{pmatrix}1&0\\1&0\end{pmatrix}\end{aligned}$$

Thus, Equation (4) can be written in the form

$$(\mathbf{A}-\lambda\mathbf{I})\left[\mathbf{C}_0+\mathbf{C}_1\lambda+\dots+\mathbf{C}_{n-1}\lambda^{n-1}\right]=p_0\mathbf{I}+p_1\lambda\mathbf{I}+\dots+(-1)^n\lambda^n\mathbf{I}. \quad (5)$$

Observe that both sides of (5) are polynomials in λ, whose coefficients are $n\times n$ matrices. Since these two polynomials are equal for all values of λ, their coefficients must agree. But if the coefficients of like powers of λ agree, then we can put in anything we want for λ and still have equality. In particular, set $\lambda=\mathbf{A}$. Then,

$$\begin{aligned}p(\mathbf{A})&=p_0\mathbf{I}+p_1\mathbf{A}+\dots+(-1)^n\mathbf{A}^n\\ &=(\mathbf{A}-\mathbf{A}\mathbf{I})\left[\mathbf{C}_0+\mathbf{C}_1\mathbf{A}+\dots+\mathbf{C}_{n-1}\mathbf{A}^{n-1}\right]=\mathbf{0}.\end{aligned}$$ □

Exercises

In each of Problems 1–4 find the general solution of the given system of differential equations.

1. $\dot{\mathbf{x}}=\begin{pmatrix}0&-1&1\\2&-3&1\\1&-1&-1\end{pmatrix}\mathbf{x}$

2. $\dot{\mathbf{x}}=\begin{pmatrix}1&1&1\\2&1&-1\\-3&2&4\end{pmatrix}\mathbf{x}$

3. $\dot{\mathbf{x}}=\begin{pmatrix}-1&-1&0\\0&-1&0\\0&0&-2\end{pmatrix}\mathbf{x}$ *Hint*: Look at Example 1 of text.

4. $\dot{\mathbf{x}}=\begin{bmatrix}2&0&-1&0\\0&2&1&0\\0&0&2&0\\0&0&-1&2\end{bmatrix}\mathbf{x}$

In each of Problems 5–8, solve the given initial-value problem

5. $\dot{\mathbf{x}}=\begin{pmatrix}-1&1&2\\-1&1&1\\-2&1&3\end{pmatrix}\mathbf{x},\quad \mathbf{x}(0)=\begin{pmatrix}1\\0\\1\end{pmatrix}$

6. $\dot{\mathbf{x}}=\begin{pmatrix}-4&-4&0\\10&9&1\\-4&-3&1\end{pmatrix}\mathbf{x},\quad \mathbf{x}(0)=\begin{pmatrix}2\\1\\-1\end{pmatrix}$

7. $\dot{\mathbf{x}}=\begin{pmatrix}1&2&-3\\1&1&2\\1&-1&4\end{pmatrix}\mathbf{x},\quad \mathbf{x}(0)=\begin{pmatrix}1\\0\\0\end{pmatrix}$

8. $\dot{\mathbf{x}}=\begin{bmatrix}3&0&0&0\\1&3&0&0\\0&0&3&0\\0&0&2&3\end{bmatrix}\mathbf{x},\quad \mathbf{x}(0)=\begin{bmatrix}1\\1\\1\\1\end{bmatrix}$

9. Let

$$\mathbf{A}=\begin{bmatrix} \lambda_1 & 0 & \dots & 0 \\ 0 & \lambda_2 & \dots & 0 \\ \vdots & \vdots & & \vdots \\ 0 & 0 & \dots & \lambda_n \end{bmatrix}.$$

Show that

$$e^{\mathbf{A}t}=\begin{bmatrix} e^{\lambda_1 t} & 0 & 0 \\ 0 & e^{\lambda_2 t} & 0 \\ \vdots & \vdots & \vdots \\ 0 & 0 & e^{\lambda_n t} \end{bmatrix}.$$

10. Let

$$\mathbf{A}=\begin{pmatrix} \lambda & 1 & 0 \\ 0 & \lambda & 1 \\ 0 & 0 & \lambda \end{pmatrix}.$$

Prove that

$$e^{\mathbf{A}t}=e^{\lambda t}\begin{bmatrix} 1 & t & \frac{1}{2}t^2 \\ 0 & 1 & t \\ 0 & 0 & 1 \end{bmatrix}.$$

Hint: Write **A** in the form

$$\mathbf{A}=\lambda\mathbf{I}+\begin{pmatrix} 0 & 1 & 0 \\ 0 & 0 & 1 \\ 0 & 0 & 0 \end{pmatrix}$$

and observe that

$$e^{\mathbf{A}t}=e^{\lambda t}\exp\left[\begin{pmatrix} 0 & 1 & 0 \\ 0 & 0 & 1 \\ 0 & 0 & 0 \end{pmatrix}t\right].$$

11. Let **A** be the $n\times n$ matrix

$$\begin{bmatrix} \lambda & 1 & 0 & \dots & 0 \\ 0 & \lambda & 1 & \dots & 0 \\ \vdots & \vdots & \vdots & & \vdots \\ 0 & 0 & 0 & \dots & 1 \\ 0 & 0 & 0 & \dots & \lambda \end{bmatrix},$$

and let **P** be the $n\times n$ matrix

$$\begin{bmatrix} 0 & 1 & 0 & \dots & 0 \\ 0 & 0 & 1 & \dots & 0 \\ \vdots & \vdots & \vdots & & \vdots \\ 0 & 0 & 0 & \dots & 1 \\ 0 & 0 & 0 & \dots & 0 \end{bmatrix}.$$

(a) Show that $\mathbf{P}^n=\mathbf{0}$.   (b) Show that $(\lambda\mathbf{I})\mathbf{P}=\mathbf{P}(\lambda\mathbf{I})$.

(c) Show that

$$e^{\mathbf{A}t}=e^{\lambda t}\left[\mathbf{I}+t\mathbf{P}+\frac{t^2\mathbf{P}^2}{2!}+\dots+\frac{t^{n-1}}{(n-1)!}\mathbf{P}^{n-1}\right].$$

12. Compute $e^{\mathbf{A}t}$ if

(a) $\mathbf{A}=\begin{bmatrix}2&1&0&0\\0&2&1&0\\0&0&2&1\\0&0&0&2\end{bmatrix}$,

(b) $\mathbf{A}=\begin{bmatrix}2&1&0&0\\0&2&1&0\\0&0&2&0\\0&0&0&2\end{bmatrix}$,   (c) $\mathbf{A}=\begin{bmatrix}2&1&0&0\\0&2&0&0\\0&0&2&0\\0&0&0&2\end{bmatrix}$.

13. (a) Show that $e^{\mathbf{T}^{-1}\mathbf{A}\mathbf{T}}=\mathbf{T}^{-1}e^{\mathbf{A}}\mathbf{T}$.

(b) Given that

$$\mathbf{T}^{-1}\mathbf{A}\mathbf{T}=\begin{pmatrix}1&0&0\\0&2&0\\0&0&-1\end{pmatrix}$$

with

$$\mathbf{T}=\begin{pmatrix}1&-1&-1\\3&-1&2\\2&2&3\end{pmatrix},$$

compute $e^{\mathbf{A}t}$.

14. Suppose that $p(\lambda)=\det(\mathbf{A}-\lambda\mathbf{I})$ has n distinct roots $\lambda_1,\dots,\lambda_n$. Prove directly that $p(\mathbf{A})\equiv(-1)^n(\mathbf{A}-\lambda_1\mathbf{I})\dots(\mathbf{A}-\lambda_n\mathbf{I})=\mathbf{0}$. *Hint*: Write any vector $\mathbf{x}$ in the form $\mathbf{x}=x_1\mathbf{v}^1+\dots+x_n\mathbf{v}^n$ where $\mathbf{v}^1,\dots,\mathbf{v}^n$ are n independent eigenvectors of $\mathbf{A}$ with eigenvalues $\lambda_1,\dots,\lambda_n$ respectively, and conclude that $p(\mathbf{A})\mathbf{x}=\mathbf{0}$ for all vectors $\mathbf{x}$.

15. Suppose that $\mathbf{A}^2=\alpha\mathbf{A}$. Find $e^{\mathbf{A}t}$.

16. Let

$$\mathbf{A}=\begin{bmatrix}1&1&1&1&1\\1&1&1&1&1\\1&1&1&1&1\\1&1&1&1&1\\1&1&1&1&1\end{bmatrix}.$$

(a) Show that $\mathbf{A}(\mathbf{A}-5\mathbf{I})=\mathbf{0}$.

(b) Find $e^{\mathbf{A}t}$.

17. Let

$$\mathbf{A}=\begin{pmatrix}0&1\\-1&0\end{pmatrix}.$$

(a) Show that $\mathbf{A}^2=-\mathbf{I}$.

(b) Show that

$$e^{\mathbf{A}t}=\begin{pmatrix} \cos t & \sin t \\ -\sin t & \cos t \end{pmatrix}.$$

In each of Problems 18–20 verify directly the Cayley–Hamilton Theorem for the given matrix **A**.

18. $\mathbf{A}=\begin{pmatrix} 1 & -1 & -1 \\ 2 & 1 & -1 \\ -1 & -1 & 1 \end{pmatrix}$ **19.** $\mathbf{A}=\begin{pmatrix} 2 & 3 & 1 \\ 2 & 3 & 1 \\ 2 & 3 & 1 \end{pmatrix}$ **20.** $\mathbf{A}=\begin{pmatrix} -1 & 0 & 1 \\ -1 & 3 & 0 \\ 2 & 4 & 6 \end{pmatrix}$

3.9 Fundamental matrix solutions; $e^{\mathbf{A}t}$

If $\mathbf{x}^1(t),\dots,\mathbf{x}^n(t)$ are n linearly independent solutions of the differential equation

$$\dot{\mathbf{x}}=\mathbf{A}\mathbf{x}, \tag{1}$$

then every solution $\mathbf{x}(t)$ can be written in the form

$$\mathbf{x}(t)=c_1\mathbf{x}^1(t)+c_2\mathbf{x}^2(t)+\dots+c_n\mathbf{x}^n(t). \tag{2}$$

Let $\mathbf{X}(t)$ be the matrix whose columns are $\mathbf{x}^1(t),\dots,\mathbf{x}^n(t)$. Then, Equation (2) can be written in the concise form $\mathbf{x}(t)=\mathbf{X}(t)\mathbf{c}$, where

$$\mathbf{c}=\begin{bmatrix} c_1 \\ \vdots \\ c_n \end{bmatrix}.$$

Definition. A matrix $\mathbf{X}(t)$ is called a *fundamental matrix solution* of (1) if its columns form a set of n linearly independent solutions of (1).

Example 1. Find a fundamental matrix solution of the system of differential equations

$$\dot{\mathbf{x}}=\begin{bmatrix} 1 & -1 & 4 \\ 3 & 2 & -1 \\ 2 & 1 & -1 \end{bmatrix}\mathbf{x}. \tag{3}$$

Solution. We showed in Section 3.6 (see Example 1) that

$$e^{t}\begin{bmatrix} -1 \\ 4 \\ 1 \end{bmatrix}, \quad e^{3t}\begin{bmatrix} 1 \\ 2 \\ 1 \end{bmatrix} \quad \text{and} \quad e^{-2t}\begin{bmatrix} -1 \\ 1 \\ 1 \end{bmatrix}$$

are three linearly independent solutions of (3). Hence

$$\mathbf{X}(t)=\begin{bmatrix} -e^{t} & e^{3t} & -e^{-2t} \\ 4e^{t} & 2e^{3t} & e^{-2t} \\ e^{t} & e^{3t} & e^{-2t} \end{bmatrix}$$

is a fundamental matrix solution of (3).

In this section we will show that the matrix $e^{\mathbf{A}t}$ can be computed directly from any fundamental matrix solution of (1). This is rather remarkable since it does not appear possible to sum the infinite series $[\mathbf{I}+\mathbf{A}t+(\mathbf{A}t)^2/2!+\ldots]$ exactly, for an arbitrary matrix $\mathbf{A}$. Specifically, we have the following theorem.

Theorem 14. *Let $\mathbf{X}(t)$ be a fundamental matrix solution of the differential equation $\dot{\mathbf{x}}=\mathbf{A}\mathbf{x}$. Then,*

$$e^{\mathbf{A}t}=\mathbf{X}(t)\mathbf{X}^{-1}(0). \tag{4}$$

In other words, the product of any fundamental matrix solution of (1) *with its inverse at $t=0$ must yield $e^{\mathbf{A}t}$.*

We prove Theorem 14 in three steps. First, we establish a simple test to determine whether a matrix-valued function is a fundamental matrix solution of (1). Then, we use this test to show that $e^{\mathbf{A}t}$ is a fundamental matrix solution of (1). Finally, we establish a connection between any two fundamental matrix solutions of (1).

Lemma 1. *A matrix $\mathbf{X}(t)$ is a fundamental matrix solution of* (1) *if, and only if, $\dot{\mathbf{X}}(t)=\mathbf{A}\mathbf{X}(t)$ and* $\det\mathbf{X}(0)\neq 0$. (*The derivative of a matrix-valued function $\mathbf{X}(t)$ is the matrix whose components are the derivatives of the corresponding components of* $\mathbf{X}(t)$.)

Proof. Let $\mathbf{x}^1(t),\ldots,\mathbf{x}^n(t)$ denote the n columns of $\mathbf{X}(t)$. Observe that

$$\dot{\mathbf{X}}(t)=\left(\dot{\mathbf{x}}^1(t),\ldots,\dot{\mathbf{x}}^n(t)\right)$$

and

$$\mathbf{A}\mathbf{X}(t)=\left(\mathbf{A}\mathbf{x}^1(t),\ldots,\mathbf{A}\mathbf{x}^n(t)\right).$$

Hence, the n vector equations $\dot{\mathbf{x}}^1(t)=\mathbf{A}\mathbf{x}^1(t),\ldots,\dot{\mathbf{x}}^n(t)=\mathbf{A}\mathbf{x}^n(t)$ are equivalent to the single matrix equation $\dot{\mathbf{X}}(t)=\mathbf{A}\mathbf{X}(t)$. Moreover, n solutions $\mathbf{x}^1(t),\ldots,\mathbf{x}^n(t)$ of (1) are linearly independent if, and only if, $\mathbf{x}^1(0),\ldots,\mathbf{x}^n(0)$ are linearly independent vectors of R^n. These vectors, in turn, are linearly independent if, and only if, $\det\mathbf{X}(0)\neq 0$. Consequently, $\mathbf{X}(t)$ is a fundamental matrix solution of (1) if, and only if, $\dot{\mathbf{X}}(t)=\mathbf{A}\mathbf{X}(t)$ and $\det\mathbf{X}(0)\neq 0$. □

Lemma 2. *The matrix-valued function $e^{\mathbf{A}t}\equiv\mathbf{I}+\mathbf{A}t+\mathbf{A}^2t^2/2!+\ldots$ is a fundamental matrix solution of* (1).

Proof. We showed in Section 3.8 that $(d/dt)e^{\mathbf{A}t}=\mathbf{A}e^{\mathbf{A}t}$. Hence $e^{\mathbf{A}t}$ is a solution of the matrix differential equation $\dot{\mathbf{X}}(t)=\mathbf{A}\mathbf{X}(t)$. Moreover, its determinant, evaluated at $t=0$, is one since $e^{\mathbf{A}t}=\mathbf{I}$. Therefore, by Lemma 1, $e^{\mathbf{A}t}$ is a fundamental matrix solution of (1). □

Lemma 3. *Let $\mathbf{X}(t)$ and $\mathbf{Y}(t)$ be two fundamental matrix solutions of* (1). *Then, there exists a constant matrix $\mathbf{C}$ such that $\mathbf{Y}(t)=\mathbf{X}(t)\mathbf{C}$.*

PROOF. By definition, the columns $\mathbf{x}^1(t),\ldots,\mathbf{x}^n(t)$ of $\mathbf{X}(t)$ and $\mathbf{y}^1(t),\ldots,\mathbf{y}^n(t)$ of $\mathbf{Y}(t)$ are linearly independent sets of solutions of (1). In particular, therefore, each column of $\mathbf{Y}(t)$ can be written as a linear combination of the columns of $\mathbf{X}(t)$; i.e., there exist constants $c_1^j,\ldots,c_n^j$ such that

$$\mathbf{y}^j(t)=c_1^j\mathbf{x}^1(t)+c_2^j\mathbf{x}^2(t)+\ldots+c_n^j\mathbf{x}^n(t),\qquad j=1,\ldots,n. \tag{5}$$

Let $\mathbf{C}$ be the matrix $(\mathbf{c}^1,\mathbf{c}^2,\ldots,\mathbf{c}^n)$ where

$$\mathbf{c}^j=\begin{pmatrix} c_1^j\\ \vdots \\ c_n^j\end{pmatrix}.$$

Then, the n equations (5) are equivalent to the single matrix equation $\mathbf{Y}(t)=\mathbf{X}(t)\mathbf{C}$. □

We are now in a position to prove Theorem 14.

PROOF OF THEOREM 14. Let $\mathbf{X}(t)$ be a fundamental matrix solution of (1). Then, by Lemmas 2 and 3 there exists a constant matrix $\mathbf{C}$ such that

$$e^{\mathbf{A}t}=\mathbf{X}(t)\mathbf{C}. \tag{6}$$

Setting $t=0$ in (6) gives $\mathbf{I}=\mathbf{X}(0)\mathbf{C}$, which implies that $\mathbf{C}=\mathbf{X}^{-1}(0)$. Hence, $e^{\mathbf{A}t}=\mathbf{X}(t)\mathbf{X}^{-1}(0)$. □

Example 1. Find $e^{\mathbf{A}t}$ if

$$\mathbf{A}=\begin{pmatrix}1&1&1\\0&3&2\\0&0&5\end{pmatrix}.$$

Solution. Our first step is to find 3 linearly independent solutions of the differential equation

$$\dot{\mathbf{x}}=\begin{pmatrix}1&1&1\\0&3&2\\0&0&5\end{pmatrix}\mathbf{x}.$$

To this end we compute

$$p(\lambda)=\det(\mathbf{A}-\lambda\mathbf{I})=\det\begin{pmatrix}1-\lambda&1&1\\0&3-\lambda&2\\0&0&5-\lambda\end{pmatrix}=(1-\lambda)(3-\lambda)(5-\lambda).$$

Thus, $\mathbf{A}$ has 3 distinct eigenvalues $\lambda=1$, $\lambda=3$, and $\lambda=5$.

(i) $\lambda=1$: Clearly,

$$\mathbf{v}^1=\begin{pmatrix}1\\0\\0\end{pmatrix}$$

is an eigenvector of $\mathbf{A}$ with eigenvalue one. Hence

$$\mathbf{x}^1(t)=e^t\begin{bmatrix}1\\0\\0\end{bmatrix}$$

is one solution of $\dot{\mathbf{x}}=\mathbf{A}\mathbf{x}$.

(ii) $\lambda=3$: We seek a nonzero solution of the equation

$$(\mathbf{A}-3\mathbf{I})\mathbf{v}=\begin{bmatrix}-2&1&1\\0&0&2\\0&0&2\end{bmatrix}\begin{bmatrix}v_1\\v_2\\v_3\end{bmatrix}=\begin{bmatrix}0\\0\\0\end{bmatrix}.$$

This implies that $v_3=0$ and $v_2=2v_1$. Hence,

$$\mathbf{v}^2=\begin{bmatrix}1\\2\\0\end{bmatrix}$$

is an eigenvector of $\mathbf{A}$ with eigenvalue 3. Consequently,

$$\mathbf{x}^2(t)=e^{3t}\begin{bmatrix}1\\2\\0\end{bmatrix}$$

is a second solution of $\dot{\mathbf{x}}=\mathbf{A}\mathbf{x}$.

(iii) $\lambda=5$: We seek a nonzero solution of the equation

$$(\mathbf{A}-5\mathbf{I})\mathbf{v}=\begin{bmatrix}-4&1&1\\0&-2&2\\0&0&0\end{bmatrix}\begin{bmatrix}v_1\\v_2\\v_3\end{bmatrix}=\begin{bmatrix}0\\0\\0\end{bmatrix}.$$

This implies that $v_2=v_3$ and $v_1=v_3/2$. Hence,

$$\mathbf{v}^3=\begin{bmatrix}1\\2\\2\end{bmatrix}$$

is an eigenvector of $\mathbf{A}$ with eigenvalue 5. Consequently,

$$\mathbf{x}^3(t)=e^{5t}\begin{bmatrix}1\\2\\2\end{bmatrix}$$

is a third solution of $\dot{\mathbf{x}}=\mathbf{A}\mathbf{x}$. These solutions are clearly linearly independent. Therefore,

$$\mathbf{X}(t)=\begin{bmatrix}e^t&e^{3t}&e^{5t}\\0&2e^{3t}&2e^{5t}\\0&0&2e^{5t}\end{bmatrix}$$

is a fundamental matrix solution. Using the methods of Section 3.5, we compute

$$\mathbf{X}^{-1}(0)=\begin{bmatrix}1&1&1\\0&2&2\\0&0&2\end{bmatrix}^{-1}=\begin{bmatrix}1&-\frac{1}{2}&0\\0&\frac{1}{2}&-\frac{1}{2}\\0&0&\frac{1}{2}\end{bmatrix}.$$

Therefore,

$$\exp\left[\begin{pmatrix}1&1&1\\0&3&2\\0&0&5\end{pmatrix}t\right]=\begin{bmatrix}e^{t}&e^{3t}&e^{5t}\\0&2e^{3t}&2e^{5t}\\0&0&2e^{5t}\end{bmatrix}\begin{bmatrix}1&-\frac{1}{2}&0\\0&\frac{1}{2}&-\frac{1}{2}\\0&0&\frac{1}{2}\end{bmatrix}$$

$$=\begin{bmatrix}e^{t}&-\frac{1}{2}e^{t}+\frac{1}{2}e^{3t}&-\frac{1}{2}e^{3t}+\frac{1}{2}e^{5t}\\0&e^{3t}&-e^{3t}+e^{5t}\\0&0&e^{5t}\end{bmatrix}.$$

Exercises

Compute $e^{\mathbf{A}t}$ for $\mathbf{A}$ equal

1. $\begin{pmatrix}1&-1&-1\\1&3&1\\-3&1&-1\end{pmatrix}$

2. $\begin{pmatrix}1&1&-1\\2&3&-4\\4&1&-4\end{pmatrix}$

3. $\begin{pmatrix}2&0&1\\1&0&1\\1&-2&0\end{pmatrix}$

4. $\begin{pmatrix}1&0&0\\2&1&-2\\3&2&1\end{pmatrix}$

5. $\begin{pmatrix}0&2&1\\-1&-3&-1\\1&1&-1\end{pmatrix}$

6. $\begin{pmatrix}1&1&1\\2&1&-1\\0&-1&1\end{pmatrix}$

7. Find $\mathbf{A}$ if

$$e^{\mathbf{A}t}=\begin{bmatrix}2e^{2t}-e^{t}&e^{2t}-e^{t}&e^{t}-e^{2t}\\e^{2t}-e^{t}&2e^{2t}-e^{t}&e^{t}-e^{2t}\\3e^{2t}-3e^{t}&3e^{2t}-3e^{t}&3e^{t}-2e^{2t}\end{bmatrix}.$$

In each of Problems 8–11, determine whether the given matrix is a fundamental matrix solution of $\dot{\mathbf{x}}=\mathbf{A}\mathbf{x}$, for some $\mathbf{A}$; if yes, find $\mathbf{A}$.

8. $\begin{bmatrix}e^{t}&e^{-t}&e^{t}+2e^{-t}\\e^{t}&-e^{-t}&e^{t}-2e^{-t}\\2e^{t}&e^{-t}&2(e^{t}+e^{-t})\end{bmatrix}$

9. $\begin{bmatrix}-5\cos 2t&-5\sin 2t&3e^{2t}\\-2(\cos 2t+\sin 2t)&2(\cos 2t-\sin 2t)&0\\\cos 2t&\sin 2t&e^{2t}\end{bmatrix}$

10. $e^t\begin{bmatrix}1 & t+1 & t^2+1\\ 1 & 2(t+1) & 4t^2\\ 1 & t+2 & 3\end{bmatrix}$ **11.** $\begin{bmatrix}e^{2t} & 2e^{-t} & e^{3t}\\ 2e^t & 2e^{-t} & e^{3t}\\ 3e^t & e^{-t} & 2e^{3t}\end{bmatrix}$

12. Let $\boldsymbol{\phi}^j(t)$ be the solution of the initial-value problem $\dot{\mathbf{x}}=\mathbf{A}\mathbf{x}$, $\mathbf{x}(0)=\mathbf{e}^j$. Show that $e^{\mathbf{A}t}=(\boldsymbol{\phi}^1,\boldsymbol{\phi}^2,\ldots,\boldsymbol{\phi}^n)$.

13. Suppose that $\mathbf{Y}(t)=\mathbf{X}(t)\mathbf{C}$, where $\mathbf{X}(t)$ and $\mathbf{Y}(t)$ are fundamental matrix solutions of $\dot{\mathbf{x}}=\mathbf{A}\mathbf{x}$, and $\mathbf{C}$ is a constant matrix. Prove that $\det\mathbf{C}\neq 0$.

14. Let $\mathbf{X}(t)$ be a fundamental matrix solution of (1), and $\mathbf{C}$ a constant matrix with $\det\mathbf{C}\neq 0$. Show that $\mathbf{Y}(t)=\mathbf{X}(t)\mathbf{C}$ is again a fundamental matrix solution of (1).

15. Let $\mathbf{X}(t)$ be a fundamental matrix solution of $\dot{\mathbf{x}}=\mathbf{A}\mathbf{x}$. Prove that the solution $\mathbf{x}(t)$ of the initial-value problem $\dot{\mathbf{x}}=\mathbf{A}\mathbf{x}$, $\mathbf{x}(t_0)=\mathbf{x}^0$ is $\mathbf{x}(t)=\mathbf{X}(t)\mathbf{X}^{-1}(t_0)\mathbf{x}^0$.

16. Let $\mathbf{X}(t)$ be a fundamental matrix solution of $\dot{\mathbf{x}}=\mathbf{A}\mathbf{x}$. Prove that $\mathbf{X}(t)\mathbf{X}^{-1}(t_0)=e^{\mathbf{A}(t-t_0)}$.

17. Here is an elegant proof of the identity $e^{\mathbf{A}t+\mathbf{B}t}=e^{\mathbf{A}t}e^{\mathbf{B}t}$ if $\mathbf{AB}=\mathbf{BA}$.

(a) Show that $\mathbf{X}(t)=e^{\mathbf{A}t+\mathbf{B}t}$ satisfies the initial-value problem $\dot{\mathbf{X}}=(\mathbf{A}+\mathbf{B})\mathbf{X}$, $\mathbf{X}(0)=\mathbf{I}$.

(b) Show that $e^{\mathbf{A}t}\mathbf{B}=\mathbf{B}e^{\mathbf{A}t}$ if $\mathbf{AB}=\mathbf{BA}$. (*Hint*: $\mathbf{A}^j\mathbf{B}=\mathbf{B}\mathbf{A}^j$ if $\mathbf{AB}=\mathbf{BA}$). Then, conclude that $(d/dt)e^{\mathbf{A}t}e^{\mathbf{B}t}=(\mathbf{A}+\mathbf{B})e^{\mathbf{A}t}e^{\mathbf{B}t}$.

(c) It follows immediately from Theorem 4, Section 3.4 that the solution $\mathbf{X}(t)$ of the initial-value problem $\dot{\mathbf{X}}=(\mathbf{A}+\mathbf{B})\mathbf{X}$, $\mathbf{X}(0)=\mathbf{I}$, is unique. Conclude, therefore, that $e^{\mathbf{A}t+\mathbf{B}t}=e^{\mathbf{A}t}e^{\mathbf{B}t}$.

3.10 The nonhomogeneous equation; variation of parameters

Consider now the nonhomogeneous equation $\dot{\mathbf{x}}=\mathbf{A}\mathbf{x}+\mathbf{f}(t)$. In this case, we can use our knowledge of the solutions of the homogeneous equation

$$\dot{\mathbf{x}}=\mathbf{A}\mathbf{x} \tag{1}$$

to help us find the solution of the initial-value problem

$$\dot{\mathbf{x}}=\mathbf{A}\mathbf{x}+\mathbf{f}(t),\qquad \mathbf{x}(t_0)=\mathbf{x}^0. \tag{2}$$

Let $\mathbf{x}^1(t),\ldots,\mathbf{x}^n(t)$ be n linearly independent solutions of the homogeneous equation (1). Since the general solution of (1) is $c_1\mathbf{x}^1(t)+\ldots+c_n\mathbf{x}^n(t)$, it is natural to seek a solution of (2) of the form

$$\mathbf{x}(t)=u_1(t)\mathbf{x}^1(t)+u_2(t)\mathbf{x}^2(t)+\ldots+u_n(t)\mathbf{x}^n(t). \tag{3}$$

This equation can be written concisely in the form $\mathbf{x}(t)=\mathbf{X}(t)\mathbf{u}(t)$ where

$\mathbf{X}(t)=(\mathbf{x}^1(t),\dots,\mathbf{x}^n(t))$ and

$$\mathbf{u}(t)=\begin{bmatrix} u_1(t) \\ \vdots \\ u_n(t) \end{bmatrix}.$$

Plugging this expression into the differential equation $\dot{\mathbf{x}}=\mathbf{A}\mathbf{x}+\mathbf{f}(t)$ gives

$$\dot{\mathbf{X}}(t)\mathbf{u}(t)+\mathbf{X}(t)\dot{\mathbf{u}}(t)=\mathbf{A}\mathbf{X}(t)\mathbf{u}(t)+\mathbf{f}(t). \tag{4}$$

The matrix $\mathbf{X}(t)$ is a fundamental matrix solution of (1). Hence, $\dot{\mathbf{X}}(t)=\mathbf{A}\mathbf{X}(t)$, and Equation (4) reduces to

$$\mathbf{X}(t)\dot{\mathbf{u}}(t)=\mathbf{f}(t). \tag{5}$$

Recall that the columns of $\mathbf{X}(t)$ are linearly independent vectors of R^n at every time t. Hence $\mathbf{X}^{-1}(t)$ exists, and

$$\dot{\mathbf{u}}(t)=\mathbf{X}^{-1}(t)\mathbf{f}(t). \tag{6}$$

Integrating this expression between t_0 and t gives

$$\begin{aligned}\mathbf{u}(t)&=\mathbf{u}(t_0)+\int_{t_0}^{t}\mathbf{X}^{-1}(s)\mathbf{f}(s)\,ds\\ &=\mathbf{X}^{-1}(t_0)\mathbf{x}^0+\int_{t_0}^{t}\mathbf{X}^{-1}(s)\mathbf{f}(s)\,ds.\end{aligned}$$

Consequently,

$$\mathbf{x}(t)=\mathbf{X}(t)\mathbf{X}^{-1}(t_0)\mathbf{x}^0+\mathbf{X}(t)\int_{t_0}^{t}\mathbf{X}^{-1}(s)\mathbf{f}(s)\,ds. \tag{7}$$

If $\mathbf{X}(t)$ is the fundamental matrix solution $e^{\mathbf{A}t}$, then Equation (7) simplifies considerably. To wit, if $\mathbf{X}(t)=e^{\mathbf{A}t}$, then $\mathbf{X}^{-1}(s)=e^{-\mathbf{A}s}$. Hence

$$\begin{aligned}\mathbf{x}(t)&=e^{\mathbf{A}t}e^{-\mathbf{A}t_0}\mathbf{x}^0+e^{\mathbf{A}t}\int_{t_0}^{t}e^{-\mathbf{A}s}\mathbf{f}(s)\,ds\\ &=e^{\mathbf{A}(t-t_0)}\mathbf{x}^0+\int_{t_0}^{t}e^{\mathbf{A}(t-s)}\mathbf{f}(s)\,ds.\end{aligned} \tag{8}$$

Example 1. Solve the initial-value problem

$$\dot{\mathbf{x}}=\begin{bmatrix} 1 & 0 & 0 \\ 2 & 1 & -2 \\ 3 & 2 & 1 \end{bmatrix}\mathbf{x}+\begin{bmatrix} 0 \\ 0 \\ e^t\cos 2t \end{bmatrix},\qquad \mathbf{x}(0)=\begin{bmatrix} 0 \\ 1 \\ 1 \end{bmatrix}.$$

Solution. We first find $e^{\mathbf{A}t}$, where

$$\mathbf{A}=\begin{pmatrix}1&0&0\\2&1&-2\\3&2&1\end{pmatrix}.$$

To this end compute

$$\det(\mathbf{A}-\lambda\mathbf{I})=\det\begin{pmatrix}1-\lambda&0&0\\2&1-\lambda&-2\\3&2&1-\lambda\end{pmatrix}=(1-\lambda)(\lambda^2-2\lambda+5).$$

Thus the eigenvalues of $\mathbf{A}$ are

$$\lambda=1\quad\text{and}\quad\lambda=\frac{2\pm\sqrt{4-20}}{2}=1\pm 2i.$$

(i) $\lambda=1$: We seek a nonzero vector $\mathbf{v}$ such that

$$(\mathbf{A}-\mathbf{I})\mathbf{v}=\begin{pmatrix}0&0&0\\2&0&-2\\3&2&0\end{pmatrix}\begin{pmatrix}v_1\\v_2\\v_3\end{pmatrix}=\begin{pmatrix}0\\0\\0\end{pmatrix}.$$

This implies that $v_1=v_3$ and $v_2=-3v_1/2$. Hence

$$\mathbf{v}=\begin{pmatrix}2\\-3\\2\end{pmatrix}$$

is an eigenvector of $\mathbf{A}$ with eigenvalue 1. Consequently,

$$\mathbf{x}^1(t)=e^t\begin{pmatrix}2\\-3\\2\end{pmatrix}$$

is a solution of the homogeneous equation $\dot{\mathbf{x}}=\mathbf{A}\mathbf{x}$.

(ii) $\lambda=1+2i$: We seek a nonzero vector $\mathbf{v}$ such that

$$\left[\mathbf{A}-(1+2i)\mathbf{I}\right]\mathbf{v}=\begin{pmatrix}-2i&0&0\\2&-2i&-2\\3&2&-2i\end{pmatrix}\begin{pmatrix}v_1\\v_2\\v_3\end{pmatrix}=\begin{pmatrix}0\\0\\0\end{pmatrix}.$$

This implies that $v_1=0$ and $v_3=-iv_2$. Hence,

$$\mathbf{v}=\begin{pmatrix}0\\1\\-i\end{pmatrix}$$

is an eigenvector of $\mathbf{A}$ with eigenvalue $1+2i$. Therefore,

$$\mathbf{x}(t)=\begin{pmatrix}0\\1\\-i\end{pmatrix}e^{(1+2i)t}$$

is a complex-valued solution of $\dot{\mathbf{x}}=\mathbf{A}\mathbf{x}$. Now,

$$\begin{pmatrix}0\\1\\-i\end{pmatrix}e^{(1+2i)t}=e^{t}(\cos 2t+i\sin 2t)\left[\begin{pmatrix}0\\1\\0\end{pmatrix}-i\begin{pmatrix}0\\0\\1\end{pmatrix}\right]$$

$$=e^{t}\left[\cos 2t\begin{pmatrix}0\\1\\0\end{pmatrix}+\sin 2t\begin{pmatrix}0\\0\\1\end{pmatrix}\right]$$

$$+ie^{t}\left[\sin 2t\begin{pmatrix}0\\1\\0\end{pmatrix}-\cos 2t\begin{pmatrix}0\\0\\1\end{pmatrix}\right].$$

Consequently,

$$\mathbf{x}^2(t)=e^{t}\begin{pmatrix}0\\ \cos 2t\\ \sin 2t\end{pmatrix}\quad\text{and}\quad \mathbf{x}^3(t)=e^{t}\begin{pmatrix}0\\ \sin 2t\\ -\cos 2t\end{pmatrix}$$

are real-valued solutions of $\dot{\mathbf{x}}=\mathbf{A}\mathbf{x}$. The solutions $\mathbf{x}^1$, $\mathbf{x}^2$, and $\mathbf{x}^3$ are linearly independent since their values at $t=0$ are clearly linearly independent vectors of R^3. Therefore,

$$\mathbf{X}(t)=\begin{pmatrix}2e^{t} & 0 & 0\\ -3e^{t} & e^{t}\cos 2t & e^{t}\sin 2t\\ 2e^{t} & e^{t}\sin 2t & -e^{t}\cos 2t\end{pmatrix}$$

is a fundamental matrix solution of $\dot{\mathbf{x}}=\mathbf{A}\mathbf{x}$. Computing

$$\mathbf{X}^{-1}(0)=\begin{pmatrix}2 & 0 & 0\\ -3 & 1 & 0\\ 2 & 0 & -1\end{pmatrix}^{-1}=\begin{pmatrix}\frac{1}{2} & 0 & 0\\ \frac{3}{2} & 1 & 0\\ 1 & 0 & -1\end{pmatrix}$$

we see that

$$e^{\mathbf{A}t}=\begin{pmatrix}2e^{t} & 0 & 0\\ -3e^{t} & e^{t}\cos 2t & e^{t}\sin 2t\\ 2e^{t} & e^{t}\sin 2t & -e^{t}\cos 2t\end{pmatrix}\begin{pmatrix}\frac{1}{2} & 0 & 0\\ \frac{3}{2} & 1 & 0\\ 1 & 0 & -1\end{pmatrix}$$

$$=e^{t}\begin{pmatrix}1 & 0 & 0\\ -\frac{3}{2}+\frac{3}{2}\cos 2t+\sin 2t & \cos 2t & -\sin 2t\\ 1+\frac{3}{2}\sin 2t-\cos 2t & \sin 2t & \cos 2t\end{pmatrix}.$$

Consequently,

$$\mathbf{x}(t)=e^{\mathbf{A}t}\begin{bmatrix}0\\1\\1\end{bmatrix}$$

$$+e^{\mathbf{A}t}\int_0^t e^{-s}\begin{bmatrix}1 & 0 & 0\\ -\frac{3}{2}+\frac{3}{2}\cos 2s-\sin 2s & \cos 2s & \sin 2s\\ 1-\frac{3}{2}\sin 2s-\cos 2s & -\sin 2s & \cos 2s\end{bmatrix}\begin{bmatrix}0\\0\\e^s\cos 2s\end{bmatrix}ds$$

$$=e^t\begin{bmatrix}0\\ \cos 2t-\sin 2t\\ \cos 2t+\sin 2t\end{bmatrix}+e^{\mathbf{A}t}\int_0^t\begin{bmatrix}0\\ \sin 2s\cos 2s\\ \cos^2 2s\end{bmatrix}ds$$

$$=e^t\begin{bmatrix}0\\ \cos 2t-\sin 2t\\ \cos 2t+\sin 2t\end{bmatrix}+e^{\mathbf{A}t}\begin{bmatrix}0\\ (1-\cos 4t)/8\\ t/2+(\sin 4t)/8\end{bmatrix}$$

$$=e^t\begin{bmatrix}0\\ \cos 2t-\sin 2t\\ \cos 2t+\sin 2t\end{bmatrix}$$

$$+e^t\begin{bmatrix}0\\ -\dfrac{t\sin 2t}{2}+\dfrac{\cos 2t-\cos 4t\cos 2t-\sin 4t\sin 2t}{8}\\ \dfrac{t\cos 2t}{2}+\dfrac{\sin 4t\cos 2t-\sin 2t\cos 4t+\sin 2t}{8}\end{bmatrix}$$

$$=e^t\begin{bmatrix}0\\ \cos 2t-\left(1+\frac{1}{2}t\right)\sin 2t\\ \left(1+\frac{1}{2}t\right)\cos 2t+\frac{5}{4}\sin 2t\end{bmatrix}.$$

As Example 1 indicates, the method of variation of parameters is often quite tedious and laborious. One way of avoiding many of these calculations is to "guess" a particular solution $\boldsymbol{\psi}(t)$ of the nonhomogeneous equation and then to observe (see Exercise 9) that every solution $\mathbf{x}(t)$ of the nonhomogeneous equation must be of the form $\boldsymbol{\phi}(t)+\boldsymbol{\psi}(t)$ where $\boldsymbol{\phi}(t)$ is a solution of the homogeneous equation.

Example 2. Find all solutions of the differential equation

$$\dot{\mathbf{x}}=\begin{bmatrix}1 & 0 & 0\\ 2 & 1 & -2\\ 3 & 2 & 1\end{bmatrix}\mathbf{x}+\begin{bmatrix}1\\0\\0\end{bmatrix}e^{ct},\qquad c\neq 1. \tag{9}$$

Solution. Let

$$\mathbf{A}=\begin{bmatrix} 1 & 0 & 0 \\ 2 & 1 & -2 \\ 3 & 2 & 1 \end{bmatrix}.$$

We "guess" a particular solution $\boldsymbol{\psi}(t)$ of the form $\boldsymbol{\psi}(t)=\mathbf{b}e^{ct}$. Plugging this expression into (9) gives

$$c\mathbf{b}e^{ct}=\mathbf{A}\mathbf{b}e^{ct}+\begin{bmatrix} 1 \\ 0 \\ 0 \end{bmatrix}e^{ct},$$

or

$$(\mathbf{A}-c\mathbf{I})\mathbf{b}=\begin{bmatrix} -1 \\ 0 \\ 0 \end{bmatrix}.$$

This implies that

$$b=\frac{-1}{1-c}\begin{bmatrix} 1 \\ \dfrac{2(c-4)}{4+(1-c)^2} \\ \dfrac{1+3c}{4+(1-c)^2} \end{bmatrix}.$$

Hence, every solution $\mathbf{x}(t)$ of (9) is of the form

$$\mathbf{x}(t)=e^{t}\left[c_1\begin{bmatrix} 2 \\ -3 \\ 2 \end{bmatrix}+c_2\begin{bmatrix} 0 \\ \cos 2t \\ \sin 2t \end{bmatrix}+c_3\begin{bmatrix} 0 \\ -\sin 2t \\ \cos 2t \end{bmatrix}\right.$$

$$\left.-\frac{1}{1-c}\begin{bmatrix} 1 \\ \dfrac{2(c-4)}{4+(1-c)^2} \\ \dfrac{1+3c}{4+(1-c)^2} \end{bmatrix}\right].$$

Remark. We run into trouble when $c=1$ because one is an eigenvalue of the matrix

$$\begin{bmatrix} 1 & 0 & 0 \\ 2 & 1 & -2 \\ 3 & 2 & 1 \end{bmatrix}.$$

More generally, the differential equation $\dot{\mathbf{x}}=\mathbf{A}\mathbf{x}+\mathbf{v}e^{ct}$ may not have a solution of the form $\mathbf{b}e^{ct}$ if c is an eigenvalue of $\mathbf{A}$. In this case we have to guess a particular solution of the form

$$\psi(t)=e^{ct}\left[\mathbf{b}_0+\mathbf{b}_1 t+\ldots+\mathbf{b}_{k-1}t^{k-1}\right]$$

for some appropriate integer k. (See Exercises 10–18).

Exercises

In each of Problems 1–6 use the method of variation of parameters to solve the given initial-value problem.

1. $\dot{\mathbf{x}}=\begin{pmatrix} 4 & 5 \\ -2 & -2 \end{pmatrix}\mathbf{x}+\begin{pmatrix} 4e^t\cos t \\ 0 \end{pmatrix},\quad \mathbf{x}(0)=\begin{pmatrix} 0 \\ 0 \end{pmatrix}$

2. $\dot{\mathbf{x}}=\begin{pmatrix} 3 & -4 \\ 1 & -1 \end{pmatrix}\mathbf{x}+\begin{pmatrix} 1 \\ 1 \end{pmatrix}e^t,\quad \mathbf{x}(0)=\begin{pmatrix} 1 \\ 1 \end{pmatrix}$

3. $\dot{\mathbf{x}}=\begin{pmatrix} 2 & -5 \\ 1 & -2 \end{pmatrix}\mathbf{x}+\begin{pmatrix} \sin t \\ \tan t \end{pmatrix},\quad \mathbf{x}(0)=\begin{pmatrix} 0 \\ 0 \end{pmatrix}$

4. $\dot{\mathbf{x}}=\begin{pmatrix} 0 & 1 \\ -1 & 0 \end{pmatrix}\mathbf{x}+\begin{pmatrix} f_1(t) \\ f_2(t) \end{pmatrix},\quad \mathbf{x}(0)=\begin{pmatrix} 0 \\ 0 \end{pmatrix}$

5. $\dot{\mathbf{x}}=\begin{pmatrix} 2 & 0 & 1 \\ 0 & 2 & 0 \\ 0 & 1 & 3 \end{pmatrix}\mathbf{x}+\begin{pmatrix} 1 \\ 0 \\ 1 \end{pmatrix}e^{2t},\quad \mathbf{x}(0)=\begin{pmatrix} 1 \\ 1 \\ 1 \end{pmatrix}$

6. $\dot{\mathbf{x}}=\begin{pmatrix} -1 & -1 & -2 \\ 1 & 1 & 1 \\ 2 & 1 & 3 \end{pmatrix}\mathbf{x}+\begin{pmatrix} 1 \\ 0 \\ 0 \end{pmatrix}e^t,\quad \mathbf{x}(0)=\begin{pmatrix} 0 \\ 0 \\ 0 \end{pmatrix}$

7. Consider the nth-order scalar differential equation

$$L[y]=\frac{d^n y}{dt^n}+a_1\frac{d^{n-1}y}{dt^{n-1}}+\ldots+a_n y=f(t). \qquad (*)$$

Let $v(t)$ be the solution of $L[y]=0$ which satisfies the initial conditions $y(0)=\ldots=y^{(n-2)}(0)=0$, $y^{(n-1)}(0)=1$. Show that

$$y(t)=\int_0^t v(t-s)f(s)\,ds$$

is the solution of (*) which satisfies the initial conditions $y(0)=\ldots=y^{(n-1)}(0)=0$. *Hint*: Convert (*) to a system of n first-order equations of the form $\dot{\mathbf{x}}=\mathbf{A}\mathbf{x}$, and show that

$$\begin{bmatrix} v(t) \\ v'(t) \\ \vdots \\ v^{(n-1)}(t) \end{bmatrix}$$

is the nth column of $e^{\mathbf{A}t}$

8. Find the solution of the initial-value problem

$$\frac{d^3y}{dt^3}+\frac{dy}{dt}=\sec t\tan t,\qquad y(0)=y'(0)=y''(0)=0.$$

9. (a) Let $\boldsymbol{\psi}(t)$ be a solution of the nonhomogeneous equation $\dot{\mathbf{x}}=\mathbf{A}\mathbf{x}+\mathbf{f}(t)$ and let $\boldsymbol{\phi}(t)$ be a solution of the homogeneous equation $\dot{\mathbf{x}}=\mathbf{A}\mathbf{x}$. Show that $\boldsymbol{\phi}(t)+\boldsymbol{\psi}(t)$ is a solution of the nonhomogeneous equation.

(b) Let $\boldsymbol{\psi}_1(t)$ and $\boldsymbol{\psi}_2(t)$ be two solutions of the nonhomogeneous equation. Show that $\boldsymbol{\psi}_1(t)-\boldsymbol{\psi}_2(t)$ is a solution of the homogeneous equation.

(c) Let $\boldsymbol{\psi}(t)$ be a particular solution of the nonhomogeneous equation. Show that any other solution $\mathbf{y}(t)$ must be of the form $\mathbf{y}(t)=\boldsymbol{\phi}(t)+\boldsymbol{\psi}(t)$ where $\boldsymbol{\phi}(t)$ is a solution of the homogeneous equation.

In each of Problems 10–14 use the method of judicious guessing to find a particular solution of the given differential equation.

10. $\dot{\mathbf{x}}=\begin{pmatrix} 2 & 1 \\ 3 & -2 \end{pmatrix}\mathbf{x}+\begin{pmatrix} 1 \\ 1 \end{pmatrix}e^{3t}$

11. $\dot{\mathbf{x}}=\begin{pmatrix} 1 & -1 \\ 1 & 3 \end{pmatrix}\mathbf{x}+\begin{pmatrix} -t^2 \\ 2t \end{pmatrix}$

12. $\dot{\mathbf{x}}=\begin{pmatrix} 1 & 3 & 2 \\ -1 & 2 & 1 \\ 4 & -1 & -1 \end{pmatrix}\mathbf{x}+\begin{pmatrix} \sin t \\ 0 \\ 0 \end{pmatrix}$

13. $\dot{\mathbf{x}}=\begin{pmatrix} 1 & 2 & -3 \\ 1 & 1 & 2 \\ 1 & -1 & 4 \end{pmatrix}\mathbf{x}+\begin{pmatrix} 1 \\ 0 \\ -1 \end{pmatrix}e^{t}$

14. $\dot{\mathbf{x}}=\begin{pmatrix} -1 & -1 & 0 \\ 0 & -4 & -1 \\ 0 & 5 & 0 \end{pmatrix}\mathbf{x}+\begin{pmatrix} 1 \\ t \\ e^{t} \end{pmatrix}$

15. Consider the differential equation

$$\dot{\mathbf{x}}=\mathbf{A}\mathbf{x}+\mathbf{v}e^{\lambda t} \qquad (*)$$

where $\mathbf{v}$ is an eigenvector of $\mathbf{A}$ with eigenvalue λ. Suppose moreover, that $\mathbf{A}$ has n linearly independent eigenvectors $\mathbf{v}^1,\mathbf{v}^2,\ldots,\mathbf{v}^n$, with distinct eigenvalues $\lambda_1,\ldots,\lambda_n$ respectively.

(a) Show that (*) has no solution $\boldsymbol{\psi}(t)$ of the form $\boldsymbol{\psi}(t)=\mathbf{a}e^{\lambda t}$. *Hint*: Write $\mathbf{a}=a_1\mathbf{v}^1+\ldots+a_n\mathbf{v}^n$.

(b) Show that (*) has a solution $\boldsymbol{\psi}(t)$ of the form

$$\boldsymbol{\psi}(t)=\mathbf{a}e^{\lambda t}+\mathbf{b}te^{\lambda t}.$$

Hint: Show that $\mathbf{b}$ is an eigenvector of $\mathbf{A}$ with eigenvalue λ and choose it so that we can solve the equation

$$(\mathbf{A}-\lambda\mathbf{I})\mathbf{a}=\mathbf{b}-\mathbf{v}.$$

In each of Problems 16–18, find a particular solution $\boldsymbol{\psi}(t)$ of the given differential equation of the form $\boldsymbol{\psi}(t)=e^{\lambda t}(\mathbf{a}+\mathbf{b}t)$.

16. $\dot{\mathbf{x}}=\begin{pmatrix}1 & 1 & -1\\ 2 & 3 & -4\\ 4 & 1 & -4\end{pmatrix}\mathbf{x}+\begin{pmatrix}1\\ 2\\ 1\end{pmatrix}e^{2t}$

17. $\dot{\mathbf{x}}=\begin{pmatrix}1 & -1 & -1\\ 1 & 3 & 1\\ -3 & 1 & -1\end{pmatrix}\mathbf{x}+\begin{pmatrix}1\\ -1\\ -1\end{pmatrix}e^{3t}$

18. $\dot{\mathbf{x}}=\begin{pmatrix}3 & 2 & 4\\ 2 & 0 & 2\\ 4 & 2 & 3\end{pmatrix}\mathbf{x}+\begin{pmatrix}2\\ 1\\ 2\end{pmatrix}e^{8t}$

3.11 Solving systems by Laplace transforms

The method of Laplace transforms introduced in Chapter 2 can also be used to solve the initial-value problem

$$\dot{\mathbf{x}}=\mathbf{A}\mathbf{x}+\mathbf{f}(t), \qquad \mathbf{x}(0)=\mathbf{x}^0. \tag{1}$$

Let

$$\mathbf{X}(s)=\begin{bmatrix}X_1(s)\\ \vdots\\ X_n(s)\end{bmatrix}=\mathcal{L}\{\mathbf{x}(t)\}=\begin{bmatrix}\int_0^\infty e^{-st}x_1(t)\,dt\\ \vdots\\ \int_0^\infty e^{-st}x_n(t)\,dt\end{bmatrix}$$

and

$$\mathbf{F}(s)=\begin{bmatrix}F_1(s)\\ \vdots\\ F_n(s)\end{bmatrix}=\mathcal{L}\{\mathbf{f}(t)\}=\begin{bmatrix}\int_0^\infty e^{-st}f_1(t)\,dt\\ \vdots\\ \int_0^\infty e^{-st}f_n(t)\,dt\end{bmatrix}.$$

Taking Laplace transforms of both sides of (1) gives

$$\begin{aligned}\mathcal{L}\{\dot{\mathbf{x}}(t)\}&=\mathcal{L}\{\mathbf{A}\mathbf{x}(t)+\mathbf{f}(t)\}=\mathbf{A}\mathcal{L}\{\mathbf{x}(t)\}+\mathcal{L}\{\mathbf{f}(t)\}\\ &=\mathbf{A}\mathbf{X}(s)+\mathbf{F}(s),\end{aligned}$$

and from Lemma 3 of Section 2.9,

$$\mathcal{L}\{\dot{\mathbf{x}}(t)\}=\begin{bmatrix}\mathcal{L}\{\dot{x}_1(t)\}\\ \vdots\\ \mathcal{L}\{\dot{x}_n(t)\}\end{bmatrix}=\begin{bmatrix}sX_1(s)-x_1(0)\\ \vdots\\ sX_n(s)-x_n(0)\end{bmatrix}$$

$$=s\mathbf{X}(s)-\mathbf{x}^0.$$

Hence,

$$s\mathbf{X}(s)-\mathbf{x}^0=\mathbf{A}\mathbf{X}(s)+\mathbf{F}(s)$$

or

$$(s\mathbf{I}-\mathbf{A})\mathbf{X}(s)=\mathbf{x}^0+\mathbf{F}(s),\qquad \mathbf{I}=\begin{bmatrix}1&0&\dots&0\\0&1&\dots&0\\ \vdots&\vdots&&\vdots\\0&0&\dots&1\end{bmatrix}. \tag{2}$$

Equation (2) is a system of n simultaneous equations for $X_1(s),\dots,X_n(s)$, and it can be solved in a variety of ways. (One way, in particular, is to multiply both sides of (2) by $(s\mathbf{I}-\mathbf{A})^{-1}$.) Once we know $X_1(s),\dots,X_n(s)$ we can find $x_1(t),\dots,x_n(t)$ by inverting these Laplace transforms.

Example 1. Solve the initial-value problem

$$\dot{\mathbf{x}}=\begin{pmatrix}1&4\\1&1\end{pmatrix}\mathbf{x}+\begin{pmatrix}1\\1\end{pmatrix}e^t,\qquad \mathbf{x}(0)=\begin{pmatrix}2\\1\end{pmatrix}. \tag{3}$$

Solution. Taking Laplace transforms of both sides of the differential equation gives

$$s\mathbf{X}(s)-\begin{pmatrix}2\\1\end{pmatrix}=\begin{pmatrix}1&4\\1&1\end{pmatrix}\mathbf{X}(s)+\frac{1}{s-1}\begin{pmatrix}1\\1\end{pmatrix}$$

or

$$(s-1)X_1(s)-4X_2(s)=2+\frac{1}{s-1}\qquad -X_1(s)+(s-1)X_2(s)=1+\frac{1}{s-1}.$$

The solution of these equations is

$$X_1(s)=\frac{2}{s-3}+\frac{1}{s^2-1},\qquad X_2(s)=\frac{1}{s-3}+\frac{s}{(s-1)(s+1)(s-3)}.$$

Now,

$$\frac{2}{s-3}=\mathcal{L}\{2e^{3t}\},\quad\text{and}\quad \frac{1}{s^2-1}=\mathcal{L}\{\sinh t\}=\mathcal{L}\left\{\frac{e^t-e^{-t}}{2}\right\}.$$

Hence,

$$x_1(t)=2e^{3t}+\frac{e^t-e^{-t}}{2}.$$

To invert $X_2(s)$, we use partial fractions. Let

$$\frac{s}{(s-1)(s+1)(s-3)}=\frac{A}{s-1}+\frac{B}{s+1}+\frac{C}{s-3}.$$

This implies that

$$A(s+1)(s-3)+B(s-1)(s-3)+C(s-1)(s+1)=s. \tag{4}$$

Setting $s=1,\ -1$, and 3 respectively in (4) gives $A=-\frac{1}{4}$, $B=-\frac{1}{8}$, and $C=\frac{3}{8}$. Consequently,

$$\begin{aligned}x_2(t)&=\mathcal{L}^{-1}\left\{-\frac{1}{4}\frac{1}{s-1}-\frac{1}{8}\frac{1}{s+1}+\frac{11}{8}\frac{1}{s-3}\right\}\\&=-\frac{1}{8}e^{-t}-\frac{1}{4}e^{t}+\frac{11}{8}e^{3t}.\end{aligned}$$

Exercises

Find the solution of each of the following initial-value problems.

1. $\dot{\mathbf{x}}=\begin{pmatrix}1&-3\\-2&2\end{pmatrix}\mathbf{x},\quad \mathbf{x}(0)=\begin{pmatrix}0\\5\end{pmatrix}$

2. $\dot{\mathbf{x}}=\begin{pmatrix}1&-1\\5&-3\end{pmatrix}\mathbf{x},\quad \mathbf{x}(0)=\begin{pmatrix}1\\2\end{pmatrix}$

3. $\dot{\mathbf{x}}=\begin{pmatrix}3&-2\\2&-2\end{pmatrix}\mathbf{x}+\begin{pmatrix}t\\3e^t\end{pmatrix},\quad \mathbf{x}(0)=\begin{pmatrix}2\\1\end{pmatrix}$

4. $\dot{\mathbf{x}}=\begin{pmatrix}1&1\\4&1\end{pmatrix}\mathbf{x}+\begin{pmatrix}2\\-1\end{pmatrix}e^t,\quad \mathbf{x}(0)=\begin{pmatrix}0\\0\end{pmatrix}$

5. $\dot{\mathbf{x}}=\begin{pmatrix}3&-4\\1&-1\end{pmatrix}\mathbf{x}+\begin{pmatrix}1\\1\end{pmatrix}e^t,\quad \mathbf{x}(0)=\begin{pmatrix}1\\1\end{pmatrix}$

6. $\dot{\mathbf{x}}=\begin{pmatrix}2&-5\\1&-2\end{pmatrix}\mathbf{x}+\begin{pmatrix}\sin t\\\tan t\end{pmatrix},\quad \mathbf{x}(0)=\begin{pmatrix}-1\\0\end{pmatrix}$

7. $\dot{\mathbf{x}}=\begin{pmatrix}4&5\\-2&-2\end{pmatrix}\mathbf{x}+\begin{pmatrix}4e^t\cos t\\0\end{pmatrix},\quad \mathbf{x}(0)=\begin{pmatrix}1\\1\end{pmatrix}$

8. $\dot{\mathbf{x}}=\begin{pmatrix}0&1\\-1&0\end{pmatrix}\mathbf{x}+\begin{pmatrix}f_1(t)\\f_2(t)\end{pmatrix},\quad \mathbf{x}(0)=\begin{pmatrix}0\\0\end{pmatrix}$

9. $\dot{\mathbf{x}}=\begin{pmatrix}2&-2\\4&-2\end{pmatrix}\mathbf{x}+\begin{pmatrix}0\\\delta(t-\pi)\end{pmatrix},\quad \mathbf{x}(0)=\begin{pmatrix}1\\0\end{pmatrix}$

10. $\dot{\mathbf{x}}=\begin{pmatrix}3&-2\\2&-2\end{pmatrix}\mathbf{x}+\begin{pmatrix}1-H_\pi(t)\\0\end{pmatrix},\quad \mathbf{x}(0)=\begin{pmatrix}0\\0\end{pmatrix}$

11. $\dot{\mathbf{x}}=\begin{pmatrix} 1 & 2 & -3\\ 1 & 1 & 2\\ 1 & -1 & 4\end{pmatrix}\mathbf{x},\quad \mathbf{x}(0)=\begin{pmatrix}1\\0\\0\end{pmatrix}$

12. $\dot{\mathbf{x}}=\begin{pmatrix} 2 & 0 & 1\\ 0 & 2 & 0\\ 0 & 0 & 3\end{pmatrix}\mathbf{x}+\begin{pmatrix}1\\0\\1\end{pmatrix}e^{2t},\quad \mathbf{x}(0)=\begin{pmatrix}1\\1\\1\end{pmatrix}$

13. $\dot{\mathbf{x}}=\begin{pmatrix} -1 & -1 & 2\\ 1 & 1 & 1\\ 2 & 1 & 3\end{pmatrix}\mathbf{x}+\begin{pmatrix}1\\0\\0\end{pmatrix}e^{t},\quad \mathbf{x}(0)=\begin{pmatrix}0\\0\\0\end{pmatrix}$

14. $\dot{\mathbf{x}}=\begin{pmatrix} 1 & 0 & 0\\ 2 & 1 & -2\\ 3 & 2 & 1\end{pmatrix}\mathbf{x}+\begin{pmatrix}0\\0\\e^{t}\cos 2t\end{pmatrix},\quad \mathbf{x}(0)=\begin{pmatrix}0\\1\\1\end{pmatrix}$

15. $\dot{\mathbf{x}}=\begin{bmatrix} 3 & 0 & 0 & 0\\ 1 & 3 & 0 & 0\\ 0 & 0 & 3 & 0\\ 0 & 0 & 2 & 3\end{bmatrix}\mathbf{x},\quad \mathbf{x}(0)=\begin{bmatrix}1\\1\\1\\1\end{bmatrix}$

4 Qualitative theory of differential equations

4.1 Introduction

In this chapter we consider the differential equation

$$\dot{\mathbf{x}} = \mathbf{f}(t, \mathbf{x}) \tag{1}$$

where

$$\mathbf{x} = \begin{bmatrix} x_1(t) \\ \vdots \\ x_n(t) \end{bmatrix},$$

and

$$\mathbf{f}(t, \mathbf{x}) = \begin{bmatrix} f_1(t, x_1, \ldots, x_n) \\ \vdots \\ f_n(t, x_1, \ldots, x_n) \end{bmatrix}$$

is a nonlinear function of $x_1, \ldots, x_n$. Unfortunately, there are no known methods of solving Equation (1). This, of course, is very disappointing. However, it is not necessary, in most applications, to find the solutions of (1) explicitly. For example, let $x_1(t)$ and $x_2(t)$ denote the populations, at time t, of two species competing amongst themselves for the limited food and living space in their microcosm. Suppose, moreover, that the rates of growth of $x_1(t)$ and $x_2(t)$ are governed by the differential equation (1). In this case, we are not really interested in the values of $x_1(t)$ and $x_2(t)$ at every time t. Rather, we are interested in the qualitative properties of $x_1(t)$ and $x_2(t)$. Specically, we wish to answer the following questions.

1. Do there exist values ξ_1 and ξ_2 at which the two species coexist together in a steady state? That is to say, are there numbers ξ_1, ξ_2 such that $x_1(t)\equiv\xi_1$, $x_2(t)\equiv\xi_2$ is a solution of (1)? Such values ξ_1, ξ_2, if they exist, are called *equilibrium points* of (1).

2. Suppose that the two species are coexisting in equilibrium. Suddenly, we add a few members of species 1 to the microcosm. Will $x_1(t)$ and $x_2(t)$ remain close to their equilibrium values for all future time? Or perhaps the extra few members give species 1 a large advantage and it will proceed to annihilate species 2.

3. Suppose that x_1 and x_2 have arbitrary values at $t=0$. What happens as t approaches infinity? Will one species ultimately emerge victorious, or will the struggle for existence end in a draw?

More generally, we are interested in determining the following properties of solutions of (1).

1. Do there exist equilibrium values

$$\mathbf{x}^0=\begin{bmatrix} x_1^0 \\ \vdots \\ x_n^0 \end{bmatrix}$$

for which $\mathbf{x}(t)\equiv\mathbf{x}^0$ is a solution of (1)?

2. Let $\boldsymbol{\phi}(t)$ be a solution of (1). Suppose that $\boldsymbol{\psi}(t)$ is a second solution with $\boldsymbol{\psi}(0)$ very close to $\boldsymbol{\phi}(0)$; that is, $\psi_j(0)$ is very close to $\phi_j(0)$, $j=1,\dots,n$. Will $\boldsymbol{\psi}(t)$ remain close to $\boldsymbol{\phi}(t)$ for all future time, or will $\boldsymbol{\psi}(t)$ diverge from $\boldsymbol{\phi}(t)$ as t approaches infinity? This question is often referred to as the problem of *stability*. It is the most fundamental problem in the qualitative theory of differential equations, and has occupied the attention of many mathematicians for the past hundred years.

3. What happens to solutions $\mathbf{x}(t)$ of (1) as t approaches infinity? Do all solutions approach equilibrium values? If they don't approach equilibrium values, do they at least approach a periodic solution?

This chapter is devoted to answering these three questions. Remarkably, we can often give satisfactory answers to these questions, even though we cannot solve Equation (1) explicitly. Indeed, the first question can be answered immediately. Observe that $\dot{\mathbf{x}}(t)$ is identically zero if $\mathbf{x}(t)\equiv\mathbf{x}^0$. Hence, $\mathbf{x}^0$ is an equilibrium value of (1), if, and only if,

$$\mathbf{f}(t,\mathbf{x}^0)\equiv\mathbf{0}. \tag{2}$$

Example 1. Find all equilibrium values of the system of differential equations

$$\frac{dx_1}{dt}=1-x_2, \qquad \frac{dx_2}{dt}=x_1^3+x_2.$$

Solution.

$$\mathbf{x}^0 = \begin{pmatrix} x_1^0 \\ x_2^0 \end{pmatrix}$$

is an equilibrium value if, and only if, $1 - x_2^0 = 0$ and $(x_1^0)^3 + x_2^0 = 0$. This implies that $x_2^0 = 1$ and $x_1^0 = -1$. Hence $\begin{pmatrix} -1 \\ 1 \end{pmatrix}$ is the only equilibrium value of this system.

Example 2. Find all equilibrium solutions of the system

$$\frac{dx}{dt} = (x-1)(y-1), \qquad \frac{dy}{dt} = (x+1)(y+1).$$

Solution.

$$\begin{pmatrix} x_0 \\ y_0 \end{pmatrix}$$

is an equilibrium value of this system if, and only if, $(x_0 - 1)(y_0 - 1) = 0$ and $(x_0 + 1)(y_0 + 1) = 0$. The first equation is satisfied if either x_0 or y_0 is 1, while the second equation is satisfied if either x_0 or y_0 is -1. Hence, $x = 1$, $y = -1$ and $x = -1$, $y = 1$ are the equilibrium solutions of this system.

The question of stability is of paramount importance in all physical applications, since we can never measure initial conditions exactly. For example, consider the case of a particle of mass one slug attached to an elastic spring of force constant 1 lb/ft, which is moving in a frictionless medium. In addition, an external force $F(t) = \cos 2t$ lb is acting on the particle. Let $y(t)$ denote the position of the particle relative to its equilibrium position. Then$(d^2y/dt^2) + y = \cos 2t$. We convert this second-order equation into a system of two first-order equations by setting $x_1 = y, x_2 = y'$. Then,

$$\frac{dx_1}{dt} = x_2, \qquad \frac{dx_2}{dt} = -x_1 + \cos 2t. \tag{3}$$

The functions $y_1(t) = \sin t$ and $y_2(t) = \cos t$ are two independent solutions of the homogeneous equation $y'' + y = 0$. Moreover, $y = -\frac{1}{3}\cos 2t$ is a particular solution of the nonhomogeneous equation. Therefore, every solution

$$\mathbf{x}(t) = \begin{pmatrix} x_1(t) \\ x_2(t) \end{pmatrix}$$

of (3) is of the form

$$\mathbf{x}(t) = c_1 \begin{pmatrix} \sin t \\ \cos t \end{pmatrix} + c_2 \begin{pmatrix} \cos t \\ -\sin t \end{pmatrix} + \begin{bmatrix} -\frac{1}{3}\cos 2t \\ \frac{2}{3}\sin 2t \end{bmatrix}. \tag{4}$$

At time $t=0$ we measure the position and velocity of the particle and obtain $y(0)=1$, $y'(0)=0$. This implies that $c_1=0$ and $c_2=\frac{4}{3}$. Consequently, the position and velocity of the particle for all future time are given by the equation

$$\begin{pmatrix} y(t) \\ y'(t) \end{pmatrix} = \begin{pmatrix} x_1(t) \\ x_2(t) \end{pmatrix} = \begin{bmatrix} \frac{4}{3}\cos t - \frac{1}{3}\cos 2t \\ -\frac{4}{3}\sin t + \frac{2}{3}\sin 2t \end{bmatrix}. \tag{5}$$

However, suppose that our measurements permit an error of magnitude 10^{-4}. Will the position and velocity of the particle remain close to the values predicted by (5)? The answer to this question had better be yes, for otherwise, Newtonian mechanics would be of no practical value to us. Fortunately, it is quite easy to show, in this case, that the position and velocity of the particle remain very close to the values predicted by (5). Let $\hat{y}(t)$ and $\hat{y}'(t)$ denote the true values of $y(t)$ and $y'(t)$ respectively. Clearly,

$$y(t)-\hat{y}(t)=\left(\tfrac{4}{3}-c_2\right)\cos t - c_1 \sin t$$

$$y'(t)-\hat{y}'(t)=-c_1\cos t-\left(\tfrac{4}{3}-c_2\right)\sin t$$

where c_1 and c_2 are two constants satisfying

$$-10^{-4}\leqslant c_1 \leqslant 10^{-4}, \qquad \tfrac{4}{3}-10^{-4}\leqslant c_2 \leqslant \tfrac{4}{3}+10^{-4}.$$

We can rewrite these equations in the form

$$y(t)-\hat{y}(t)=\left[c_1^2+\left(\tfrac{4}{3}-c_2\right)^2\right]^{1/2}\cos(t-\delta_1), \qquad \tan\delta_1=\frac{c_1}{c_2-\frac{4}{3}}$$

$$y'(t)-\hat{y}'(t)=\left[c_1^2+\left(\tfrac{4}{3}-c_2\right)^2\right]^{1/2}\cos(t-\delta_2), \qquad \tan\delta_2=\frac{\frac{4}{3}-c_2}{c_1}.$$

Hence, both $y(t)-\hat{y}(t)$ and $y'(t)-\hat{y}'(t)$ are bounded in absolute value by $[c_1^2+(\frac{4}{3}-c_2)^2]^{1/2}$. This quantity is at most $\sqrt{2}\,10^{-4}$. Therefore, the true values of $y(t)$ and $y'(t)$ are indeed close to the values predicted by (5).

As a second example of the concept of stability, consider the case of a particle of mass m which is supported by a wire, or inelastic string, of length l and of negligible mass. The wire is always straight, and the system is free to vibrate in a vertical plane. This configuration is usually referred to as a simple pendulum. The equation of motion of the pendulum is

$$\frac{d^2y}{dt^2}+\frac{g}{l}\sin y=0$$

where y is the angle which the wire makes with the vertical line A0 (see

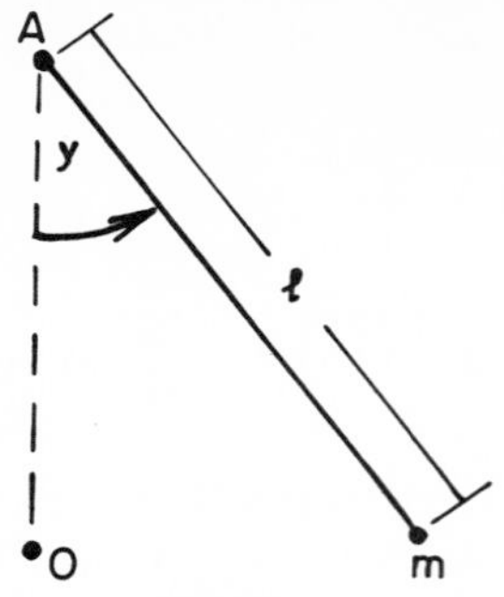

Figure 1

Figure 1). Setting $x_1 = y$ and $x_2 = dy/dt$ we see that

$$\frac{dx_1}{dt} = x_2, \qquad \frac{dx_2}{dt} = -\frac{g}{l}\sin x_1. \tag{6}$$

The system of equations (6) has equilibrium solutions $x_1 = 0$, $x_2 = 0$, and $x_1 = \pi$, $x_2 = 0$. (If the pendulum is suspended in the upright position $y = \pi$ with zero velocity, then it will remain in this upright position for all future time.) These two equilibrium solutions have very different properties. If we disturb the pendulum slightly from the equilibrium position $x_1 = 0$, $x_2 = 0$, by either displacing it slightly, or giving it a small velocity, then it will execute small oscillations about $x_1 = 0$. On the other hand, if we disturb the pendulum slightly from the equilibrium position $x_1 = \pi$, $x_2 = 0$, then it will either execute very large oscillations about $x_1 = 0$, or it will rotate around and around ad infinitum. Thus, the slightest disturbance causes the pendulum to deviate drastically from its equilibrium position $x_1 = \pi$, $x_2 = 0$. Intuitively, we would say that the equilibrium value $x_1 = 0$, $x_2 = 0$ of (6) is stable, while the equilibrium value $x_1 = \pi$, $x_2 = 0$ of (6) is unstable.

The question of stability is usually very difficult to resolve, because we cannot solve (1) explicitly. The only case which is manageable is when $\mathbf{f}(t, \mathbf{x})$ does not depend explicitly on t; that is, $\mathbf{f}$ is a function of $\mathbf{x}$ alone. Such differential equations are called *autonomous*. And even for autonomous differential equations, there are only two instances, generally, where we can completely resolve the stability question. The first case is when $\mathbf{f}(\mathbf{x}) = \mathbf{A}\mathbf{x}$, and the second case is when we are only interested in the stability of an equilibrium solution of $\dot{\mathbf{x}} = \mathbf{f}(\mathbf{x})$. These two cases are treated fully in Sections 4.2 and 4.3 in the unabridged version of this text.

Question 3 is extremely important in many applications since an answer to this question is a prediction concerning the long time evolution of the system under consideration. A partial answer to this question is given in Section 4.2 where we study differential equations from a *geometric* point of view. A more complete answer is given in Sections 4.6–4.12 in the unabridged version of this text, where we also present many exciting applications.

EXERCISES

In each of Problems 1–8, find all equilibrium values of the given system of differential equations.

1. $\dfrac{dx}{dt} = x - x^2 - 2xy$

$\dfrac{dy}{dt} = 2y - 2y^2 - 3xy$

2. $\dfrac{dx}{dt} = -\beta xy + \mu$

$\dfrac{dy}{dt} = \beta xy - \gamma y$

3. $\dfrac{dx}{dt} = ax - bxy$

$\dfrac{dy}{dt} = -cy + dxy$

$\dfrac{dz}{dt} = z + x^2 + y^2$

4. $\dfrac{dx}{dt} = -x - xy^2$

$\dfrac{dy}{dt} = -y - yx^2$

$\dfrac{dz}{dt} = 1 - z + x^2$

5. $\dfrac{dx}{dt} = xy^2 - x$

$\dfrac{dy}{dt} = x\sin\pi y$

6. $\dfrac{dx}{dt} = \cos y$

$\dfrac{dy}{dt} = \sin x - 1$

7. $\dfrac{dx}{dt} = -1 - y - e^x$

$\dfrac{dy}{dt} = x^2 + y(e^x - 1)$

$\dfrac{dz}{dt} = x + \sin z$

8. $\dfrac{dx}{dt} = x - y^2$

$\dfrac{dy}{dt} = x^2 - y$

$\dfrac{dz}{dt} = e^z - x$

9. Consider the system of differential equations

$$\frac{dx}{dt} = ax + by, \qquad \frac{dy}{dt} = cx + dy. \qquad (*)$$

(i) Show that $x=0$, $y=0$ is the only equilibrium point of (*) if $ad - bc \neq 0$.
(ii) Show that (*) has a line of equilibrium points if $ad - bc = 0$.

10. Let $x = x(t)$, $y = y(t)$ be the solution of the initial-value problem

$$\frac{dx}{dt} = -x - y, \qquad \frac{dy}{dt} = 2x - y, \qquad x(0) = y(0) = 1.$$

Suppose that we make an error of magnitude 10^{-4} in measuring $x(0)$ and $y(0)$. What is the largest error we make in evaluating $x(t)$, $y(t)$ for $0 \leqslant t < \infty$?

11. (a) Verify that

$$\mathbf{x} = \begin{pmatrix} 1 \\ 0 \\ 0 \end{pmatrix} e^{-t}$$

is the solution of the initial-value problem

$$\dot{\mathbf{x}} = \begin{pmatrix} 1 & 1 & 1 \\ 2 & 1 & -1 \\ -3 & 2 & 4 \end{pmatrix} \mathbf{x} - \begin{pmatrix} 2 \\ 2 \\ -3 \end{pmatrix} e^{-t}, \qquad \mathbf{x}(0) = \begin{pmatrix} 1 \\ 0 \\ 0 \end{pmatrix}.$$

(b) Let $\mathbf{x}=\psi(t)$ be the solution of the above differential equation which satisfies the initial condition

$$\mathbf{x}(0)=\mathbf{x}^0\neq\begin{pmatrix}1\\0\\0\end{pmatrix}.$$

Show that each component of $\psi(t)$ approaches infinity, in absolute value, as $t\to\infty$.

4.2 The phase-plane

In this section we begin our study of the "geometric" theory of differential equations. For simplicity, we will restrict ourselves, for the most part, to the case $n=2$. Our aim is to obtain as complete a description as possible of all solutions of the system of differential equations

$$\frac{dx}{dt}=f(x,y),\qquad \frac{dy}{dt}=g(x,y). \tag{1}$$

To this end, observe that every solution $x=x(t)$, $y=y(t)$ of (1) defines a curve in the three-dimensional space t,x,y. That is to say, the set of all points $(t,x(t),y(t))$ describe a curve in the three-dimensional space t,x,y. For example, the solution $x=\cos t$, $y=\sin t$ of the system of differential equations

$$\frac{dx}{dt}=-y,\qquad \frac{dy}{dt}=x$$

describes a helix (see Figure 1) in (t,x,y) space.

The geometric theory of differential equations begins with the important observation that every solution $x=x(t)$, $y=y(t)$, $t_0\leqslant t\leqslant t_1$, of (1) also defines a curve in the $x-y$ plane. To wit, as t runs from t_0 to t_1, the set of points $(x(t),y(t))$ trace out a curve C in the $x-y$ plane. This curve is called the *orbit*, or *trajectory*, of the solution $x=x(t)$, $y=y(t)$, and the $x-y$ plane is called the *phase-plane* of the solutions of (1). Equivalently, we can think of the orbit of $x(t)$, $y(t)$ as the path that the solution traverses in the $x-y$ plane.

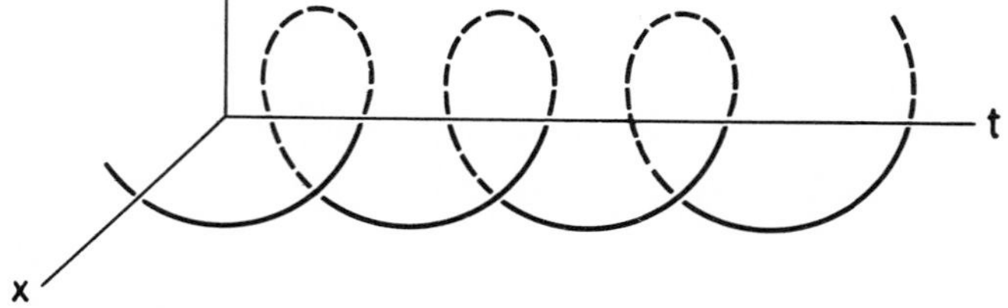

Figure 1. Graph of the solution $x=\cos t$, $y=\sin t$

Example 1. It is easily verified that $x=\cos t$, $y=\sin t$ is a solution of the system of differential equations $\dot{x}=-y$, $\dot{y}=x$. As t runs from 0 to 2π, the set of points $(\cos t, \sin t)$ trace out the unit circle $x^2+y^2=1$ in the $x-y$ plane. Hence, the unit circle $x^2+y^2=1$ is the orbit of the solution $x=\cos t$, $y=\sin t$, $0 \leq t \leq 2\pi$. As t runs from 0 to ∞, the set of points $(\cos t, \sin t)$ trace out this circle infinitely often.

Example 2. It is easily verified that $x=e^{-t}\cos t$, $y=e^{-t}\sin t$, $-\infty<t<\infty$, is a solution of the system of differential equations $dx/dt=-x-y$, $dy/dt=x-y$. As t runs from $-\infty$ to ∞, the set of points $(e^{-t}\cos t, e^{-t}\sin t)$ trace out a spiral in the $x-y$ plane. Hence, the orbit of the solution $x=e^{-t}\cos t$, $y=e^{-t}\sin t$ is the spiral shown in Figure 2.

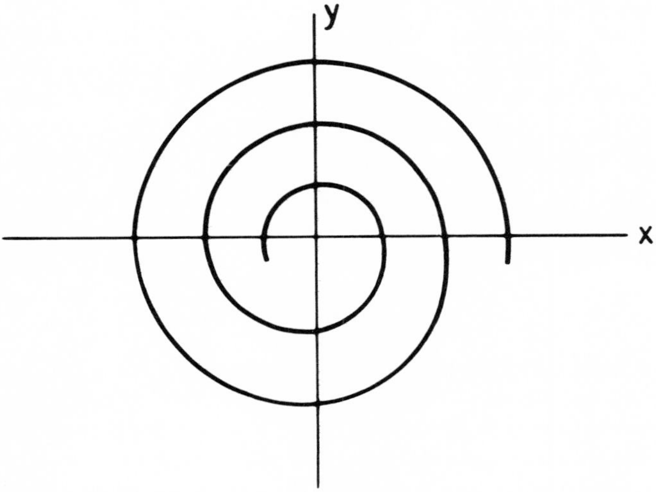

Figure 2. Orbit of $x=e^{-t}\cos t$, $y=e^{-t}\sin t$

Example 3. It is easily verified that $x=3t+2$, $y=5t+7$, $-\infty<t<\infty$ is a solution of the system of differential equations $dx/dt=3$, $dy/dt=5$. As t runs from $-\infty$ to ∞, the set of points $(3t+2, 5t+7)$ trace out the straight line through the point $(2,7)$ with slope $\frac{5}{3}$. Hence, the orbit of the solution $x=3t+2$, $y=5t+7$ is the straight line $y=\frac{5}{3}(x-2)+7$, $-\infty<x<\infty$.

Example 4. It is easily verified that $x=3t^2+2$, $y=5t^2+7$, $0 \leq t<\infty$ is a solution of the system of differential equations

$$\frac{dx}{dt}=6\left[(y-7)/5\right]^{1/2}, \qquad \frac{dy}{dt}=10\left[(x-2)/3\right]^{1/2}.$$

All of the points $(3t^2+2, 5t^2+7)$ lie on the line through $(2,7)$ with slope $\frac{5}{3}$. However, x is always greater than or equal to 2, and y is always greater than or equal to 7. Hence, the orbit of the solution $x=3t^2+2$, $y=5t^2+7$, $0\leqslant t<\infty$, is the straight line $y=\frac{5}{3}(x-2)+7$, $2\leqslant x<\infty$.

Example 5. It is easily verified that $x=3t+2$, $y=5t^2+7$, $-\infty<t<\infty$, is a solution of the system of differential equations

$$\frac{dx}{dt}=y-\frac{5}{9}(x-2)^2-4,\qquad \frac{dy}{dt}=\frac{10}{3}(x-2).$$

The orbit of this solution is the set of all points $(x,y)=(3t+2,5t^2+7)$. Solving for $t=\frac{1}{3}(x-2)$, we see that $y=\frac{5}{9}(x-2)^2+7$. Hence, the orbit of the solution $x=3t+2$, $y=5t^2+7$ is the parabola $y=\frac{5}{9}(x-2)^2+7$, $|x|<\infty$.

One of the advantages of considering the orbit of the solution rather than the solution itself is that is is often possible to obtain the orbit of a solution without prior knowledge of the solution. Let $x=x(t)$, $y=y(t)$ be a solution of (1). If $x'(t)$ is unequal to zero at $t=t_1$, then we can solve for $t=t(x)$ in a neighborhood of the point $x_1=x(t_1)$ (see Exercise 4). Thus, for t near t_1, the orbit of the solution $x(t)$, $y(t)$ is the curve $y=y(t(x))$. Next, observe that

$$\frac{dy}{dx}=\frac{dy}{dt}\frac{dt}{dx}=\frac{dy/dt}{dx/dt}=\frac{g(x,y)}{f(x,y)}.$$

Thus, the orbits of the solutions $x=x(t)$, $y=y(t)$ of (1) are the solution curves of the first-order scalar equation

$$\frac{dy}{dx}=\frac{g(x,y)}{f(x,y)}.\tag{2}$$

Therefore, it is not necessary to find a solution $x(t)$, $y(t)$ of (1) in order to compute its orbit; we need only solve the single first-order scalar differential equation (2).

Remark. From now on, we will use the phrase "the orbits of (1)" to denote the totality of orbits of solutions of (1).

Example 6. The orbits of the system of differential equations

$$\frac{dx}{dt}=y^2,\qquad \frac{dy}{dt}=x^2\tag{3}$$

are the solution curves of the scalar equation $dy/dx=x^2/y^2$. This equation is separable, and it is easily seen that every solution is of the form $y(x)=(x^3-c)^{1/3}$, c constant. Thus, the orbits of (3) are the set of all curves $y=(x^3-c)^{1/3}$.

Example 7. The orbits of the system of differential equations

$$\frac{dx}{dt}=y(1+x^2+y^2),\qquad \frac{dy}{dt}=-2x(1+x^2+y^2) \tag{4}$$

are the solution curves of the scalar equation

$$\frac{dy}{dx}=-\frac{2x(1+x^2+y^2)}{y(1+x^2+y^2)}=-\frac{2x}{y}.$$

This equation is separable, and all solutions are of the form $\frac{1}{2}y^2+x^2=c^2$. Hence, the orbits of (4) are the families of ellipses $\frac{1}{2}y^2+x^2=c^2$.

Warning. A solution curve of (2) is an orbit of (1) only if dx/dt and dy/dt are not zero simultaneously along the solution. If a solution curve of (2) passes through an equilibrium point of (1), then the entire solution curve is not an orbit. Rather, it is the union of several distinct orbits. For example, consider the system of differential equations

$$\frac{dx}{dt}=y(1-x^2-y^2),\qquad \frac{dy}{dt}=-x(1-x^2-y^2). \tag{5}$$

The solution curves of the scalar equation

$$\frac{dy}{dx}=\frac{dy/dt}{dx/dt}=-\frac{x}{y}$$

are the family of concentric circles $x^2+y^2=c^2$. Observe, however, that every point on the unit circle $x^2+y^2=1$ is an equilibrium point of (5). Thus, the orbits of this system are the circles $x^2+y^2=c^2$, for $c\neq 1$, and all points on the unit circle $x^2+y^2=1$. Similarly, the orbits of (3) are the curves $y=(x^3-c)^{1/3}$, $c\neq 0$; the half-lines $y=x$, $x>0$, and $y=x$, $x<0$; and the point $(0,0)$.

It is not possible, in general, to explicitly solve Equation (2). Hence, we cannot, in general, find the orbits of (1). Nevertheless, it is still possible to obtain an accurate description of all orbits of (1). This is because the system of differential equations (1) sets up a *direction field* in the $x-y$ plane. That is to say, the system of differential equations (1) tells us how fast a solution moves along its orbit, and in what direction it is moving. More precisely, let $x=x(t)$, $y=y(t)$ be a solution of (1). As t increases, the point $(x(t),y(t))$ moves along the orbit of this solution. Its velocity in the x-direction is dx/dt; its velocity in the y-direction is dy/dt; and the magnitude of its velocity is $[(dx(t)/dt)^2+(dy(t)/dt)^2]^{1/2}$. But $dx(t)/dt=f(x(t),y(t))$, and $dy(t)/dt=g(x(t),y(t))$. Hence, at each point (x,y) in the phase plane of (1) we know (i), the tangent to the orbit through (x,y) (the line through (x,y) with direction numbers $f(x,y)$, $g(x,y)$ respectively) and (ii), the speed $[f^2(x,y)+g^2(x,y)]^{1/2}$ with which the solution is traversing its orbit. This information can often be used to deduce important properties of the orbits of (1).

The notion of orbit can easily be extended to the case $n>2$. Let $\mathbf{x}=\mathbf{x}(t)$ be a solution of the vector differential equation

$$\dot{\mathbf{x}}=\mathbf{f}(\mathbf{x}), \qquad \mathbf{x}=\begin{bmatrix} x_1 \\ \vdots \\ x_n \end{bmatrix}, \qquad \mathbf{f}(\mathbf{x})=\begin{bmatrix} f_1(x_1,\dots,x_n) \\ \vdots \\ f_n(x_1,\dots,x_n) \end{bmatrix} \tag{6}$$

on the interval $t_0 \leqslant t \leqslant t_1$. As t runs from t_0 to t_1, the set of points $(x_1(t),\dots,x_n(t))$ trace out a curve C in the n-dimensional space $x_1,x_2,\dots,x_n$. This curve is called the orbit of the solution $\mathbf{x}=\mathbf{x}(t)$, for $t_0 \leqslant t \leqslant t_1$, and the n-dimensional space $x_1,\dots,x_n$ is called the phase-space of the solutions of (6).

Exercises

In each of Problems 1–3, verify that $x(t), y(t)$ is a solution of the given system of equations, and find its orbit.

1. $\dot{x}=1, \quad \dot{y}=2(1-x)\sin(1-x)^2$
$x(t)=1+t, \quad y(t)=\cos t^2$

2. $\dot{x}=e^{-x}, \quad \dot{y}=e^{e^x-1}$
$x(t)=\ln(1+t), \quad y(t)=e^t$

3. $\dot{x}=1+x^2, \quad \dot{y}=(1+x^2)\sec^2 x$
$x(t)=\tan t, \quad y(t)=\tan(\tan t)$

4. Suppose that $x'(t_1)\neq 0$. Show that we can solve the equation $x=x(t)$ for $t=t(x)$ in a neighborhood of the point $x_1=x(t_1)$. *Hint*: If $x'(t_1)\neq 0$, then $x(t)$ is a strictly monotonic function of t in a neighborhood of $t=t_1$.

Find the orbits of each of the following systems.

5. $\dot{x}=y,$
$\dot{y}=-x$

6. $\dot{x}=y(1+x^2+y^2),$
$\dot{y}=-x(1+x^2+y^2)$

7. $\dot{x}=y(1+x+y),$
$\dot{y}=-x(1+x+y)$

8. $\dot{x}=y+x^2y,$
$\dot{y}=3x+xy^2$

9. $\dot{x}=xye^{-3x},$
$\dot{y}=-2xy^2$

10. $\dot{x}=4y,$
$\dot{y}=x+xy^2$

11. $\dot{x}=ax-bxy,$
$\dot{y}=cx-dxy$
$(a,b,c,d$ positive)

12. $\dot{x}=x^2+\cos y,$
$\dot{y}=-2xy$

13. $\dot{x}=2xy,$
$\dot{y}=x^2-y^2$

14. $\dot{x}=y+\sin x,$
$\dot{y}=x-y\cos x$

4.3 Lanchester's combat models and the battle of Iwo Jima

During the first World War, F. W. Lanchester [4] pointed out the importance of the concentration of troops in modern combat. He constructed mathematical models from which the expected results of an engagement could be obtained. In this section we will derive two of these models, that of a conventional force versus a conventional force, and that of a conventional force versus a guerilla force. We will then solve these models, or equations, and derive "Lanchester's square law," which states that the *strength* of a combat force is proportional to the square of the number of combatants entering the engagement. Finally, we will fit one of these models, with astonishing accuracy, to the battle of Iwo Jima in World War II.

(a) *Construction of the models*

Suppose that an "x-force" and a "y-force" are engaged in combat. For simplicity, we define the strengths of these two forces as their number of combatants. (See Howes and Thrall [3] for another definition of combat strength.) Thus let $x(t)$ and $y(t)$ denote the number of combatants of the x and y forces, where t is measured in days from the start of the combat. Clearly, the rate of change of each of these quantities equals its *reinforcement rate* minus its *operational loss rate* minus its *combat loss rate*.

The operational loss rate. The operational loss rate of a combat force is its loss rate due to non-combat mishaps; i.e., desertions, diseases, etc. Lanchester proposed that the operational loss rate of a combat force is proportional to its strength. However, this does not appear to be very realistic. For example, the desertion rate in a combat force depends on a host of psychological and other intangible factors which are difficult even to describe, let alone quantify. We will take the easy way out here and consider only those engagements in which the operational loss rates are negligible.

The combat loss rate. Suppose that the x-force is a conventional force which operates in the open, comparatively speaking, and that every member of this force is within "kill range" of the enemy y. We also assume that as soon as the conventional force suffers a loss, fire is concentrated on the remaining combatants. Under these "ideal" conditions, the combat loss rate of a conventional force x equals $ay(t)$, for some positive constant a. This constant is called the *combat effectiveness coefficient* of the y-force.

The situation is very different if x is a guerilla force, invisible to its opponent y and occupying a region R. The y-force fires into R but cannot know when a kill has been made. It is certainly plausible that the combat loss rate for a guerilla force x should be proportional to $x(t)$, for the larger $x(t)$, the greater the probability that an opponent's shot will kill. On the other hand, the combat loss rate for x is also proportional to $y(t)$, for the larger y, the greater the number of x-casualties. Thus, the combat loss rate for a guerilla force x equals $cx(t)y(t)$, where the constant c is called the *combat effectiveness coefficient* of the opponent y.

The reinforcement rate. The reinforcement rate of a combat force is the rate at which new combatants enter (or are withdrawn from) the battle. We denote the reinforcement rates of the x- and y-forces by $f(t)$ and $g(t)$ respectively.

Under the assumptions listed above, we can now write down the following two Lanchestrian models for conventional–guerilla combat.

$$\text{Conventional combat:} \quad \begin{cases} \dfrac{dx}{dt} = -ay + f(t) \\ \dfrac{dy}{dt} = -bx + g(t) \end{cases} \tag{1a}$$

$$\begin{array}{c}\text{Conventional–guerilla combat:} \\ (x = \text{guerilla})\end{array} \quad \begin{cases} \dfrac{dx}{dt} = -cxy + f(t) \\ \dfrac{dy}{dt} = -dx + g(t) \end{cases} \tag{1b}$$

The system of equations (1a) is a linear system and can be solved explicitly once a, b, $f(t)$, and $g(t)$ are known. On the other hand, the system of equations (1b) is nonlinear, and its solution is much more difficult. (Indeed, it can only be obtained with the aid of a digital computer.)

It is very instructive to consider the special case where the reinforcement rates are zero. This situation occurs when the two forces are "isolated." In this case (1a) and (1b) reduce to the simpler systems

$$\frac{dx}{dt} = -ay, \qquad \frac{dy}{dt} = -bx \tag{2a}$$

and

$$\frac{dx}{dt} = -cxy, \qquad \frac{dy}{dt} = -dx. \tag{2b}$$

Conventional combat: The square law. The orbits of system (2a) are the solution curves of the equation

$$\frac{dy}{dx} = \frac{bx}{ay} \quad \text{or} \quad ay\frac{dy}{dx} = bx.$$

Integrating this equation gives

$$ay^2 - bx^2 = ay_0^2 - bx_0^2 = K. \tag{3}$$

The curves (3) define a family of hyperbolas in the x–y plane and we have indicated their graphs in Figure 1. The arrowheads on the curves indicate the direction of changing strengths as time passes.

Let us adopt the criterion that one force wins the battle if the other force vanishes first. Then, y wins if $K>0$ since the x-force has been annihilated by the time $y(t)$ has decreased to $\sqrt{K/a}$. Similarly, x wins if $K<0$.

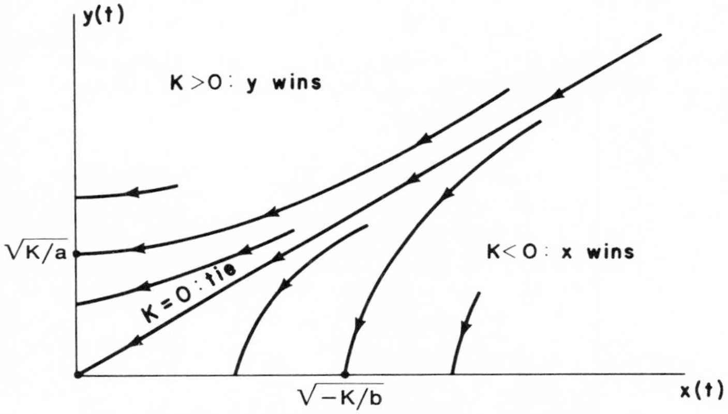

Figure 1. The hyperbolas defined by (3)

Remark 1. Equation (3) is often referred to as "Lanchester's square law," and the system (2a) is often called the square law model, since the strengths of the opposing forces appear *quadratically* in (3). This terminology is rather anomolous since the system (2a) is actually a linear system.

Remark 2. The y-force always seeks to establish a setting in which $K>0$. That is to say, the y-force wants the inequality

$$ay_0^2 > bx_0^2$$

to hold. This can be accomplished by increasing a; i.e. by using stronger and more accurate weapons, or by increasing the initial force y_0. Notice though that a doubling of a results in a doubling of ay_0^2 while a doubling of y_0 results in a *four-fold* increase of ay_0^2. This is the essence of Lanchester's square law of conventional combat.

Conventional–guerilla combat. The orbits of system (2b) are the solution curves of the equation

$$\frac{dy}{dx} = \frac{dx}{cxy} = \frac{d}{cy}. \tag{4}$$

Multiplying both sides of (4) by cy and integrating gives

$$cy^2 - 2dx = cy_0^2 - 2dx_0 = M. \tag{5}$$

The curves (5) define a family of parabolas in the x–y plane, and we have indicated their graphs in Figure 2. The y-force wins if $M>0$, since the x-force has been annihilated by the time $y(t)$ has decreased to $\sqrt{M/c}$. Similarly, x wins if $M<0$.

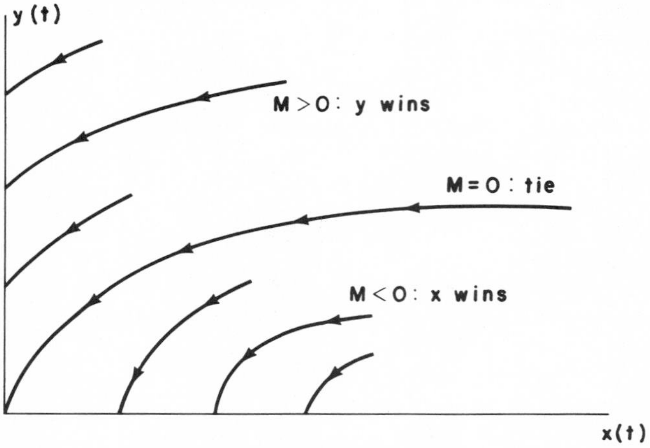

Figure 2. The parabolas defined by (5)

Remark. It is usually impossible to determine, a priori, the numerical value of the combat coefficients a, b, c, and d. Thus, it would appear that Lanchester's combat models have little or no applicability to real-life engagements. However, this is not so. As we shall soon see, it is often possible to determine suitable values of a and b (or c and d) using data from the battle itself. Once these values are established for one engagement, they are known for all other engagements which are fought under similar conditions.

(b) *The battle of Iwo Jima*

One of the fiercest battles of World War II was fought on the island of Iwo Jima, 660 miles south of Tokyo. Our forces coveted Iwo Jima as a bomber base close to the Japanese mainland, while the Japanese needed the island as a base for fighter planes attacking our aircraft on their way to bombing missions over Tokyo and other major Japanese cities. The American invasion of Iwo Jima began on February 19, 1945, and the fighting was intense throughout the month long combat. Both sides suffered heavy casualties (see Table 1). The Japanese had been ordered to fight to the last man, and this is exactly what they did. The island was declared "secure" by the American forces on the 28th day of the battle, and all active combat ceased on the 36th day. (The last two Japanese holdouts surrendered in 1951!)

Table 1. Casualties at Iwo Jima

Total United States casualties at Iwo Jima				
	Killed, missing or died of wounds	Wounded	Combat Fatigue	Total
Marines	5,931	17,272	2,648	25,851
Navy units:				
Ships and air units	633	1,158		1,791
Medical corpsmen	195	529		724
Seabees	51	218		269
Doctors and dentists	2	12		14
Army units in battle	9	28		37
Grand totals	6,821	19,217	2,648	28,686
Japanese casualties at Iwo Jima				
Defense forces (Estimated)		Prisoners		Killed
21,000		Marine 216		20,000
		Army 867		
		Total 1,083		

(Newcomb [6], page 296)

The following data is available to us from the battle of Iwo Jima.

1. *Reinforcement rates*. During the conflict Japanese troops were neither withdrawn nor reinforced. The Americans, on the other hand, landed 54,000 troops on the first day of the battle, none on the second, 6,000 on the third, none on the fourth and fifth, 13,000 on the sixth day, and none thereafter. There were no American troops on Iwo Jima prior to the start of the engagement.

2. *Combat losses*. Captain Clifford Morehouse of the United States Marine Corps (see Morehouse [5]) kept a daily count of all American combat losses. Unfortunately, no such records are available for the Japanese forces. Most probably, the casualty lists kept by General Kuribayashi (commander of the Japanese forces on Iwo Jima) were destroyed in the battle itself, while whatever records were kept in Tokyo were consumed in the fire bombings of the remaining five months of the war. However, we can infer from Table 1 that approximately 21,500 Japanese forces were on Iwo Jima at the start of the battle. (Actually, Newcomb arrived at the figure of 21,000 for the Japanese forces, but this is a little low since he apparently did not include some of the living and dead found in the caves in the final days.)

3. *Operational losses*. The operational losses on both sides were negligible.

Now, let $x(t)$ and $y(t)$ denote respectively, the active American and Japanese forces on Iwo Jima t days after the battle began. The data above suggests the following Lanchestrian model for the battle of Iwo Jima:

$$\begin{aligned}\frac{dx}{dt}&=-ay+f(t)\\\frac{dy}{dt}&=-bx\end{aligned}\tag{6}$$

where a and b are the combat effectiveness coefficients of the Japanese and American forces, respectively, and

$$f(t)=\begin{cases}54{,}000 & 0\leqslant t<1\\0 & 1\leqslant t<2\\6{,}000 & 2\leqslant t<3\\0 & 3\leqslant t<5\\13{,}000 & 5\leqslant t<6\\0 & t\geqslant 6\end{cases}$$

Using the method of variation of parameters developed in Section 3.12 or the method of elimination in Section 2.14, it is easily seen that the solution of (6) which satisfies $x(0)=0$, $y(0)=y_0=21{,}500$ is given by

$$x(t)=-\sqrt{\frac{a}{b}}\,y_0\cosh\sqrt{ab}\,t+\int_0^t\cosh\sqrt{ab}\,(t-s)f(s)\,ds\tag{7a}$$

and

$$y(t)=y_0\cosh\sqrt{ab}\,t-\sqrt{\frac{b}{a}}\int_0^t\sinh\sqrt{ab}\,(t-s)f(s)\,ds\tag{7b}$$

where

$$\cosh x\equiv(e^x+e^{-x})/2\quad\text{and}\quad\sinh x\equiv(e^x-e^{-x})/2.$$

The problem before us now is this: Do there exist constants a and b so that (7a) yields a good fit to the data compiled by Morehouse? This is an extremely important question. An affirmative answer would indicate that Lanchestrian models do indeed describe real life battles, while a negative answer would shed a dubious light on much of Lanchester's work.

As we mentioned previously, it is extremely difficult to compute the combat effectiveness coefficients a and b of two opposing forces. However, it is often possible to determine suitable values of a and b once the data for the battle is known, and such is the case for the battle of Iwo Jima.

The calculation of a and b. Integrating the second equation of (6) between 0 and s gives

$$y(s)-y_0=-b\int_0^s x(t)\,dt$$

so that

$$b = \frac{y_0 - y(s)}{\int_0^s x(t)\,dt}. \tag{8}$$

In particular, setting $s = 36$ gives

$$b = \frac{y_0 - y(36)}{\int_0^{36} x(t)\,dt} = \frac{21{,}500}{\int_0^{36} x(t)\,dt}. \tag{9}$$

Now the integral on the right-hand side of (9) can be approximated by the Riemann sum

$$\int_0^{36} x(t)\,dt \cong \sum_{i=1}^{36} x(i)$$

and for $x(i)$ we enter the number of effective American troops on the ith day of the battle. Using the data available from Morehouse, we compute for b the value

$$b = \frac{21{,}500}{2{,}037{,}000} = 0.0106. \tag{10}$$

Remark. We would prefer to set $s = 28$ in (8) since that was the day the island was declared secure, and the fighting was only sporadic after this day. However, we don't know $y(28)$. Thus, we are forced here to take $s = 36$.

Next, we integrate the first equation of (6) between $t = 0$ and $t = 28$ and obtain that

$$\begin{aligned} x(28) &= -a\int_0^{28} y(t)\,dt + \int_0^{28} f(t)\,dt \\ &= -a\int_0^{28} y(t)\,dt + 73{,}000. \end{aligned}$$

There were 52,735 effective American troops on the 28th day of the battle. Thus

$$a = \frac{73{,}000 - 52{,}735}{\int_0^{28} y(t)\,dt} = \frac{20{,}265}{\int_0^{28} y(t)\,dt}. \tag{11}$$

Finally, we approximate the integral on the right-hand side of (11) by the Riemann sum

$$\int_0^{28} y(t)\,dt \cong \sum_{j=1}^{28} y(j)$$

and we approximate $y(j)$ by

$$y(j) = y_0 - b\int_0^j x(t)\,dt$$

$$\cong 21{,}500 - b\sum_{i=i}^{j} x(i).$$

Again, we replace $x(i)$ by the number of effective American troops on the ith day of the battle. The result of this calculation is (see Engel [2])

$$a = \frac{20{,}265}{372{,}500} = 0.0544. \tag{12}$$

Figure 3 below compares the actual American troop strength with the values predicted by Equation (7a) (with $a = 0.0544$ and $b = 0.0106$). The fit is remarkably good. Thus, it appears that a Lanchestrian model does indeed describe real life engagements.

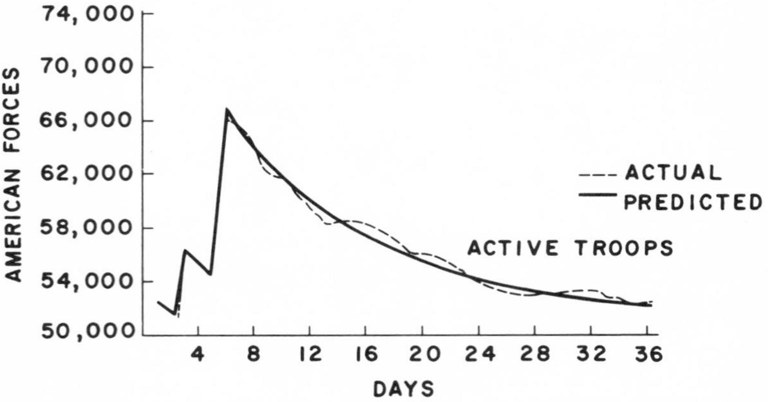

Figure 3. Comparison of actual troop strength with predicted troop strength

Remark. The figures we have used for American reinforcements include *all* the personnel put ashore, both combat troops and support troops. Thus the numbers a and b that we have computed should be interpreted as the *average* effectiveness per man ashore.

References

1. Coleman, C. S., Combat Models, MAA Workshop on Modules in Applied Math, Cornell University, Aug. 1976.
2. Engel, J. H., A verification of Lanchester's law, *Operations Research*, **2**, (1954), 163–171.
3. Howes, D. R., and Thrall, R. M., A theory of ideal linear weights for heterogeneous combat forces, *Naval Research Logistics Quarterly*, vol. 20, 1973, pp. 645–659.
4. Lanchester, F. W., *Aircraft in Warfare, the Dawn of the Fourth Arm*. Tiptree, Constable and Co., Ltd., 1916.

5. Morehouse, C. P., *The Iwo Jima Operation*, USMCR, Historical Division, Hdqr. USMC, 1946.

6. Newcomb, R. F., *Iwo Jima*. New York: Holt, Rinehart, and Winston, 1965.

EXERCISES

1. Derive Equations (7a) and (7b).

2. The system of equations

$$\begin{aligned}\dot{x} &= -ay \\ \dot{y} &= -by - cxy\end{aligned} \tag{13}$$

is a Lanchestrian model for conventional–guerilla combat, in which the operational loss rate of the guerilla force y is proportional to $y(t)$.
(a) Find the orbits of (13).
(b) Who wins the battle?

3. The system of equations

$$\begin{aligned}\dot{x} &= -ay \\ \dot{y} &= -bx - cxy\end{aligned} \tag{14}$$

is a Lanchestrian model for conventional–guerilla combat, in which the operational loss rate of the guerilla force y is proportional to the strength of the conventional force x. Find the orbits of (14).

4. The system of equations

$$\begin{aligned}\dot{x} &= -cxy \\ \dot{y} &= -dxy\end{aligned} \tag{15}$$

is a Lanchestrian model for guerilla–guerilla combat in which the operational loss rates are negligible.
(a) Find the orbits of (15).
(b) Who wins the battle?

5. The system of equations

$$\begin{aligned}\dot{x} &= -ay - cxy \\ \dot{y} &= -bx - dxy\end{aligned} \tag{16}$$

is a Lanchestrian model for guerilla–guerilla combat in which the operational loss rate of each force is proportional to the strength of its opponent. Find the orbits of (16).

6. The system of equations

$$\begin{aligned}\dot{x} &= -ax - cxy \\ \dot{y} &= -by - dxy\end{aligned} \tag{17}$$

is a Lanchestrian model for guerilla–guerilla combat in which the operational loss rate of each force is proportional to its strength.
(a) Find the orbits of (17).
(b) Show that the x and y axes are both orbits of (17).
(c) Using the fact that two orbits of (17) cannot intersect, show that there is no clear-cut winner in this battle. *Hint*: Show that $x(t)$ and $y(t)$ can never become zero in finite time. (It can be shown that both $x(t)$ and $y(t)$ approach zero as $t \to \infty$.)

Appendix A

Some simple facts concerning functions of several variables

1. A function $f(x,y)$ is continuous at the point (x_0,y_0) if for every $\varepsilon>0$ there exists $\delta(\varepsilon)$ such that

$$|f(x,y)-f(x_0,y_0)|<\varepsilon \quad \text{if} \quad |x-x_0|+|y-y_0|<\delta(\varepsilon).$$

2. The partial derivative of $f(x,y)$ with respect to x is the ordinary derivative of f with respect to x, assuming that y is constant. In other words

$$\frac{\partial f(x_0,y_0)}{\partial x}=\lim_{h\to 0}\frac{f(x_0+h,y_0)-f(x_0,y_0)}{h}.$$

3. (a) A function $f(x_1,\dots,x_n)$ is said to be differentiable if

$$f(x_1+\Delta x_1,\dots,x_n+\Delta x_n)-f(x_1,\dots,x_n)=\frac{\partial f}{\partial x_1}\Delta x_1+\dots+\frac{\partial f}{\partial x_n}\Delta x_n+e$$

where $e/[|\Delta x_1|+\dots+|\Delta x_n|]$ approaches zero as $|\Delta x_1|+\dots+|\Delta x_n|$ approaches zero. (b) A function $f(x_1,\dots,x_n)$ is differentiable in a region R if $\partial f/\partial x_1,\dots,\partial f/\partial x_n$ are continuous in R.

4. Let $f=f(x_1,\dots,x_n)$ and $x_j=g_j(y_1,\dots,y_m)$, $j=1,\dots,n$. If f and g are differentiable, then

$$\frac{\partial f}{\partial y_k}=\sum_{j=1}^{n}\frac{\partial f}{\partial x_j}\frac{\partial g_j}{\partial y_k}.$$

This is the chain rule of partial differentiation.

5. If all the partial derivatives of order two of f are continuous in a region R, then

$$\frac{\partial^2 f}{\partial x_j \partial x_k} = \frac{\partial^2 f}{\partial x_k \partial x_j}; \qquad j,k = 1,\ldots,n.$$

6. The general term in the Taylor series expansion of f about $x_1 = x_1^0, \ldots, x_n = x_n^0$ is

$$\frac{1}{j_1! \ldots j_n!} \frac{\partial^{j_1 + \cdots + j_n} f(x_1^0, \ldots, x_n^0)}{\partial x_1^{j_1} \ldots \partial x_n^{j_n}} (x_1 - x_1^0)^{j_1} \ldots (x_n - x_n^0)^{j_n}.$$

Appendix B

Sequences and series

1. A sequence of numbers a_n, $n=1,2,\ldots$ is said to converge to the limit a if the numbers a_n get closer and closer to a as n approaches infinity. More precisely, the sequence (a_n) converges to a if for every $\varepsilon>0$ there exists $N(\varepsilon)$ such that

$$|a-a_n|<\varepsilon \quad \text{if } n\geqslant N(\varepsilon).$$

2. **Theorem 1.** *If $a_n\to a$ and $b_n\to b$, then $a_n\pm b_n\to a\pm b$.*

PROOF. Let $\varepsilon>0$ be given. Choose $N_1(\varepsilon),N_2(\varepsilon)$ such that

$$|a-a_n|<\frac{\varepsilon}{2} \quad \text{for } n\geqslant N_1(\varepsilon), \quad \text{and} \quad |b-b_n|<\frac{\varepsilon}{2} \quad \text{for } n\geqslant N_2(\varepsilon).$$

Let $N(\varepsilon)=\max\{N_1(\varepsilon),N_2(\varepsilon)\}$. Then, for $n\geqslant N(\varepsilon)$,

$$|a_n\pm b_n-(a\pm b)|\leqslant|a_n-a|+|b_n-b|<\frac{\varepsilon}{2}+\frac{\varepsilon}{2}=\varepsilon.$$ □

3. **Theorem 2.** *Suppose that $a_{n+1}\geqslant a_n$, and there exists a number K such that $|a_n|\leqslant K$, for all n. Then, the sequence (a_n) has a limit.*

PROOF. Exactly the same as the proof of Lemma 1, Section 4.8. □

4. The infinite series $a_1+a_2+\ldots=\sum a_n$ is said to converge if the sequence of partial sums

$$s_1=a_1,\quad s_2=a_1+a_2,\quad \ldots,\quad s_n=a_1+a_2+\ldots+a_n,\ldots$$

has a limit.

5. The sum and difference of two convergent series are also convergent. This follows immediately from Theorem 1.

6. **Theorem 3.** *Let $a_n \geqslant 0$. The series $\sum a_n$ converges if there exists a number K such that $a_1 + \ldots + a_n \leqslant K$ for all n.*

PROOF. The sequence of partial sums $s_n = a_1 + \ldots + a_n$ satisfies $s_{n+1} \geqslant s_n$. Since $s_n \leqslant K$, we conclude from Theorem 2 that $\sum a_n$ converges. □

7. **Theorem 4.** *The series $\sum a_n$ converges if there exists a number K such that $|a_1| + \ldots + |a_n| \leqslant K$ for all n.*

PROOF. From Theorem 3, $\sum |a_n|$ converges. Let $b_n = a_n + |a_n|$. Clearly, $0 \leqslant b_n \leqslant 2|a_n|$. Thus, $\sum b_n$ also converges. This immediately implies that the series

$$\sum a_n = \sum [b_n - |a_n|]$$

also converges. □

8. **Theorem 5** (Cauchy ratio test). *Suppose that the sequence $|a_{n+1}/a_n|$ has a limit λ. Then, the series $\sum a_n$ converges if $\lambda < 1$ and diverges if $\lambda > 1$.*

PROOF. Suppose that $\lambda < 1$. Then, there exists $\rho < 1$, and an index N, such that $|a_{n+1}| \leqslant \rho |a_n|$ for $n \geqslant N$. This implies that $|a_{N+p}| \leqslant \rho^p |a_N|$. Hence,

$$\sum_{n=N}^{N+K} |a_n| \leqslant (1 + \rho + \ldots + \rho^K)|a_N| \leqslant \frac{|a_N|}{1-\rho},$$

and $\sum a_n$ converges.

If $\lambda > 1$, then $|a_{N+p}| \geqslant \rho^p |a_N|$, with $\rho > 1$. Thus $|a_n| \to \infty$ as $n \to \infty$. But $|a_n|$ must approach zero if $\sum a_n$ converges to s, since

$$|a_{n+1}| = |s_{n+1} - s_n| \leqslant |s_{n+1} - s| + |s_n - s|$$

and both these quantities approach zero as n approaches infinity. Thus, $\sum a_n$ diverges. □

Answers to odd-numbered exercises

Chapter 1

Section 1.2

1. $y(t)=ce^{-\sin t}$ 3. $y(t)=\dfrac{t+c}{1+t^2}$ 5. $y(t)=\exp(-\frac{1}{3}t^3)[\int \exp(\frac{1}{3}t^3)\,dt+c]$

7. $y(t)=\dfrac{c+\int (1+t^2)^{1/2}(1+t^4)^{1/4}\,dt}{(1+t^2)^{1/2}(1+t^4)^{1/4}}$ 9. $y(t)=\exp\left(-\int_0^t \sqrt{1+s^2}\,e^{-s}\,ds\right)$

11. $y(t)=\dfrac{3e^{t^2}-1}{2}$ 13. $y(t)=e^{-t}\left[2e+\int_1^t \dfrac{e^s}{1+s^2}\,ds\right]$

15. $y(t)=\left(\dfrac{t^5}{5}+\dfrac{2t^3}{3}+t+c\right)(1+t^2)^{-1/2}$ 17. $y(t)=\begin{cases} 2(1-e^{-t}), & 0\leqslant t<1 \\ 2(e-1)e^{-t}, & t>1 \end{cases}$

21. Each solution approaches a distinct limit as $t\to 0$.

23. All solutions approach zero as $t\to\pi/2$.

Section 1.3

3. 127,328 5. (a) $N_{238}(t)=N_{238}(0)2^{-10^{-9}t/4.5}$; (b) $N_{235}(t)=N_{235}(0)2^{-10^{-9}t/.707}$

7. About 13,550 B.C.

Section 1.4

1. $y(t)=\dfrac{t+c}{1-ct}$ 3. $y(t)=\tan(t-\frac{1}{2}t^2+c)$ 5. $y(t)=\arcsin(c\sin t)$

7. $y(t)=\left[9+2\ln\left(\dfrac{1+t^2}{5}\right)\right]^{1/2}, \quad -\infty<t<\infty$

9. $y(t)=1-[4+2t+2t^2+t^3]^{1/2}, \quad -2<t<\infty$

11. $y(t)=\dfrac{a^2kt}{1+akt}, \quad \dfrac{-1}{ak}<t<\infty$, if $a=b$

$y(t)=\dfrac{ab[1-e^{k(b-a)t}]}{a-be^{k(b-a)t}}, \quad \dfrac{1}{k(b-a)}\ln\dfrac{a}{b}<t<\infty; \quad a\neq b$

13. (b) $y(t)=-t$ and $y(t)=\dfrac{ct^2}{1-ct}$ **15.** $y(t)=\dfrac{t^2-1}{2}$

17. $y=c^2e^{-2\sqrt{t/y}}, \quad t>0; \; y=-c^2e^{2\sqrt{t/y}}, \quad t<0$ **19.** $t+ye^{t/y}=c$

21. (b) $|(b+c)(at+by)+an+bm|=ke^{(b+c)(at-cy)}$

23. $(t+2y)^2-(t+2y)=c-7t$

Section 1.5

3. $a>0.4685$

5. (a) $\dfrac{dp}{dt}=0.003p-0.001p^2-0.002$; (b) $p(t)=\dfrac{1{,}999{,}998-999{,}998e^{-0.001t}}{999{,}999-999{,}998e^{-0.001t}}$, $p(t)\to 2$ as $t\to\infty$

Section 1.6

1. $V(t)=\dfrac{W-B}{c}[1-e^{(-cg/W)t}]$

7. $V(t)=\sqrt{322}\left[\dfrac{K_0e^{(64.4)(t-5)/\sqrt{322}}+1}{K_0e^{(64.4)(t-5)/\sqrt{322}}-1}\right], \quad K_0=\dfrac{\sqrt{322}\left(1-\dfrac{1}{\sqrt{e}}\right)+1}{\sqrt{322}\left(1-\dfrac{1}{\sqrt{e}}\right)-1}$

9. (a) $\sqrt{V}-\sqrt{V_0}+\dfrac{mg}{c}\ln\dfrac{mg-c\sqrt{V}}{mg-c\sqrt{V_0}}=\dfrac{-ct}{2m}$; (b) $V_T=\dfrac{m^2g^2}{c^2}$

11. (a) $v\dfrac{dv}{dy}=\dfrac{-gR^3}{(y+R)^2}$; (b) $V_0=\sqrt{2gR}$

Section 1.7

3. $\dfrac{2\ln 5}{\ln 2}$ yrs. **5.** 10^4e^{10} **7.** (a) $c(t)=1-e^{-0.001t}$; (b) $1000\ln\frac{5}{4}$ min.

9. $c(t)=\frac{1}{2}(1-e^{-0.02t})+\dfrac{Q_0}{150}e^{-0.02t}$ **11.** 48% **13.** $xy=c$

15. $x^2+y^2=cy$ **17.** $y^2(\ln y^2-1)=2(c-x^2)$

Section 1.8

3. $t^2\sin y+y^3e^t=c$ **5.** $t+y+te^{ty}=c$ **7.** $y(t)=t^{-2/3}$

9. $y(t)=-t^2+\sqrt{t^4-(t^3-1)}$ **11.** $t^2-3y+(\cos 2t)e^{ty}=1$

13. $a=-2$; $y(t)=\pm\left[\dfrac{t(2t-1)}{2(1+ct)}\right]^{1/2}$ **19.** $a=1$; $y(t)=\arcsin[(c-t)e^t]$

SECTION 1.9

1. $y_n(t)=t^2+\frac{t^4}{2!}+\frac{t^6}{3!}+\ldots+\frac{t^{2n}}{n!}$

3. $y_1(t)=e^t-1;\quad y_2(t)=t-e^t+\frac{1+e^{2t}}{2}$

$$y_3(t)=-\frac{107}{48}+\frac{t}{4}+\frac{t^2}{2}+\frac{t^3}{3}+2(1-t)e^t+\frac{(1+t)e^{2t}}{2}-\frac{e^{3t}}{3}+\frac{e^{4t}}{16}$$

19. $y(t)=\sin\frac{t^2}{2}$

SECTION 1.10

1. $y_n=(-7)^n-\frac{1}{4}[(-7)^n-1]$ **3.** (a) $E_n\leqslant\frac{1}{2}(3^n-1)$; (b) $E_n\leqslant 2(2^n-1)$

5. $y_n=a_1\ldots a_{n-1}\left[y_1+\sum_{j=1}^{n-1}\frac{b_j}{a_1\ldots a_j}\right]$ **7.** $y_{25}=\frac{1}{25}\left[1+\sum_{j=1}^{24}2^j\right]$

9. (a) $x=\$251.75$; (b) $x=\$289.50$

Chapter 2

SECTION 2.1

1. (a) $(4-3t)e^t$; (b) $3\sqrt{3}\,t\sin\sqrt{3}\,t$; (c) $2(4-3t)e^t+12\sqrt{3}\,t\sin\sqrt{3}\,t$; (d) $2-3t^2$; (e) $5(2-3t^2)$; (f) 0; (g) $2-3t^2$

5. (b) $W=\frac{-3}{2t^{3/2}}$; (d) $y(t)=2\sqrt{t}$

7. (a) $-(b\sin at\sin bt+a\cos at\cos bt)$; (b) 0; (c) $(b-a)e^{(a+b)t}$; (d) e^{2at}; (e) t; (f) $-be^{2at}$

13. $W=\frac{1}{t}$

SECTION 2.2

1. $y(t)=c_1e^t+c_2e^{-t}$ **3.** $y(t)=e^{3t/2}\left[c_1e^{\sqrt{5}\,t/2}+c_2e^{-\sqrt{5}\,t/2}\right]$

5. $y(t)=\frac{1}{5}(e^{4t}+4e^{-t})$ **7.** $y(t)=\frac{\sqrt{5}}{3}e^{-t/2}\left[e^{3\sqrt{5}\,t/10}-e^{-3\sqrt{5}\,t/10}\right]$

9. $V\geqslant -3$ **11.** $y(t)=c_1t+c_2/t^5$

Section 2.2.1

1. $y(t)=e^{-t/2}\left[c_1\cos\frac{\sqrt{3}\,t}{2}+c_2\sin\frac{\sqrt{3}\,t}{2}\right]$

3. $y(t)=e^{-t}(c_1\cos\sqrt{2}\,t+c_2\sin\sqrt{2}\,t)$

5. $y(t)=e^{-t/2}\left[\cos\frac{\sqrt{7}\,t}{2}-\frac{3}{\sqrt{7}}\sin\frac{\sqrt{7}\,t}{2}\right]$

9. $y(t)=e^{(t-2)/3}\left[\cos\frac{\sqrt{11}\,(t-2)}{3}-\frac{4}{\sqrt{11}}\sin\frac{\sqrt{11}\,(t-2)}{3}\right]$

11. $y_1(t)=\cos\omega t,\ y_2(t)=\sin\omega t$

15. $\sqrt{i}=\frac{\pm(1+i)}{\sqrt{2}};\ \sqrt{1+i}=\frac{\pm1}{\sqrt{2}}\left[\sqrt{\sqrt{2}+1}+i\sqrt{\sqrt{2}-1}\right]$

$\sqrt{-i}=\frac{\pm(1-i)}{\sqrt{2}};\ \sqrt{\sqrt{i}}=\pm\left[\sqrt{\sqrt{2}-1}+i\sqrt{\sqrt{2}+1}\right],\quad (\sqrt{i}=e^{i\pi/4})$

19. $y(t)=\frac{1}{\sqrt{t}}\left[c_1\cos\frac{\sqrt{7}}{2}\ln t+c_2\sin\frac{\sqrt{7}}{2}\ln t\right]$

Section 2.2.2

1. $y(t)=(c_1+c_2t)e^{3t}$ **3.** $y(t)=(1+\frac{1}{3}t)e^{-t/3}$ **7.** $y(t)=2(t-\pi)e^{2(t-\pi)/3}$

11. $y(t)=(c_1+c_2t)e^{t^2}$ **13.** $y(t)=c_1t^2+c_2(t^2-1)$ **15.** $y(t)=c_1(t+1)+c_2e^{2t}$

17. $y(t)=c_1(1+3t)+c_2e^{3t}$ **19.** $y(t)=\frac{c_1+c_2\ln t}{t}$

Section 2.3

1. $y(t)=c_1+c_2e^{2t}+t^2$ **3.** $y(t)=e^{t^2}+2e^{t}-2e^{-t^3}$

Section 2.4

1. $y(t)=(c_1+\ln\cos t)\cos t+(c_2+t)\sin t$

3. $y(t)=c_1e^{t/2}+[c_2+9t-2t^2+\frac{1}{3}t^3]e^{t}$

5. $y(t)=\frac{1}{13}(2\cos t-3\sin t)e^{-t}-e^{-t}+\frac{24}{13}e^{-t/3}$

7. $y(t)=\int_0^t\sqrt{s+1}\,[e^{2(t-s)}-e^{(t-s)}]\,ds$ **9.** $y(t)=c_1t^2+\frac{c_2}{t}+\frac{t^2\ln t}{3}$

11. $y(t)=\sqrt{t}\left[c_1+c_2\ln t+\int_0^t f\sqrt{s}\,\cos s[\ln t-\ln s]\,ds\right]$

Section 2.5

1. $\psi(t)=\frac{t^3-2t-1}{3}$ **3.** $\psi(t)=t(\frac{1}{4}-\frac{1}{4}t+\frac{1}{6}t^2)e^{t}$ **5.** $\psi(t)=\frac{t^2}{2}e^{-t}$

7. $\psi(t)=\frac{1}{16}t[\sin 2t-2t\cos 2t]$ **9.** $\psi(t)=\frac{1}{5}+\frac{1}{17}(\cos 2t-4\sin 2t)$

11. $\psi(t)=\frac{-1}{50}(\cos t+7\sin t)+\left(\frac{t}{2}-\frac{1}{5}\right)\frac{te^{2t}}{5}$ **13.** $\psi(t)=t(e^{2t}-e^{t})$

15. $\psi(t)=-\frac{1}{16}\cos 3t+\frac{1}{4}t\sin t$

17. (b) $\psi(t)=0.005+\frac{1}{32{,}000{,}018}(15\sin 30t-20{,}000\cos 30t)+\frac{15}{2}\sin 10t$

SECTION 2.6

1. Amplitude $=\frac{1}{4}$, period $=2\pi/\sqrt{64.4}$, frequency $=\sqrt{64.4}$

7. $\alpha \geqslant \beta$, where $\left[1+(1-\beta)^2\right]e^{-2\beta}=10^{-6}$

9. $y(t)=\dfrac{e^{-(t-\pi)}}{2}\left[\left(\dfrac{1}{2}+\dfrac{(\pi+1)}{2}e^{-\pi}\right)\cos(t-\pi)+\dfrac{\pi}{2}e^{-\pi}\sin(t-\pi)\right]$

11. $\pi/2$ seconds

Section 2.6.2

3. $Q(1)=\dfrac{12}{10^6}\left[1-\left(\cos 500\sqrt{3}+\dfrac{1}{\sqrt{3}}\sin 500\sqrt{3}\right)e^{-500}\right]$

Steady state charge $=\dfrac{3}{250{,}000}$

7. $\omega=\left(\dfrac{1}{LC}-\dfrac{R^2}{2L^2}\right)^{1/2}$

SECTION 2.8

1. $y(t)=a_0e^{-t^2/2}+a_1\sum\limits_{n=0}^{\infty}\dfrac{(-1)^n t^{2n+1}}{3\cdot 5\ldots(2n+1)}$

3. $y(t)=a_0\left[1+\dfrac{3t^2}{2^2}+\dfrac{3t^4}{2^4\cdot 2!}-\dfrac{3t^6}{2^6\cdot 3!}+\dfrac{3\cdot 3t^8}{2^8\cdot 4!}-\dfrac{3\cdot 3\cdot 5t^{10}}{2^{10}\cdot 5!}+\ldots\right]+a_1\left(t+\dfrac{t^3}{3}\right)$

5. $y(t)=\sum\limits_{n=0}^{\infty}(n+1)(t-1)^{2n}$

7. $y(t)=-2[t+\dfrac{t^6}{5\cdot 6}+\dfrac{t^{11}}{5\cdot 6\cdot 10\cdot 11}+\dfrac{t^{16}}{5\cdot 6\cdot 10\cdot 11\cdot 15\cdot 16}+\ldots]$

9. (a) $y_1(t)=1-\dfrac{\lambda t^2}{2!}-\dfrac{\lambda(4-\lambda)t^4}{4!}-\dfrac{\lambda(4-\lambda)(8-\lambda)t^6}{6!}+\ldots$

(b) $y_2(t)=t+(2-\lambda)\dfrac{t^3}{3!}+(2-\lambda)(6-\lambda)\dfrac{t^5}{5!}+(2-\lambda)(6-\lambda)(10-\lambda)\dfrac{t^7}{7!}+\ldots$

11. (a) $y_1(t)=1-\alpha^2\dfrac{t^2}{2!}-\alpha^2(2^2-\alpha^2)\dfrac{t^4}{4!}$

$-\alpha^2(2^2-\alpha^2)(4^2-\alpha^2)\dfrac{t^6}{6!}+\ldots$

$y_2(t)=t+(1^2-\alpha^2)\dfrac{t^3}{3!}+(1^2-\alpha^2)(3^2-\alpha^2)\dfrac{t^5}{5!}+\ldots$

Section 2.8.1

1. Yes **3.** Yes **5.** No

7. $y(t)=c_1t^{5/2}\left[1+\dfrac{t^2}{2\cdot 5}+\dfrac{t^4}{2\cdot 4\cdot 5\cdot 7}+\ldots\right]$

$+c_2t^{-1/2}\left[1-\dfrac{t^2}{2}-\dfrac{t^4}{2\cdot 4}-\dfrac{t^6}{2\cdot 4\cdot 6\cdot 8}+\ldots\right]$

9. $y(t)=c_1e^{t^2/2}+c_2\left[t^3+\frac{t^5}{5}+\frac{t^7}{5\cdot7}+\frac{t^9}{5\cdot7\cdot9}+\dots\right]$

11. $y(t)=\frac{c_1}{t}\left[1-t-\frac{t^2}{2!}-\frac{t^3}{3\cdot3!}-\frac{t^4}{3\cdot5\cdot4!}-\dots\right]$
$+c_2t^{1/2}\left[1+\frac{t}{5}+\frac{t^2}{5\cdot7\cdot2!}+\frac{t^3}{5\cdot7\cdot9\cdot3!}+\dots\right]$

13. $y(t)=c_1\left(1+2t+\frac{t^2}{3}\right)+c_2t^{1/2}\left[1+\frac{t}{2}+\frac{t^2}{2^2\cdot2!\cdot5}-\frac{t^3}{2^3\cdot3!\cdot5\cdot7}+\dots\right]$

15. $y(t)=c_1\left[1+\frac{3t}{1\cdot3}+\frac{3^2t^2}{3\cdot7\cdot2!}+\frac{3^3t^3}{3\cdot7\cdot11\cdot3!}+\dots\right]$
$+c_2t^{1/4}\left[1+\frac{3t}{1\cdot5}+\frac{3^2t^2}{5\cdot9\cdot2!}+\frac{3^3t^3}{5\cdot9\cdot13\cdot3!}+\dots\right]$

17. $y(t)=c_1\left(\frac{1}{t}-1\right)+\frac{c_2}{t}\left(e^{-t/2}+\frac{t}{2}-1\right)$

19. $y(t)=(1-t^2+\frac{1}{8}t^4)\left[c_1+c_2\int\frac{e^{t^2/2}}{t[1-t^2+t^4/8]^2}dt\right]$

21. (b) $y_1(t)=1+2t+t^2+\frac{4t^3}{15}+\frac{t^4}{14}+\dots$
$y_2(t)=t^{1/2}\left[1+\frac{5t}{6}+\frac{17t^2}{60}+\frac{89t^3}{60(21)}+\frac{941t^4}{36(21)(60)}+\dots\right]$

25. (b) $y(t)=1-\frac{\lambda t}{(1!)^2}-\frac{\lambda(1-\lambda)t^2}{(1!)^2(2!)^2}+\dots+\frac{(-\lambda)(1-\lambda)\dots(n-1-\lambda)}{(n!)^2}t^n+\dots$

Section 2.9

1. $\frac{1}{s^2}$ **3.** $\frac{s-a}{(s-a)^2+b^2}$ **5.** $\frac{1}{2}\left[\frac{1}{s}+\frac{s}{s^2+4a^2}\right]$

7. $\frac{1}{2}\left[\frac{a+b}{s^2+(a+b)^2}+\frac{a-b}{s^2+(a-b)^2}\right]$ **9.** $\sqrt{\frac{\pi}{s}}$

15. $y(t)=2e^t-\frac{1}{2}e^{4t}-\frac{1}{2}e^{2t}$ **17.** $Y(s)=\frac{s^2+6s+6}{(s+1)^3}$

19. $Y(s)=\frac{2s^2+s+2}{(s^2+3s+7)(s^2+1)}$ **23.** $y(t)=-\frac{1}{6}e^t+\frac{1}{2}e^{2t}-\frac{1}{2}e^{3t}+\frac{1}{6}e^{4t}$

Section 2.10

1. $\frac{n!}{s^{n+1}}$ **3.** $\frac{2as}{(s^2+a^2)^2}$ **5.** $\frac{15}{8s^3}\sqrt{\frac{\pi}{s}}$

7. (a) $\frac{\pi}{2}-\arctan s$; (b) $\ln\frac{s}{\sqrt{s^2+a^2}}$; (c) $\ln\frac{s-b}{s-a}$ **9.** $-\frac{5}{3}+2e^{-t}+\frac{2}{3}e^{-3t}$

11. $\left[\cosh\frac{\sqrt{57}\,t}{2}+\frac{3}{\sqrt{57}}\sinh\frac{\sqrt{57}\,t}{2}\right]e^{3t/2}$ **13.** $\frac{1}{2}(3-t)t^2e^{-t}$

15. $\frac{1}{2}(\cos t - te^{-t})$ **19.** $y(t)=\cos t+\frac{5}{2}\sin t-\frac{1}{2}t\cos t$ **21.** $y(t)=\frac{1}{6}t^3e^t$

23. $y(t)=1+e^{-t}+\left[\cos\frac{\sqrt{3}\,t}{2}-\frac{7}{\sqrt{3}}\sin\frac{\sqrt{3}\,t}{2}\right]e^{-t/2}$

Section 2.11

1. $y(t)=(2+3t)e^{-t}+2H_3(t)\left[(t-5)+(t-1)e^{-(t-3)}\right]$

3. $y(t)=3\cos 2t-\sin 2t+\frac{1}{4}(1-\cos 2t)-\frac{1}{4}(1-\cos 2(t-4))H_4(t)$

5. $y(t)=3\cos t-\sin t+\frac{1}{2}t\sin t+\frac{1}{2}H_{\pi/2}(t)\left[\left(t-\frac{1}{2}\pi\right)\sin t-\cos t\right]$

7. $$y(t)=\frac{1}{49}\left[(7t-1)+\left(\cos\frac{\sqrt{27}\,t}{2}-\frac{13}{\sqrt{27}}\sin\frac{\sqrt{27}\,t}{2}\right)e^{-t/2}\right.$$
$$-H_2(t)(7t-1)-H_2(t)\left[13\cos\sqrt{27}\,\frac{(t-2)}{2}\right.$$
$$\left.\left.+\sqrt{27}\,\sin\frac{\sqrt{27}\,(t-2)}{2}\right]e^{-(t-2)/2}\right]$$

9. $y(t)=te^t+H_1(t)\left[2+t+(2t-5)e^{t-1}\right]-H_2(t)\left[1+t+(2t-7)e^{t-2}\right]$

Section 2.12

3. (a) $y(t)=(\cosh\frac{1}{2}t-3\sinh\frac{1}{2}t)e^{3t/2}-2H_2(t)\sinh\frac{1}{2}(t-2)e^{3(t-2)/2}$

5. $y(t)=\frac{1}{2}(\sin t-t\cos t)-H_\pi(t)\sin t$

7. $y(t)=3te^{-t}+\frac{1}{2}t^2e^{-t}+3H_3(t)(t-3)e^{-(t-3)}$

Section 2.13

1. $\dfrac{e^{bt}-e^{at}}{b-a}$ **3.** $\dfrac{a\sin at-b\sin bt}{a^2-b^2}$ **5.** $\dfrac{\sin at-at\cos at}{2a}$ **7.** $t-\sin t$

9. $\frac{1}{2}t\sin t$ **11.** $(t-2)+(t+2)e^{-t}$ **13.** $y(t)=t+\frac{3}{2}\sin 2t$ **15.** $y(t)=\frac{1}{2}t^2$

17. $y(t)=\frac{1}{2}\sin t+(1-\frac{3}{2}t)e^{-t}$

Section 2.14

1. $x(t)=c_1e^{3t}+3c_2e^{4t},\quad y(t)=c_1e^{3t}+2c_2e^{4t}$

3. $x(t)=2c_1+2c_2\sin t\; e^{-2t},\quad y(t)=c_1(1-\sin t)+c_2(\cos t+\sin t)e^{-2t}$

5. $x(t)=\frac{7}{4}e^{3t}+\frac{1}{4}e^{-t},\quad y(t)=\frac{7}{8}e^{3t}-\frac{1}{8}e^{-t}$

7. $x(t)=\cos t\; e^{-t},\quad y(t)=(2\cos t+\sin t)e^{-t}$

9. $x(t)=2(t\cos t+3t\sin t+\sin t)e^t,\quad y(t)=-2t\sin t\; e^t$

11. $x(t)=-4\sin t-\frac{1}{2}t\sin t-t\cos t+5\cos t\ln(\sec t+\tan t),$
$y(t)=-t\sin t-t\cos t-\frac{1}{2}\sin^2 t\cos t-5\sin t\cos t+5\sin^2 t$
$-\frac{1}{2}\sin^3 t+5(\cos t-\sin t)\ln(\sec t+\tan t)$

Section 2.15

1. $y(t)=c_1e^t+c_2e^{-t}+c_3e^{2t}$ **3.** $y(t)=(c_1+c_2t+c_3t^2)e^{2t}+c_4e^{-t}$ **5.** $y(t)=0$

7. $y(t)=-3-2t-\frac{1}{2}t^2+(3-t)e^t$ **9.** $\psi(t)=1-\cos t-\ln\cos t-\sin t\ln(\sec t+\tan t)$

11. $$\psi(t)=\frac{1}{\sqrt{2}}\int_0^t\Big[\sin(t-s)/\sqrt{2}\,\cosh(t-s)/\sqrt{2}\\ -\cos(t-s)/\sqrt{2}\,\sinh(t-s)/\sqrt{2}\,\Big]g(s)\,ds$$

13. $\psi(t)=\frac{1}{4}t(e^{-2t}-1)-\frac{1}{5}\sin t$ **15.** $\psi(t)=\frac{1}{4}t^2\left[\left(\frac{1}{2}-\frac{1}{3}t\right)\cos t+\left(\frac{3}{4}+\frac{1}{6}t-\frac{1}{12}t^2\right)\sin t\right]$

17. $\psi(t)=t-1+\frac{1}{2}te^{-t}$

Chapter 3

Section 3.1

1. $\dot{x}_1=x_2$, $\dot{x}_2=x_3$, $\dot{x}_3=-x_2^2$ **3.** $\dot{x}_1=x_2$, $\dot{x}_2=x_3$, $\dot{x}_3=x_4$, $\dot{x}_4=1-x_3$ **7.** $\dot{\mathbf{x}}=\begin{pmatrix}5 & 5\\ -1 & 7\end{pmatrix}\mathbf{x}$, $\mathbf{x}(3)=\begin{pmatrix}0\\6\end{pmatrix}$

9. $\dot{\mathbf{x}}=\begin{pmatrix}0 & 0 & -1\\ 1 & 0 & 0\\ 0 & -1 & 0\end{pmatrix}\mathbf{x}$, $\mathbf{x}(-1)=\begin{pmatrix}2\\3\\4\end{pmatrix}$

11. (a) $\begin{pmatrix}1\\3\\-1\end{pmatrix}$; (b) $\begin{pmatrix}2\\0\\-1\end{pmatrix}$; (c) $\begin{pmatrix}-1\\4\\2\end{pmatrix}$

13. (a) $\begin{pmatrix}11\\16\\7\end{pmatrix}$; (b) $\begin{pmatrix}-7\\2\\-3\end{pmatrix}$; (c) $\begin{pmatrix}5\\10\\1\end{pmatrix}$; (d) $\begin{pmatrix}-1\\9\\1\end{pmatrix}$ **15.** $\mathbf{A}=\begin{pmatrix}\frac{1}{2} & \frac{7}{2}\\ -2 & 4\end{pmatrix}$

Section 3.2

1. Yes **3.** No **5.** No **7.** Yes **9.** No **11.** Yes

Section 3.3

1. Linearly dependent **3.** Linearly independent

5. (a) Linearly dependent; (b) Linearly dependent

7. (b) $y_1(t)=e^t$, $y_2(t)=e^{-t}$ **9.** $p_1(t)=t-1$, $p_2(t)=t^2-6$

11. $\mathbf{x}^1=\begin{pmatrix}-1\\-1\\1\end{pmatrix}$ **17.** $x_1=y_1$, $x_2=\frac{1}{\sqrt{2}}(y_2-y_3)$, $x_3=\frac{1}{\sqrt{2}}(y_2+y_3)$

Section 3.4

1. $\mathbf{x}^1(t)=\begin{bmatrix} \cos\sqrt{3}\,t/2\ e^{-t/2} \\ \left(-\frac{1}{2}\cos\sqrt{3}\,t/2-\frac{1}{2}\sqrt{3}\ \sin\sqrt{3}\ t/2\right)e^{-t/2} \end{bmatrix}$

$\mathbf{x}^2(t)=\begin{bmatrix} \sin\sqrt{3}\,t/2\ e^{-t/2} \\ \left(\frac{3}{2}\cos\sqrt{3}\,t/2-\frac{1}{2}\sin\sqrt{3}\ t/2\right)e^{-t/2} \end{bmatrix}$

3. $\mathbf{x}^1(t)=\begin{pmatrix} 0 \\ e^t \end{pmatrix}$, $\mathbf{x}^2(t)=\begin{pmatrix} e^t \\ 2te^t \end{pmatrix}$ **5.** Yes **7.** Yes **9.** No **11.** (b) No

Section 3.5

1. -48 **3.** 0 **5.** -97 **9.** No solutions

11. $\begin{bmatrix} x_1 \\ x_2 \\ x_3 \\ x_4 \end{bmatrix}=\begin{bmatrix} 1 \\ 0 \\ 0 \\ 0 \end{bmatrix}$ **13.** $\begin{bmatrix} x_1 \\ x_2 \\ x_3 \\ x_4 \end{bmatrix}=c\begin{bmatrix} -1 \\ -2 \\ 1 \\ 4 \end{bmatrix}$

15. $\mathbf{AB}=\begin{pmatrix} 8 & 5 \\ 20 & 13 \end{pmatrix}$, $\mathbf{BA}=\begin{pmatrix} 13 & 20 \\ 5 & 8 \end{pmatrix}$

17. $\mathbf{AB}=\begin{bmatrix} 10 & 10 & 10 & 10 \\ 10 & 10 & 10 & 10 \\ 10 & 10 & 10 & 10 \\ 10 & 10 & 10 & 10 \end{bmatrix}$, $\mathbf{BA}=\begin{bmatrix} 4 & 4 & 4 & 4 \\ 8 & 8 & 8 & 8 \\ 12 & 12 & 12 & 12 \\ 16 & 16 & 16 & 16 \end{bmatrix}$

21. $\frac{1}{18}\begin{pmatrix} -3 & -4 & 9 \\ 0 & -12 & 18 \\ 6 & 14 & -18 \end{pmatrix}$ **23.** $\frac{i}{2}\begin{pmatrix} 2i & 0 & -2i \\ 1-2i & -1 & -1+i \\ 1-2i & 1 & 1+i \end{pmatrix}$

25. $\frac{i}{2}\begin{pmatrix} 0 & -2i & 0 \\ -1 & 1 & 1-i \\ 1 & -1 & -1-i \end{pmatrix}$ **33.** $\mathbf{x}=\begin{pmatrix} 0 \\ 0 \\ 0 \end{pmatrix}$ **35.** $\mathbf{x}=\begin{bmatrix} 0 \\ 0 \\ 0 \\ 0 \end{bmatrix}$ **37.** $\mathbf{x}=c\begin{bmatrix} -5 \\ -43 \\ 17 \\ 10 \end{bmatrix}$

Section 3.6

1. $\mathbf{x}(t)=c_1\begin{pmatrix} 1 \\ 1 \end{pmatrix}e^{3t}+c_2\begin{pmatrix} 3 \\ 2 \end{pmatrix}e^{4t}$ **3.** $\mathbf{x}(t)=c_1\begin{pmatrix} 1 \\ 0 \\ -1 \end{pmatrix}e^{-t}+c_2\begin{pmatrix} 0 \\ 2 \\ -1 \end{pmatrix}e^{-t}+c_3\begin{pmatrix} 2 \\ 1 \\ 2 \end{pmatrix}e^{8t}$

5. $\mathbf{x}(t)=c_1\begin{pmatrix} 2 \\ 0 \\ -1 \end{pmatrix}e^{-10t}+\left[c_2\begin{pmatrix} 0 \\ 1 \\ 0 \end{pmatrix}+c_3\begin{pmatrix} 1 \\ 0 \\ 2 \end{pmatrix}\right]e^{5t}$

7. $\mathbf{x}(t)=\frac{7}{4}\begin{pmatrix} 1 \\ 2 \end{pmatrix}e^{3t}+\frac{1}{4}\begin{pmatrix} 1 \\ -2 \end{pmatrix}e^{-t}$

9. $\mathbf{x}(t)=\begin{pmatrix}1\\-2\\-1\end{pmatrix}e^{2t}$ **11.** $\mathbf{x}(t)=\begin{pmatrix}-2\\0\\3\end{pmatrix}e^{-2t}$

13. (b) $\mathbf{x}(t)=9\begin{pmatrix}1\\0\\1\end{pmatrix}e^{t-1}+\frac{4}{3}\begin{pmatrix}-7\\2\\-13\end{pmatrix}e^{-(t-1)}+\frac{4}{3}\begin{pmatrix}1\\1\\1\end{pmatrix}e^{2(t-1)}$

Section 3.7

1. $\mathbf{x}(t)=\left[c_1\begin{pmatrix}2\\1-\sin t\end{pmatrix}+c_2\begin{pmatrix}2\sin t\\\cos t+\sin t\end{pmatrix}\right]e^{-2t}$

3. $\mathbf{x}(t)=e^t\left[c_1\begin{pmatrix}2\\-2\\3\end{pmatrix}+c_2\begin{pmatrix}0\\\cos 2t\\\sin 2t\end{pmatrix}+c_3\begin{pmatrix}0\\\sin 2t\\-\cos 2t\end{pmatrix}\right]$ **5.** $\mathbf{x}(t)=\begin{pmatrix}\cos t\\2\cos t+\sin t\end{pmatrix}e^{-t}$

7. $\mathbf{x}(t)=\begin{pmatrix}2\\-2\\1\end{pmatrix}e^{-2t}+\begin{bmatrix}-\sqrt{2}\,\sin\sqrt{2}\,t-\sqrt{2}\,\cos\sqrt{2}\,t\\\cos\sqrt{2}\,t-\sqrt{2}\,\sin\sqrt{2}\,t\\-3\cos\sqrt{2}\,t\end{bmatrix}$ **9.** $\mathbf{x}(0)=\begin{pmatrix}x_1\\0\\x_3\end{pmatrix}$

Section 3.8

1. $\mathbf{x}(t)=c_1\begin{pmatrix}0\\1\\0\end{pmatrix}e^{-2t}+\left[c_2\begin{pmatrix}1\\1\\0\end{pmatrix}+c_3\begin{pmatrix}t\\t\\1\end{pmatrix}\right]e^{-t}$

3. $\mathbf{x}(t)=c_1\begin{pmatrix}0\\0\\1\end{pmatrix}e^{-2t}+\left[c_2\begin{pmatrix}1\\0\\0\end{pmatrix}+c_3\begin{pmatrix}-t\\1\\0\end{pmatrix}\right]e^{-t}$

5. $\mathbf{x}(t)=\begin{pmatrix}1\\0\\1\end{pmatrix}e^{t}$ **7.** $\mathbf{x}(t)=\begin{pmatrix}1-t\\t\\t\end{pmatrix}e^{2t}$

13. (b) $\frac{1}{10}\begin{bmatrix}7e^t-5e^{2t}+8e^{-t} & -e^t+5e^{2t}-4e^{-t} & 3e^t-5e^{2t}-2e^{-t}\\21e^t-5e^{2t}-16e^{-t} & -3e^t+5e^{2t}+8e^{-t} & 9e^t-5e^{2t}-4e^{-t}\\14e^t+10e^{2t}-24e^{-t} & -2e^t-10e^{2t}+12e^{-t} & 6e^t+10e^{2t}-6e^{-t}\end{bmatrix}$

15. $\mathbf{I}+\frac{\mathbf{A}}{\alpha}(e^{\alpha t}-1)$

Section 3.9

1. $\frac{1}{5}\begin{bmatrix}e^{-2t}+5e^{2t}-e^{3t} & 5e^{2t}-5e^{3t} & e^{-2t}-e^{3t}\\-e^{-2t}+e^{3t} & 5e^{3t} & -e^{-2t}+e^{3t}\\4e^{-2t}-5e^{2t}+e^{3t} & -5e^{2t}+5e^{3t} & 4e^{-2t}+e^{3t}\end{bmatrix}$

3. $\frac{1}{5}\begin{bmatrix}6e^{2t}-\cos t-2\sin t & -2e^{2t}+2\cos t+2\sin t & 2e^{2t}-2\cos t+\sin t\\3e^{2t}-3\cos t-\sin t & -e^{2t}+6\cos t+2\sin t & e^{2t}-\cos t+3\sin t\\5\sin t & -10\sin t & 5\cos t\end{bmatrix}$

5. $\begin{bmatrix} (t+1)e^{-t} & (t+1)e^{-t}-e^{-2t} & e^{-t}-e^{-2t} \\ -te^{-t} & -te^{-t}+e^{-2t} & e^{-2t}-e^{-t} \\ te^{-t} & te^{-t} & e^{-t} \end{bmatrix}$

7. $\mathbf{A}=\begin{pmatrix} 3 & 1 & -1 \\ 1 & 3 & -1 \\ 3 & 3 & -1 \end{pmatrix}$ **9.** $\mathbf{A}=\frac{1}{13}\begin{pmatrix} 16 & -25 & 30 \\ 8 & -6 & -24 \\ 0 & 13 & 26 \end{pmatrix}$ **11.** No

SECTION 3.10

1. $\mathbf{x}(t)=2e^{t}\begin{pmatrix} t\cos t+3t\sin t+\sin t \\ -t\sin t \end{pmatrix}$

3. $\mathbf{x}(t)=\begin{bmatrix} -4\sin t-\frac{1}{2}t\sin t-t\cos t+5\cos t\ln(\sec t+\tan t) \\ -t\sin t-t\cos t+\sin t\cos^2 t-\frac{1}{2}\sin^2 t\cos t \\ -5\sin t\cos t+5\sin^2 t-\frac{1}{2}\sin^3 t \\ +5(\cos t-\sin t)\ln(\sec t+\tan t) \end{bmatrix}$

5. $\mathbf{x}(t)=\begin{bmatrix} 3e^{3t}-2e^{2t}-te^{2t} \\ e^{2t} \\ 3e^{3t}-2e^{2t} \end{bmatrix}$ **11.** $\psi(t)=\begin{pmatrix} \frac{3}{4}t^2+\frac{1}{2}t+\frac{1}{8} \\ -\frac{1}{4}t^2-t+\frac{1}{8} \end{pmatrix}$

13. $\psi(t)=\begin{pmatrix} 2 \\ -2 \\ -1 \end{pmatrix}e^{t}$ **17.** $\psi(t)=\begin{pmatrix} 1 \\ -1 \\ -1 \end{pmatrix}te^{3t}$

SECTION 3.11

1. $\mathbf{x}(t)=e^{-t}\begin{pmatrix} 3 \\ 2 \end{pmatrix}+e^{4t}\begin{pmatrix} -3 \\ 3 \end{pmatrix}$

3. $\mathbf{x}(t)=-\frac{4}{3}\begin{pmatrix} 1 \\ 2 \end{pmatrix}e^{-t}+\frac{1}{6}\begin{pmatrix} 2 \\ 1 \end{pmatrix}e^{2t}+\frac{1}{2}\begin{pmatrix} 0 \\ 1 \end{pmatrix}-t\begin{pmatrix} 1 \\ 1 \end{pmatrix}+3\begin{pmatrix} 1 \\ 1 \end{pmatrix}e^{t}$

5. $\mathbf{x}(t)=e^{t}\begin{pmatrix} 1-t-t^2 \\ 1-\frac{3}{2}t^2 \end{pmatrix}$ **7.** $\mathbf{x}(t)=2e^{t}\begin{pmatrix} t\cos t+3t\sin t+\sin t \\ -t\sin t \end{pmatrix}$

9. $\mathbf{x}(t)=\begin{cases} \begin{pmatrix} \cos 2t+\sin 2t \\ 2\sin 2t \end{pmatrix}, & t<\pi \\ \begin{pmatrix} \cos 2t \\ \sin 2t+\cos 2t \end{pmatrix}, & t>\pi \end{cases}$ **11.** $\mathbf{x}(t)=\begin{pmatrix} 1-t \\ t \\ t \end{pmatrix}e^{2t}$

13. $\mathbf{x}(t)=\begin{bmatrix} t-t^2-\frac{1}{6}t^3-\frac{1}{2}t^4+\frac{1}{12}t^5 \\ -\frac{1}{2}t^2 \\ t^2+\frac{1}{6}t^3 \end{bmatrix}$ **15.** $\mathbf{x}(t)=\begin{bmatrix} 1 \\ 1+t \\ 1 \\ 1+2t \end{bmatrix}e^{3t}$

Chapter 4

Section 4.1

1. $x=0, y=0;\quad x=0, y=1;\quad x=1, y=0;\quad x=\frac{1}{2}, y=\frac{1}{4}$

3. $x=0, y=0, z=0;\quad x=\frac{c}{d}, y=\frac{a}{b}, z=-\left(\frac{c^2}{d^2}+\frac{a^2}{b^2}\right)$

5. $x=0, y=y_0;\quad y_0$ arb.
$x=x_0, y=1;\quad x_0$ arb.
$x=x_0, y=-1;\quad x_0$ arb.

7. $x=0, y=-2, z=n\pi$

Section 4.2

1. $y=\cos(x-1)^2$ **3.** $y=\tan x$ **5.** $x^2+y^2=c^2$

7. The points $x=x_0, y=y_0$ with $x_0+y_0=-1$, and the circles $x^2+y^2=c^2$, minus these points.

9. The points $x=0, y=y_0;\quad x=x_0, y=0$, and the curves $y=ce^{-(2/3)e^{3x}}, x>0$; $y=ce^{-(2/3)e^{3x}}, x<0$.

11. The points $x=0, y=y_0$, and the curves

$$(by-dx)-\frac{(ad-bc)}{d}\ln|c-dy|=k,\quad x>0$$

$$(by-dx)-\frac{(ad-bc)}{d}\ln|c-dy|=k,\quad x<0$$

13. The point $x=0, y=0$; and the curves $xy^2-\frac{1}{3}x^3=c$.

Section 4.3

3. $\frac{y}{c}-\frac{b}{c^2}\ln(b+cy)=\frac{x^2}{2a}+k$ **5.** $\frac{y}{d}-\frac{x}{c}=\frac{b}{d^2}\ln(b+dy)-\frac{a}{c^2}\ln(a+cx)+k$

Index

Adjoint matrix 233
Analytic functions 138
Atomic waste problem 37
Autonomous systems 284

Bernoulli equation 57
Bessel equation 86
  order ν 153
  order zero 151
Bessel function
  $J_\nu(t)$ 153
  $J_0(t)$ 152

Carbon dating 19
Cauchy ratio test 137
Cayley–Hamilton theorem 259
Characteristic equation 87
  complex roots 90
  real and equal roots 94
  real and unequal roots 87
Characteristic polynomial 241
  complex roots 249
  real distinct roots 242
  repeated roots 253
Combat models 291
Complex exponentials 90
Compound interest 19
Convolution integral 182

Damping force 115
Detecting art forgeries 11
Detection of diabetes 127
Difference equations 72
Differential operators 78
Dirac, P. A. M. 173
Dirac delta function 173
  Laplace transform of 176
Direction field 289
Discontinuous forcing function 168

Eigenvalues of
  matrix 241
    multiplicity of 255
Eigenvectors of a matrix 241
  linear independence of 242
Electrical circuits 124
Equilibrium points 281
Escape velocity 42
Euler equation 94, 99
Exact equations 48, 50
Existence and uniqueness theorems for
  first-order equations 57
  second-order equations 78
  systems of first-order equations 221

First-order differential equations 1, 2
  exact 48, 50
  existence and uniqueness theorem for 57
  general solutions of 4, 21, 50
  homogeneous 3
  initial-value problem 5, 7, 21
  integrating factor for 8, 53
  linear 3
  nonlinear 3
  separable 21
  several variables 194
  systems of 195
Frobenius method 146
Fundamental matrix 263
Fundamental set of solutions 82

General solutions
 first-order linear equations 4, 21, 50
 higher-order equations 189
 second-order equations 81, 91, 97
 systems of first-order equations 242
Generalized functions 179
Gompertzian relation 43

Hermite equation 145
Hermite polynomial 145
Higher-order equations 188
 general solution 189
 variation of parameters 192
Homogeneous equations 25
Hypergeometric equations 154

Identity matrix 232
Impedance 126
Indicial equation 150
Initial-value problem 5, 7, 21, 78, 188, 194
Integral equation 59
Integrating factor 8, 53
Inverse matrix 234

Jump discontinuities 157, 168
Judicious guessing of particular solutions 106

Kirchoff's second law 125

Laguerre equation 154
Laplace transform 154
 of the convolution 182
 definition 155
 of derivatives 158
 of systems 276
 of the Dirac delta function 176
 existence of 157
 inverse of 159
Legendre equation 145
Legendre polynomials 145
Linear algebra 201, 226
Linear independence 85, 211
 of eigenvectors 242
 of vector functions 223
 of vectors 211
Linear operators 80
Linear differential equations, *see* First-order differential equations, Second-order differential equations, or Systems of first-order differential equations.
Logistic curves 30
Logistic law of population growth 29

Malthusian law of population growth 27
Matrices 197
 addition of 206
 adjoint 233
 characteristic polynomial 241
 determinant of 227
 diagonal 227
  lower 227
  upper 228
 eigenvalues of 241
 eigenvectors of 241
 fundamental 263
 identity 232
 inverse 234
 product of 231
Mechanical vibrations 114
 damped forced vibrations 118
 damped free vibrations 116
 forced free vibrations 119
 free vibrations 115
 resonant frequency 122
Method of elimination 186
Mixing problems 44

Natural frequency 120
Neumann function
 $Y_0(t)$ 152
 $Y_\nu(t)$ 154
Newtonian mechanics 37, 76, 115
Nonlinear differential equations
 autonomous system 284

Ohm's law 125
Orbits 286, 290
Order of a differential equation 1
Ordinary differential equations
 definition 1

Particular solution 100, 272
Phase plane 286
Picard iterates 60
Piecewise continuous functions 157
Population models 27
 Logistic law of population growth 29
 Malthusian law of population growth 27
Power series 137

Qualitative theory 280

Radioactive dating 12
Reduction of order 96
Reduction to systems of equations 195
Regular singular points 148
Resonant frequency 122
Resonance 119
Reynolds number 41

Schwartz, Laurent 174, 178
Second-order differential equations 76, 87
 characteristic equation 87
  complex roots 90

real and equal roots 94
real and unequal roots 87
Euler equation 94, 99
existence and uniqueness theorem 78
fundamental set of solutions 82
general solutions 81, 91, 97
homogeneous equation 77
judicious guessing of particular solution 106
nonhomogeneous equation 100
discontinuous function 168
particular solution 100
reduction of order 96
series solution 134
singular points 146
variation of parameters 104
Separable equations 20
Series solution 134
recurrence formula 136
when roots of the indicial equation are equal 150
when roots of the indicial equation differ by an integer 150
Singular points 146
Solutions, definition 1
first-order equations 4, 21, 50
higher-order equations 189
second-order equations 81, 91, 97
systems of first-order equations 242
Spring-mass dashpot system 114
damped vibrations 116, 118
forced vibrations 118, 119
free vibrations 115
Stability 281
Systems of algebraic equations 226
Systems of linear first-order equations 195
characteristic polynomial 241
complex roots 249
real distinct roots 242
repeated roots 253
definition 195
existence and uniqueness 221
fundamental matrix 263
general solution of 242
nonhomogeneous 268
reduction to 195
stability of 281
variation of parameters 268

Tacoma Bridge disaster 122
Taylor series 138
Tchebycheff equation 145
Tchebycheff polynomials 145
Tumor growth 43

Variation of parameters 104
for systems of equations 268
Vectors 196
linear independence of 211
solutions of systems of equations 226
valued functions 196
Vector spaces 203
dimension of 213
basis 217

Wronskian 82